Molecular Biotechnology:
Principles and Practices

Molecular Biotechnology: Principles and Practices

Contributors
───────
D. N. Lázaro-Silva, J. C. P. De Mattos et al.

www.aurisreference.com

Molecular Biotechnology: Principles and Practices

Contributors: D. N. Lázaro-Silva, J. C. P. De Mattos et al.

Published by Auris Reference Limited
www.aurisreference.com

United Kingdom

Copyright 2016
Printed in 2017 for Sale in the Indian Subcontinent

The information in this book has been obtained from highly regarded resources. The copyrights for individual articles remain with the authors, as indicated. All chapters are distributed under the terms of the Creative Commons Attribution License, which permit unrestricted use, distribution, and reproduction in any medium, provided the original author and source are credited.

Notice

Contributors, whose names have been given on the book cover, are not associated with the Publisher. The editors and the Publisher have attempted to trace the copyright holders of all material reproduced in this publication and apologise to copyright holders if permission has not been obtained. If any copyright holder has not been acknowledged, please write to us so we may rectify.

Reasonable efforts have been made to publish reliable data. The views articulated in the chapters are those of the individual contributors, and not necessarily those of the editors or the Publisher. Editors and/or the Publisher are not responsible for the accuracy of the information in the published chapters or consequences from their use. The Publisher accepts no responsibility for any damage or grievance to individual(s) or property arising out of the use of any material(s), instruction(s), methods or thoughts in the book.

Molecular Biotechnology: Principles and Practices

ISBN: 978-1-78154-962-9

British Library Cataloguing in Publication Data
A CIP record for this book is available from the British Library

Printed in the United Kingdom

Exclusively distributed by CBS Publishers & Distributors Pvt. Ltd.

Sales & Distribution Rights only for India, Pakistan, Bangladesh, Sri Lanka, Nepal and Bhutan. This book is not to be sold outside these territories.

Contents

List of Abbreviations vii
List of Contributors.. ix
Preface... xvii

Chapter 1	**The Use of DNA Extraction for Molecular Biology and Biotechnology Training: A Practical and Alternative Approach**............. 1	

D. N. Lázaro-Silva, J. C. P. De Mattos, Helena C. Castro,
G. G. Alves, and L. M. F. Amorim

Chapter 2	**Molecular Cloning and Expression of a Family 6 Cellobiohydrolase Gene cbhII from Penicillium funiculosum NCL1**................. 19	

Vanitha Chinnathambi, Meera Balasubramanium, Ramani Gurusamy,
and Gunasekaran Paramasamy

Chapter 3	**Molecular Cloning of a Chitinase Gene from the Ovotestis of Kuroda's Sea Hare Aplysia kurodai**.. 35	

Gaku Matsunaga, Syuuji Karasuda, Ryo Nishino,
Hideto Fukushima, and Masahiro Matsumiya

Chapter 4	**Whole-genome Molecular Haplotyping of Single Cells**...................... 49	

H Christina Fan, Jianbin Wang, Anastasia Potanina,
and Stephen R Quake

Chapter 5	**Project-Based Learning to Promote Effective Learning in Biotechnology Courses**....................................... 81	

Farahnaz Movahedzadeh, Ryan Patwell, Jenna E. Rieker, and
Trinidad Gonzalez

Chapter 6	**Carotenoids from Haloarchaea and Their Potential in Biotechnology**... 97	

Montserrat Rodrigo-Baños, Inés Garbayo, Carlos Vílchez,
María José Bonete, and Rosa María Martínez-Espinosa

Chapter 7	**The Bifunctional Dihydrofolate Reductase Thymidylate Synthase of tetrahymena Thermophila Provides A Tool For Molecular and Biotechnology Applications**.. 131	

Lutz Herrmann, Ulrike Bockau, Arno Tiedtke,
Marcus WW Hartmann and Thomas Weide

Chapter 8	**Function and Biotechnology of Extremophilic Enzymes in Low Water Activity** 149	
	Ram Karan, Melinda D Capes and Shiladitya DasSarma	
Chapter 9	**Marine Microbial Biodiversity, Bioinformatics and Biotechnology (M2B3) Data Reporting and Service Standards** 187	
	Petra ten Hoopen, Stéphane Pesant, Renzo Kottmann, Anna Kopf, Mesude Bicak, Simon Claus, Klaas Deneudt, Catherine Borremans, Peter Thijsse, Stefanie Dekeyzer, Dick MA Schaap, Chris Bowler, Frank Oliver Glöckner and Guy Cochrane	
Chapter 10	**Advances in Biotechnology and Genomics of Switchgrass** 205	
	Madhugiri Nageswara-Rao, Jaya R Soneji, Charles Kwit and C Neal Stewart Jr	
Chapter 11	**A Novel, Highly Selective RT-QPCR Method for Quantification of MSRV Using PNA Clamping Syncytin-1 (ERVWE1)** 241	
	Grzegorz Machnik, Estera Skudrzyk, Łukasz Bułdak, Krzysztof Łabuzek, Jarosław Ruczyn´ski, Magdalena Alenowicz, Piotr Rekowski, Piotr Jan Nowak, Bogusław Okopien	
Chapter 12	**Microfluidics in Biotechnology** 267	
	Richard Barrycorresponding and Dimitri Ivanov	
	Citations 275	
	Index 277	

List of Abbreviations

AFLP	Amplified Fragment Length Polymorphism
BAC	Bacterial Artificial Chromosome
BES	BAC-End Sequence
CAD	Cinnamyl Alcohol Dehydrogenase
CBDs	Chitin Binding Domains
CDI	Common Data Index
CNS	Central Nervous System
COMT	Caffeic acid O-methyltransferase
DArT	Diversity Array Technology
ENA	European Nucleotide Archive
EST	Expressed Sequence Tags
EurOBIS	European Ocean Biogeographic Information System
GH	Glycoside Hydrolase
HIV	Human Immunodeficiency Virus
ICoMM	International Census Of Marine Microbes
M2B3	Marine Microbial Biodiversity, Bioinformatics and Biotechnology
Micro B3 IS	Micro B3 Information System
Micro B3	Marine Microbial Biodiversity, Bioinformatics, Biotechnology
NGS	Next generation sequencing
OGC	Open Geospatial Consortium
OSD	Ocean Sampling Day
PCR	Polymerase Chain Reaction
RAPD	Randomly Amplified Polymorphic DNA
RFLP	Restriction Fragment Length Polymorphism
SDRFs	Single dose Restriction Fragments
SNPs	Single Nucleotide Polymorphisms
SPL	Squamosa Promoter Binding Like
SSRs	Simple Sequence Repeats
STSs	Sequence-tagged sites
TILLING	Targeting Induced Local Lesions In Genomes
UFF	Fluminense Federal University
WFS	Web Feature Service
WGS	Whole Genome Sequencing.

List of Contributors

D. N. Lázaro-Silva
Post-Graduation Program of Science and Biotechnology, Biology Institute, Fluminense Federal University, Niterói, Brazil

J. C. P. De Mattos
Biophysics and Biometry Departament, Roberto Alcantara Gomes Biolology Institute, Rio de Janeiro State University, Rio de Janeiro, Brazil

Helena C. Castro
Post-Graduation Program of Science and Biotechnology, Biology Institute, Fluminense Federal University, Niterói, Brazil

G. G. Alves
Post-Graduation Program of Science and Biotechnology, Biology Institute, Fluminense Federal University, Niterói, Brazil

L. M. F. Amorim
Post-Graduation Program of Science and Biotechnology, Biology Institute, Fluminense Federal University, Niterói, Brazil

Vanitha Chinnathambi
Center for Marine Bioprospecting, AMET University Kanathur, Chennai, India

Meera Balasubramanium
Department of Genetics, Center for Excellence in Genomic Sciences, School of Biological Sciences, Madurai Kamaraj University, Madurai, India

Ramani Gurusamy
Department of Genetics, Center for Excellence in Genomic Sciences, School of Biological Sciences, Madurai Kamaraj University, Madurai, India

Gunasekaran Paramasamy
Department of Genetics, Center for Excellence in Genomic Sciences, School of Biological Sciences, Madurai Kamaraj University, Madurai, India

Gaku Matsunaga
Department of Marine Science and Resources, College of Bioresource Sciences, Nihon University, Kanagawa, Japan

Syuuji Karasuda
Department of Marine Science and Resources, College of Bioresource Sciences, Nihon University, Kanagawa, Japan

Ryo Nishino
Department of Marine Science and Resources, College of Bioresource Sciences, Nihon University, Kanagawa, Japan

Hideto Fukushima
Department of Marine Science and Resources, College of Bioresource Sciences, Nihon University, Kanagawa, Japan

Masahiro Matsumiya
Department of Marine Science and Resources, College of Bioresource Sciences, Nihon University, Kanagawa, Japan

H Christina Fan
Department of Bioengineering, Stanford University, Stanford, California, USA

Jianbin Wang
Department of Bioengineering, Stanford University, Stanford, California, USA

Anastasia Potanina
Howard Hughes Medical Institute, Stanford University, Stanford, California, USA

Stephen R Quake
Department of Bioengineering, Stanford University, Stanford, California, USA
Department of Applied Physics, Stanford University, Stanford, California, USA

Farahnaz Movahedzadeh
Department of Biological Sciences, Harold Washington College, 30 East Lake Street, Chicago, IL 60601, USA

Ryan Patwell
Department of Biological Sciences, Harold Washington College, 30 East Lake Street, Chicago, IL 60601, USA

Jenna E. Rieker
Department of Biological Sciences, Harold Washington College, 30 East Lake Street, Chicago, IL 60601, USA

Trinidad Gonzalez
Department of Biological Sciences, Harold Washington College, 30 East Lake Street, Chicago, IL 60601, USA

Montserrat Rodrigo-Baños
Biochemistry and Molecular Biology Division, Agrochemistry and Biochemistry Department, Faculty of Sciences, University of Alicante, Ap. 99, E-03080 Alicante, Spain

Inés Garbayo
Algal Biotechnology Group, University of Huelva and Marine International Campus of Excellence (CEIMAR), CIDERTA and Faculty of Sciences, 21071 Huelva, Spain

Carlos Vílchez
Algal Biotechnology Group, University of Huelva and Marine International Campus of Excellence (CEIMAR), CIDERTA and Faculty of Sciences, 21071 Huelva, Spain

María José Bonete
Biochemistry and Molecular Biology Division, Agrochemistry and Biochemistry Department, Faculty of Sciences, University of Alicante, Ap. 99, E-03080 Alicante, Spain

Rosa María Martínez-Espinosa
Biochemistry and Molecular Biology Division, Agrochemistry and Biochemistry Department, Faculty of Sciences, University of Alicante, Ap. 99, E-03080 Alicante, Spain

Lutz Herrmann
Cilian AG, Johann-Krane Weg 42, D-48149 Münster, Germany

Ulrike Bockau
Cilian AG, Johann-Krane Weg 42, D-48149 Münster, Germany
Institut für allgemeine Zoologie und Genetik, Universität Münster, Schloßplatz 5, D-48149 Münster, Germany

Arno Tiedtke
Institut für allgemeine Zoologie und Genetik, Universität Münster, Schloßplatz 5, D-48149 Münster, Germany

Marcus WW Hartmann
Cilian AG, Johann-Krane Weg 42, D-48149 Münster, Germany

Thomas Weide
Cilian AG, Johann-Krane Weg 42, D-48149 Münster, Germany

Ram Karan
Department of Microbiology and Immunology, University of Maryland School of Medicine
Institute of Marine and Environmental Technology, University System of Maryland

Melinda D Capes
Department of Microbiology and Immunology, University of Maryland School of Medicine
Institute of Marine and Environmental Technology, University System of Maryland

Shiladitya DasSarma
Department of Microbiology and Immunology, University of Maryland School of Medicine
Institute of Marine and Environmental Technology, University System of Maryland

Petra ten Hoopen
European Nucleotide Archive, EMBL-EBI, Wellcome Trust Genome Campus Hinxton

Stéphane Pesant
European Nucleotide Archive, EMBL-EBI, Wellcome Trust Genome Campus Hinxton

Renzo Kottmann
European Nucleotide Archive, EMBL-EBI, Wellcome Trust Genome Campus Hinxton

Anna Kopf
European Nucleotide Archive, EMBL-EBI, Wellcome Trust Genome Campus Hinxton

Mesude Bicak
European Nucleotide Archive, EMBL-EBI, Wellcome Trust Genome Campus Hinxton

Simon Claus
European Nucleotide Archive, EMBL-EBI, Wellcome Trust Genome Campus Hinxton

Klaas Deneudt
European Nucleotide Archive, EMBL-EBI, Wellcome Trust Genome Campus Hinxton

Catherine Borremans
European Nucleotide Archive, EMBL-EBI, Wellcome Trust Genome Campus Hinxton

Peter Thijsse
European Nucleotide Archive, EMBL-EBI, Wellcome Trust Genome Campus Hinxton

Stefanie Dekeyzer
European Nucleotide Archive, EMBL-EBI, Wellcome Trust Genome Campus Hinxton

Dick MA Schaap
European Nucleotide Archive, EMBL-EBI, Wellcome Trust Genome Campus Hinxton

Chris Bowler
European Nucleotide Archive, EMBL-EBI, Wellcome Trust Genome Campus Hinxton

Frank Oliver Glöckner
European Nucleotide Archive, EMBL-EBI, Wellcome Trust Genome Campus Hinxton

Guy Cochrane
European Nucleotide Archive, EMBL-EBI, Wellcome Trust Genome Campus Hinxton

Madhugiri Nageswara-Rao
Department of Plant Sciences, The University of Tennessee, 252 Ellington Plant Sciences, 2431 Joe Johnson Dr., Knoxville, TN 37996, USA
Department of Biological Sciences, Polk State College, Winter Haven, FL 33881, USA

Jaya R Soneji
Department of Biological Sciences, Polk State College, Winter Haven, FL 33881, USA

Charles Kwit
Department of Plant Sciences, The University of Tennessee, 252 Ellington Plant Sciences, 2431 Joe Johnson Dr., Knoxville, TN 37996, USA

C Neal Stewart Jr
Department of Plant Sciences, The University of Tennessee, 252 Ellington Plant Sciences, 2431 Joe Johnson Dr., Knoxville, TN 37996, USA
Department of Biological Sciences, Polk State College, Winter Haven, FL 33881, USA

Irshad Ul Haq
NUST Center of Virology & Immunology (NCVI), National University of Sciences & Technology (NUST), H-12, Islamabad 44000, Pakistan

Waqas Nasir Chaudhry
NUST Center of Virology & Immunology (NCVI), National University of Sciences & Technology (NUST), H-12, Islamabad 44000, Pakistan

Maha Nadeem Akhtar
NUST Center of Virology & Immunology (NCVI), National University of Sciences & Technology (NUST), H-12, Islamabad 44000, Pakistan

Saadia Andleeb
NUST Center of Virology & Immunology (NCVI), National University of Sciences & Technology (NUST), H-12, Islamabad 44000, Pakistan

Ishtiaq Qadri
NUST Center of Virology & Immunology (NCVI), National University of Sciences & Technology (NUST), H-12, Islamabad 44000, Pakistan

Grzegorz Machnik
Department of Pharmacology, Medical University of Silesia, Medyko´w 18, 40-752 Katowice, Poland

Estera Skudrzyk
Department of Pharmacology, Medical University of Silesia, Medyko´w 18, 40-752 Katowice, Poland

Łukasz Bułdak
Department of Pharmacology, Medical University of Silesia, Medyków 18, 40-752 Katowice, Poland

Krzysztof Łabuzek
Department of Pharmacology, Medical University of Silesia, Medyków 18, 40-752 Katowice, Poland

Jarosław Ruczyński
Faculty of Chemistry, University of Gdańsk, Wita Stwosza 63, 80-308 Gdańsk, Poland

Magdalena Alenowicz
Faculty of Chemistry, University of Gdańsk, Wita Stwosza 63, 80-308 Gdańsk, Poland

Piotr Rekowski
Faculty of Chemistry, University of Gdańsk, Wita Stwosza 63, 80-308 Gdańsk, Poland

Piotr Jan Nowak
Department of Nephrology, Hypertension, and Kidney Transplantation, Medical University of Łódź, Pomorska 251, 92-213 Łódź, Poland

Bogusław Okopień
Department of Pharmacology, Medical University of Silesia, Medyków 18, 40-752 Katowice, Poland

Preface

Molecular biotechnology is the use of laboratory techniques to study and modify nucleic acids and proteins for applications in areas such as human and animal health, agriculture, and the environment. The text *Molecular Biotechnology: Principles and Practices* provides comprehensive and up-to-date coverage of key concepts in molecular biotechnology. In first chapter, we create a dynamic and practical course that receives positive evaluation by the students of biological sciences and biomedicine Second chapter deals with expression and purification of cellobiohydrolase gene from Penicillium funiculosum NCL1. In third chapter, we report that we successfully cloned and sequenced a chitinase gene from the ovotestis of Kuroda's sea hare Aplysia kurodai. Fourth chapter focuses on whole-genome molecular haplotyping of single cells. Fifth chapter demonstrates the benefits of redesigning a standard lab-based molecular biology course to create a more effective learning environment. Sixth chapter deals with carotenoids from haloarchaea and their potential in biotechnology. The cloning, characterization and functional analysis of *Tetrahymena thermophila's* DHFR-TS has been presented in seventh chapter. In eighth chapter, we summarize the current state of knowledge of extremophilic enzymes functioning in high salinity and cold temperatures, focusing on their strategy for function at low water activity. Ninth chapter presents marine microbial biodiversity, bioinformatics and biotechnology (M2B3) standards for "reporting" and "serving" data. Advances in biotechnology and genomics of switchgrass have been focused in tenth chapter. A novel, highly selective RT-QPCR method for quantification of MSRV using PNA clamping syncytin-1 (ERVWE1) has been discussed in eleventh chapter. Last chapter deals with microfluidics in biotechnology.

Chapter 1

THE USE OF DNA EXTRACTION FOR MOLECULAR BIOLOGY AND BIOTECHNOLOGY TRAINING: A PRACTICAL AND ALTERNATIVE APPROACH

D. N. Lázaro-Silva[1], J. C. P. De Mattos[2], Helena C. Castro[1], G. G. Alves[1], and L. M. F. Amorim[1]

[1]Post-Graduation Program of Science and Biotechnology, Biology Institute, Fluminense Federal University, Niterói, Brazil

[2]Biophysics and Biometry Departament, Roberto Alcantara Gomes Biology Institute, Rio de Janeiro State University, Rio de Janeiro, Brazil

ABSTRACT

DNA extraction and polymerase chain reaction are molecular biology techniques widely used in research, diagnostics and biotechnology industry. In this work, we report a practical course designed for undergraduate students of health, biological sciences and biotechnology areas. The course is five days long and emphasizes basic procedures in practical activities. The students were challenged to use three different protocols for DNA isolation (one using commercial silica column and two, in-house) using no lab animals but themselves. They evaluated the DNA yield, and potential amplification of the extracted DNA by PCR using their own cells genetic material. Students were stimulated to adopt a critical view during the course to develop skills including critical analysis of primary results and possible errors, scientific report writing and a problem solving profile.

INTRODUCTION

Biological Sciences and Biotechnology are inherently complex and highly abstract. Therefore, professionals of health and biological sciences and biotechnology are challenged with techniques such blood collection and extraction of DNA, evaluation by spectrophotometry and agarose gel electrophoresis as well as polymerase chain reaction (PCR) amplification of the DNA. However, these complex topics are often presented to undergraduate students during theoretical classes or with practical activities involving classical protocols of DNA isolation from fruit (Miller, 1994, Tibell and Rundgren, 2010, Sitaraman, 2012). Despite these strategies are very easy, low-cost and adequate to teach principles of molecular biology, they have limited resemblance to the actual professional activities on the environment of health science careers. In addition, some skills are necessary to perform these basic molecular procedures and should be obtained during graduation as endorsed by reviews of biology curricula like "BIO2010" (Slonczewsk, 2004).

Isolating high-quality nucleic acids from blood is a part of the many diagnostics with forensic and/or scientific applications (Poh, 2014). Despite the concern to use noninvasive sources of DNA (Hearn, 2010), the use of peripheral blood still remains very common in research and diagnosis, due to its reliability as a source of genetic material (Martin, 1992). Furthermore, blood is the main target of study in clinical research and diagnosis to effectively detect pathogens (Schijman, 2011) as it allows to access the genetic material of pathogens such as Pneumococcal pneumonia (Marchese, 2011), Trypanosoma cruzi (González, 2010) and several viruses (Mello, 2007). Furthermore, regardless of the relative degree of invasiveness, blood collection can often be used to avoid techniques that are even more invasive. It allows prenatal diagnosis, and the detection of genetic disorders such as hemophilia in fetuses (without the use of amniotic fluid) (Tsui, 2011) or even sex determination (Mortarino, 2011).

The use of blood from laboratory animals for in vivo experiments in education is still looking for alternatives (Cardoso and Almeida, 2014). However, it is important that the students of health science careers be prepared for the issues involving this topic in their future professional lives (Badyal and Desai 2014). This includes questions involving collecting, manipulating, evaluating and comparing human blood. Therefore, alternatives for the use of blood from lab animals should be developed for substitution purpose without loosing the applied perspective. In addition, besides approaching the technique, the alternative option should still allow the discussion in different levels, from research to the professional application with a dynamic view for every student.

Thus, the aim of this work is to propose a practical summer course to undergraduate students of biomedicine, biotechnology and biological sciences, containing basic procedures of using human blood DNA extraction, including spectrophotometry, agarose gels and Biosafety on blood manipulation.

Therefore, the students are challenged to evaluate their own blood by using three different DNA extraction techniques (one from commercial sources and two, in-house), getting first-hand contact with the equipments and procedures of a real molecular biology laboratory. The course requires and works several skills of these students, including: data interpretation, problem solving, experimental design, scientific report writing and collaborative work, with critical analysis of primary results and possible errors.

METHODOLOGY

Subjects

A group of ten undergraduate students of both genders (7 females and 3 males), age 21.4 ± 1.43 (20 to 24 years), 100% single, 100% student occupation, 100% auto-declared white skin color, from São Paulo (10%), Minas Gerais (10%) and Rio de Janeiro (80%) on the fourth period (2nd year) of Biology (n = 6) or Biomedicine (n = 4) courses from the Fluminense Federal University (UFF), Rio de Janeiro—Brazil, were invited to participate in the molecular biology summer course offered at the Molecular Oncology Laboratory of Biology Institute of UFF. All students signed the informed consent to collect blood, perform DNA extractions and donate DNA to the laboratory's DNA collection (Antônio Pedro Hospital Ethics Committee CONEP 02509 No. 0012.0258000-09, 2009).

Summer Course Design

The laboratory classes were divided into five meetings with a maximum of 7 hours (Table 1). The DNA extraction was distributed into the first three days (Figure 1). The basic procedures executed were:

- Preparation of solutions, dilutions and sterilization;
- Preparation and use of agarose gels to separate genomic DNA and PCR products;
- Interpretation and trouble shooting of agarose gel electrophoresis;
- PCR reaction, the importance of controls; interpretation and troubleshooting of PCR reaction;
- The use of the laboratory equipments as: autoclave, micropipetters, centrifuges, balance, horizontal electro- phoresis system, UV transilluminator, termocycler and photodocumentation system.

Table 1. Schedule of activities of the molecular biology summer course.

Day	Approximate time used	Content
1	1 h	Introductory remarks on: biosafety in the handling of chemical and biological materials; procedures on DNA extraction protocols
	2:30 h	DNA extraction solutions preparation and autoclaving
	30 min	Blood sample collection
	1:30 h	Protocol 1 starting. Hemolysis and nucleus lysis, overnight
	1:30 h	Protocol 2 starting. Hemolysis and nucleus lysis, overnight
2	3:30 h	DNA extraction using Protocol 3
	2 h	Protein extraction and DNA precipitation of samples from protocols 1 and 2
3	2 h	DNA recover, wash and solubilization of samples from protocols 1 and 2
	1:30 h	Spectrophotometric quantification of DNA extracted from protocols 1, 2 and 3
	2:30 h	Agarose gel electrophoresis of Genomic DNA
4	4 h	PCR reaction
	2 h	Agarose gel electrophoresis of PCR products
5	4 h	Graphics and Statistical analysis of the results to answer the questions above: I. Using spectrophotometer quantification calculate mean and standard deviation of DNA solutions: a. Concentrations b. 260/280 ratio c. Yield per protocol d. Yield per blood volume used II. Construct graphics using the calculated values for a, b, c and d items III. Identify the statistical difference among the three DNA extraction techniques on a, b, c and d items

On the last day of the course they were asked to write a final text, including graphics, to describe and illustrate the obtained results.

BioSafety

Students were instructed about laboratory safety, such as the use of individual protection equipments. They wore laboratory coat, gloves and safety glasses all time when in the laboratory and handling blood samples. Any residual blood or contaminated vial was disposed in a Biohazard waste container that was autoclaved prior to discard. Students used ultraviolet protective glasses and gloves when viewing and handling gels with ethidium bromide (EtBr) on the UV transilluminator. To minimize exposure to EtBr, only the dilute staining solution was available to them. Besides, a restricted area of the laboratory was reserved to gel staining, and students were instructed to leave all EtBr-contaminated materials in this area only. All solutions and remaining buffer were treated as hazardous waste and disposed in adequate containers to be discarded.

DNA Extraction Protocols

Nine milliliters of blood were collected from the each student by venous puncture with Ethylenediaminetetraacetic acid (EDTA)—containing tubes, and aliquoted to be submitted to three different DNA extraction protocols (Figure 1). For safety reasons, students manipulated only their own blood samples.

Protocol 1 (MILLER, 1988)

Red blood cells in 4 mL of blood were lysed with 8 mL of of haemolysis buffer (155 mM NH_4Cl, 10 mM $KHCO_3$, 1 mM EDTA) in an ice bath per thirty minutes. After centrifugation (500 ×g/10 min) the leukocytes nuclei pellet was washed with 8 mL of hemolytic buffer until disappearance of red color, indicating the absence of hemoglobin.

Leukocytes were lysed with 3 mL of nuclei lysis buffer (10 mM Tris-HCl, 400 mM NaCl and 2 mM Na_2EDTA, pH 8.2) containing 200 µL of 20% SDS and 20 µL of proteinase K (25 mg/mL) at 56°C overnight. On the next day, the digestion was centrifuged (500 ×g/10 min) and the supernatant passed to another tube. Proteins were precipitated with 1 mL of NaCl (6 M) and after centrifugation (12,000 ×g/10 min), the aqueous phase was transferred to a new tube. Two volumes of ice cold ethanol were added and DNA precipitation visualized. DNA was collected (14,000 ×g/10 min) and washed with 1 mL of 70% ethanol to remove excess of salt. This step must be repeated until the precipitate becomes translucent. Dried samples were eluted with 100 µL of TE

(10 mM Tris-HCl, 0.2 mM Na$_2$EDTA, pH 7.5) for 2 hours at 37°C and stored at ?20°C.

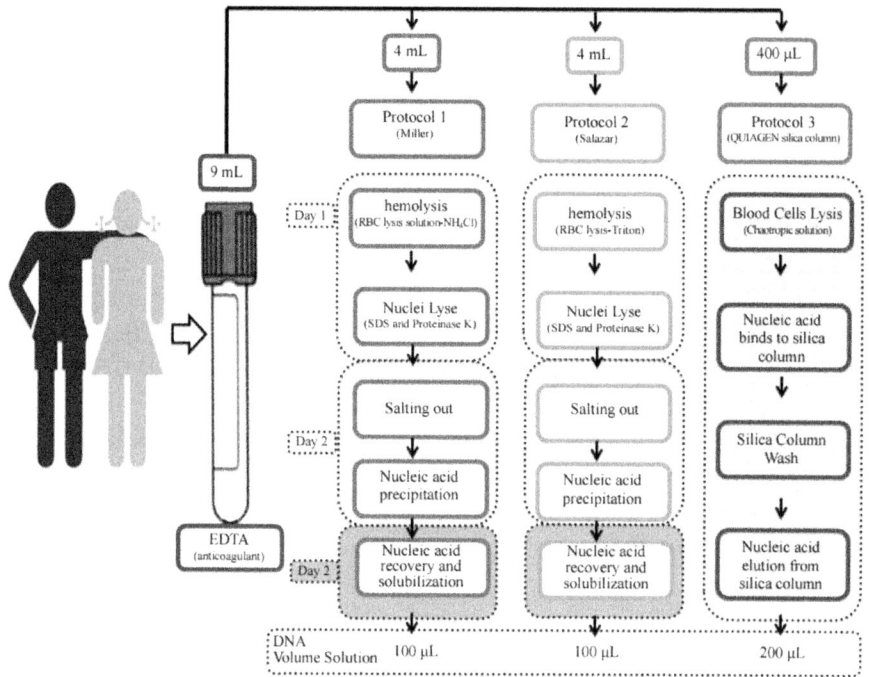

Figure 1. Detailed schedule of the procedures used in the practical summer course applied to undergraduate students of biomedicine and biological sciences at Universidade Federal Fluminense, containing basic procedures using human blood, and three different protocols.

Protocol 2 SALAZAR (1998)

Red blood cells in 4 mL of blood were lysed with 8 mL of TM1-TRITON (10 mM Tris-HCl, pH 8.0, 10 mM KCl, 10 mM MgCl$_2$, 2 mM EDTA, pH 8.0, and 25 mL/L Triton X-100) for 15 minutes at room temperature. After centrifugation (500 ×g/10 min), the pellet was washed with 6 mL of TM1 (10 mM Tris-HCl, pH 8.0, 10 mM KCl, 10 mM MgCl$_2$, 2 mM EDTA, pH 8.0) until hemoglobin removal. The nuclei of leukocytes was lysed with 500 µL of TM2 (10 mM Tris-HCl, pH 8.0, 10 mM KCl, 10 mM MgCl$_2$, 2 mM EDTA, pH 8.0, 0.4 M NaCl) and 50 µL of SDS 10% at 37°C overnight. The proteins were removed by saline precipitation with 80 µL of NaCl (5 M) and centrifugation (10,000 ×g/5 min). The supernatant was transferred to a new tube, DNA precipitated with 2 volumes of ice cold ethanol and washed, for salt removing, with ethanol 70%. The elution was made in 100 µL of TE (10 mM Tris-HCl, 0.2 mM Na$_2$EDTA, pH 7.5) in a water bath for 15 minutes at 60°C. The DNA was stored at ?20°C until further use.

Protocol 3 QIAamp DNA Blood Mini Kit Silica Column

The extraction using QIAamp DNA Blood Mini Kit (Qiagen, Germany) was performed following manufacturer's instructions. Briefly, 30 µL of proteinase K (100 mg/mL), 400 µL of blood and 400 µL of AL buffer were added, in this order, to a microcentrifuge tube and mixed by inversion. After incubation at 56°C for 10 minutes, 200 µL of absolute ethanol was added and the tube mixed again by pulse-vortexing for 15 seconds. The mixture was transferred to the spin column and centrifuged (6000 ×g/1 min). The column was then placed in a clean 2 mL tube, 500 µL of AW1 buffer was added and after centrifugation (6000 ×g/1 min) the column was placed into a new collection tube. The AW2 buffer (500 µL) was added, and the column was centrifuged (20,000 ×g/3 min). The column was then placed in a new 1.5 mL microcentrifuge tube, 200 µL double-distilled water was added, incubated at room temperature for 5 minutes and finally centrifuged (8000 ×g/3 min) to recover DNA. The eluted DNA was stored at ?20°C until further use.

Genomic DNA Quantification and Purity Measurements

The concentration and purity of genomic DNA were determined by measuring the absorbance at 260 nm and A_{260}:A_{280} absorbance ratios respectively, using a spectrophotometer NanoDrop ND-1000.

GENOMIC DNA INTEGRITY AND PCR PRODUCT VERIFICATION

Agarose gel electrophoresis was used to determine the integrity of the DNA. Briefly, 5 µL of genomic DNA mixed with 1 µl of loading buffer (50% of Glycerol, 0.125% of Bromophenol blue, 0.125% of Xylene Cyanol, TE pH 8.0) was loaded onto 1% agarose gel with ethidium bromide (0.5 µg/mL) and electrophoresed for 45 minutes at 80 mV followed by visualization under UV light (Sambrook, 1998).

The size of the product (197 bp) was confirmed by electrophoresis on a 2% agarose gel containing 8 µL of the PCR products mixed with 1 µL of loading buffer and using a 100 base pair marker as a reference (Fermentas Life Sciences, Lithuania).

PCR Reaction

In order to evaluate DNA quality for amplification by PCR-RFLP, DNA samples of each protocol were assayed as previously described by Nomura (2007) with modifications. Briefly, the PCR reaction was carried out in a 50 µL mix containing 100 ng genomic DNA, Taq Buffer (Fermentas, MD, USA), 0.2 mM of dNTPs (Life Tech., Brazil), 0.2 µM of each primer (Forward: 5'-CTCCTCCTCCTCTGCTCCTC-3'; Reverse: 5'-GGGGCTAGCT-CGGGACTC-3'-Operon, USA), 5% of DMSO and 1.25 U Taq DNA Polymerase (Fermentas, MD, USA). Cycling conditions comprised an initial denaturation step for 10 minutes at 95°C, followed by 35 cycles of three steps: 95°C for 30 seconds, 60°C for 30 seconds, and 72°C for 1 minute. A final extension step was performed at 72°C for 10 minutes.

Assessment of Student Acceptance

A survey with 5 questions regarding the students' opinions about the summer course was applied on the last day for the 10 participants. The questions involved the evaluation of the class dynamics and how the course reached their expectation, as detailed in Table 2.

In order to assess if there is a public interest in attending the summer course, which justifies the assembly of it, ninety undergraduate students of biology and medicine were asked if they would like to perform a summer curse of molecular biology. They were previously informed about the content of the course (DNA extraction from blood, electrophoresis and PCR) and answered this question after a practical class of DNA extraction from fruits in the discipline of biochemistry.

RESULTS AND DISCUSSION

According to Voet et al. (2003), research experience is an essential part of any molecular biology curriculum, and may be obtained through well planned laboratory courses. Current literature reports about the lack of disciplines that not only get the themes in a practical perspective, but also involves data interpretation, problem solving, experimental design, scientific report writing and collaborative work, with critical analysis of primary results and possible errors (Coil, 2010; Azer, 2013).

Table 2. Survey of students' opinions about the summer course.

#	Question	Mark the answer with X (1 = completely disagree => 5 = completely agree)				
Q1	What have you learned during the Course? (Answer below)					
Q2	You understand each step and the role of each solution on the 3 protocols used for DNA extraction	1	2	3	4	5
Q3	You approve the dynamic used to theme comprehension	1	2	3	4	5
Q4	Your expectations were met	1	2	3	4	5
Q5	Classify the summer course	1	2	3	4	5
Q6	What did you like best and or would modify in the course? (Answer below)					
Q7	Would you like to participate in another summer course of molecular biology	Yes			No	

In the present work, the proposed Summer Course in Molecular Biology challenged the students from biomedicine and biology undergraduate courses to isolate and manipulate DNA from human blood samples. Thus, these future specialists could get first-hand contact with the professional laboratory environment, techniques and equipment. Herein we report the results of the students' experience and reports about the comparison of three DNA extraction techniques performed during this summer course.

DNA Yield and Purity

DNA extraction for DNA collection samples is a critical procedure as it is time-consuming and may involve ethical aspects. The three protocols proposed here included two in-house procedures of low cost (Aidar, 2007) and low-toxicity as it does not use phenol-chloroform extraction (Lum, 1998;Rogers, 2007).

According to the students' assay reports, the concentration of DNA samples were 260.1 ± 44.9 ng/µL (26.3 ? 529.6 ng/µL), 268.3 ± 43.7 ng/µL (131.4 - 463.5 ng/µL) and 33.8 ± 3.4 ng/µL (16.5 - 46.4 ng/µL), with a total DNA mass of 26.0, 26.8 and 6.8 µg to Protocols 1, 2 and 3 respectively (Figure 2(a) and Figure 2(b)). The two in-house procedures extracted 4 more times DNA than the commercial silica column ($p < 0.001$). However, considering the DNA mass corrected per blood volume used on extractions (6.503 ± 1.122 ng, 6.707 ± 1.092 ng and 33.75 ± 3.319 ng/ml of blood to Protocols 1, 2 and

3 respectively) commercial silica column was the most profitable protocol (p < 0.001; Figure 2(c)). The 260/280 ratio is used to assess the purity of DNA and RNA. A ratio of ~1.8 is generally accepted as "pure" for DNA (Sambrook, 1989). The OD 260/280 ratio was 1.838 ± 0.014 (1.74 - 1.87), 1.831 ± 0.028 (1.60 - 1.89) and 1.779 ± 0.047 (1.55 ? 2.02) to Protocols 1, 2 and 3 respectively. The student data showed that the in-house procedures provided DNA with good purity with no statistical difference of the commercial silica column (p > 0.05, Figure 2(d)). Although silica columns involved a four step procedure: 1) blood cells lysis, 2) DNA bind to the silica, 3) silica wash and 4) DNA elution from silica (Figure 1(a)), they were performed in only one day. The students' results showed that the in-house methods were able to extract large quantities of DNA in three days (Figure 1(a)). The comparison of the students revealed that the salt-precipitation method is economically more viable while the technique described by Salazar (1998)can be performed in a smaller number of days since nuclei lysis could be carried out at 65°C for 30 minutes. The students detected these protocols advantages and disadvantages and informed in their reports.

DNA Integrity

The student assays reports described all DNA extraction procedures (Figure 2). They described that some samples obtained by using protocol 2 presented a smearing, indicating degradation (Figure 2(e)). It is also possible to see a pattern of bands, characteristic of apoptosis, in samples 1S and 3S. According to Takayama (2001), it is common to observe apoptotic DNA fragments in samples stored at room temperature or stored for up to 2 days at 4°C, with no implications for the PCR. The blood samples were collected at the beginning of classes and were kept at least 3 hours at room temperature prior to be extracted.

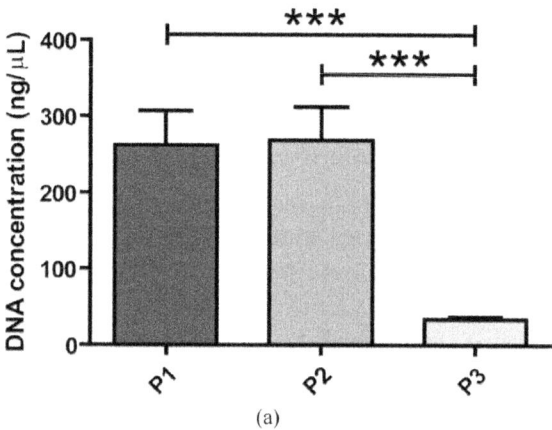

(a)

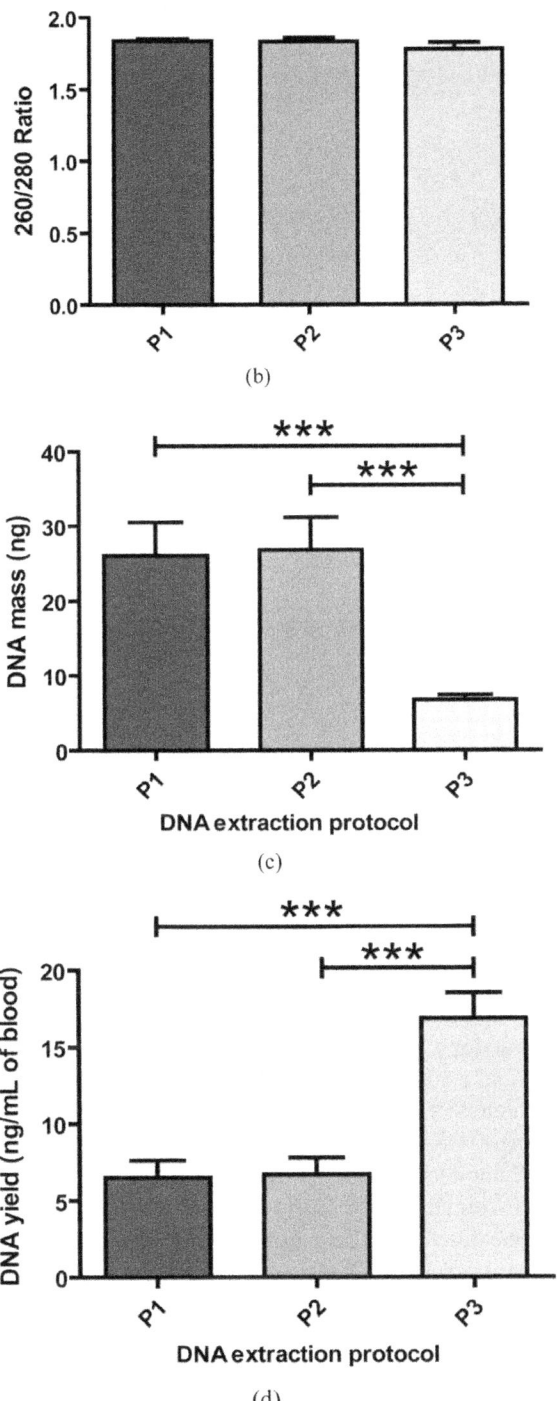

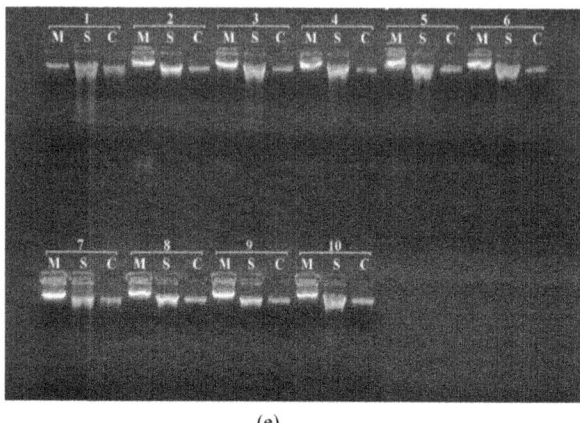

(e)

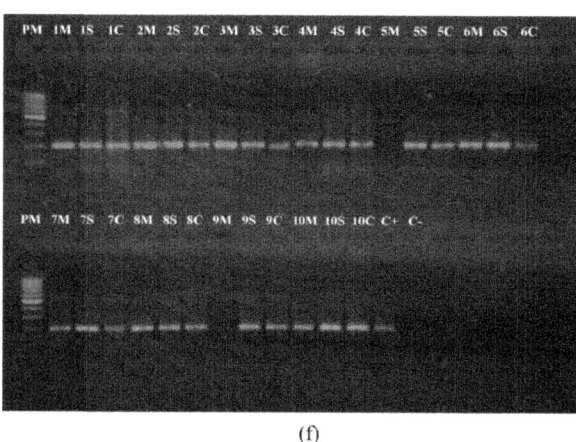

(f)

Figure 2. DNA concentration (a), purity (b), mass (c) and yield per milliliters of blood (d) obtained on each protocol of DNA extraction assay performed by the students (***$p < 0.0001$). (e) Agarose gel electrophoresis of genomic DNA obtained by three different DNA extractions. Samples (5 µL) were separated in 1% agarose gel for 45 minutes at 80 mA and stained with ethidium bromide. The numbers refer to the students DNAs and the letters to the extraction techniques used (M—Protocol 1- Miller, S—Protocol 2-Salazar, C—Protocol 3-silica column). (f) Agarose gel electrophoresis of PCR products. Samples (8 µL) were separated in 2% agarose gel for 45 minutes at 80 mA and stained with ethidium bromide. The numbers refer to the students DNAs and the letters to the extraction techniques used (PM—molecular weight, M—Protocol 1-Miller, S—Protocol 2-Salazar, C—Protocol 3-Silica column).

To avoid this, the students suggested that in the next course, the blood collection be carried out after the preparation of DNA extraction solutions (Table 1). The Protocol 2 seems to promote precipitation of fragmented DNA as it was detected apoptotic DNA fragments from apoptotic leukocytes (Figure 2(e)).

PCR Amplification

Polymerase chain reaction (PCR) is an important tool in molecular biology, genetics, forensics, and other fields. This technique is used to manipulate DNA for cloning and mutagenesis, and it can be used to detect genetic disorders, DNA fingerprinting, and identification of criminals from small samples of bodily fluids and hair (Sambrook, 1989). The presence of contaminants, DNA degradation or low quantity of the DNA could inhibit PCR reaction (Gryson, 2010).

The students of biomedicine and biological sciences were able to analyze their own blood in this course. PCR reactions using DNA samples obtained from protocol 2 and commercial kit were successfully performed on all samples, and in 80% of the samples extracted by protocol 1 (Figure 2(f)). According to the students' observation, the failure in amplifying the two DNA samples from protocol 1 was due to the students handling error since samples 5 M and 9 M were successfully amplified in a second reaction.

Assessment of Student Learning, Acceptance and Demand

Student learning was assessed by: instructor observation of the student laboratory work and a questionnaire presented in Table 2. The results of the answers of the 10 participants are presented in Figure 3.

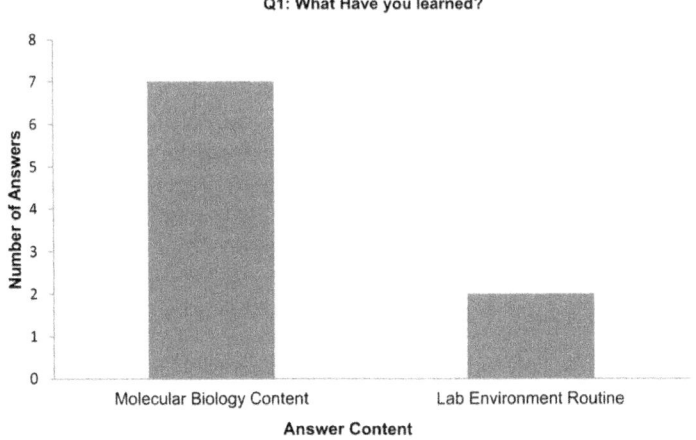

(a)

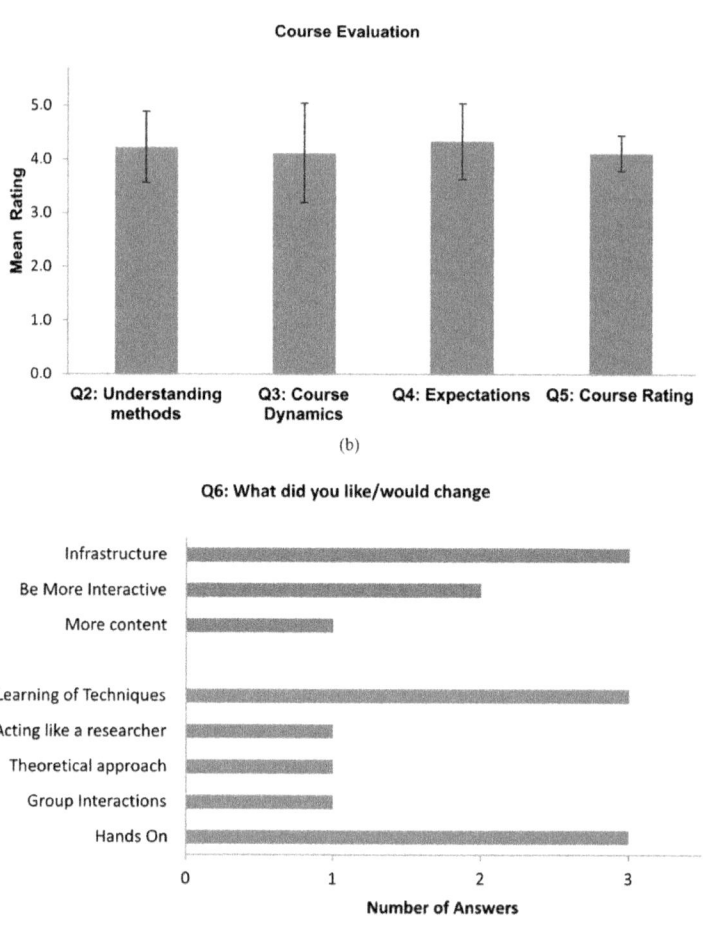

Figure 3. Evaluation of the summer course by the students of biomedicine and biological sciences. (a) Perception of Learning as inferred from the answers to the question 1 (Q1) of the questionnaire (n = 9 answers). (b) Mean Rating on a Lickert 1 - 5 Scale (1 = completely disagree, 5 = completely agree) to the understanding of content, dynamics of the course, reaching initial expectations and quality of course (results as mean ± SD, n = 10 answers). (c) Evaluation of the features that the students liked the most (blue bars), or would change (red bars) on the Summer Course (n = 9 answers).

According to the answers to the first question (Q1), most students felt that they learned molecular biology-related content (e.g. PCR, DNA isolation and amplification), indicating that the course somehow reached its main objective. In addition, some answers also stated that they learned practices pertaining to the daily routine of a laboratory (Figure 3(a)). This kind of statement reveals the potential of this dynamic as a practical course involving real equipment and laboratory techniques approximating the undergraduate students to their future professional environment.

The rates of the course quality and dynamics (1 - 5 Likert scale) determined by the students were all similar (~4) to both understanding of content and dynamics, reaching the initial expectations and course quality (Figure 3(b)). Overall, these results indicate a very good acceptance of the Course format, even though indicating that some changes could be performed to be totally approved by students, according to Q6.

Figure 3(c) shows what the students liked the most, or would prefer to change on the Course. The fact that they were able to learn the techniques by hands-on activities led most of them to enjoy the course, while at least one statement reported the simulation of the researchers' activities and group interactions. Once again, students enjoyed and identified the importance of the practical activities where they were in first-hand contact with professional "researcher laboratory" activities.

Regarding to the suggestions of changes (red bars on Figure 3), students asked for improvements on the infrastructure (3 students), more interactive activities (2 students) and more molecular biology content (1 student). In fact, one important concern in practical courses is the demand for adequate infrastructure, especially concerning the availability of individual equipments, to avoid adopting a "demonstrative" profile for these activities.

Nevertheless, in agreement with the good acceptance revealed in Figure 3(b), all students stated that they would like to attend other editions of the Summer Course, when answering the Question 7. When asking other 90 students from medicine and biology courses (47% biology and 53% medicine) that had had the class using fruit as DNA source about attending a summer course regarding molecular biology, many of them answered positively (62%) about this possibility (48% medicine and 52% biology), which pointed to the need of knowing more about this subject and possibly for more applied classes.

FINAL CONSIDERATIONS

According to different authors, there is an urgent need for different approaches with an applied perspective for teaching biotechnology and molecular biology. In this work, we create a dynamic and practical course that receives positive evaluation by the students of biological sciences and biomedicine. The students' reports reveal that the DNA extraction protocols evaluated here are simple and suitable for using PCR. The use of in-house procedures that are lower-cost methods and that can be performed in just one day (e.g. Protocol 2) allows the inclusion of this practical class in the disciplines of undergraduate courses of

Health and Biological Sciences areas. Moreover, another important feature of the parallel evaluation of three different methods during the summer course includes that the students are challenged with 3 different options, and they have to consider other parameters such as costs, time-consumption, availability of infrastructure, evaluation and comparison of results, quality assessment and other advanced issues usually not covered in classical courses. In this regard, students are able to conclude that even though the use of commercial kit shortens the experimental time, it is much more expensive than the in-house version, even though with similar or even better results. Such kind of stimuli may contribute to the potential development of critical thinking and problem-solving skills, both highly praised on cell and molecular biology curricula (Voet et al., 2003), and extremely related to the reality of the laboratory/work environment of these professionals.

ACKNOWLEDGEMENTS

We thank to FAPERJ, UFF-FOPESQ, CAPES, CNPq and CNPq/PIBIC for the fellowships and financial support.

REFERENCES

1. Aidar, M., & Line, S. R. (2007). A Simple and Cost-Effective Protocol for DNA Isolation from Buccal Epithelial Cells. Brazilian Dental Journal, 18, 148-152.
2. Azer, S. A., Guerrero, A. P., & Walsh, A. (2013). Enhancing Learning Approaches: Practical Tips for Students and Teachers. Medical Teacher, 35, 433-443.
3. Badyal, D. K., & Desai, C. (2014). Animal Use in Pharmacology Education and Research: The Changing Scenario. Indian Journal of Pharmacology, 46, 257-265.
4. Cardoso, C. V. P., & Almeida, A. E. C. C. (2014). Laboratory Animal: Biological Reagent or Living Being? Brazilian Journal of Medical and Biological Research, 47, 19-23.
5. Coil, D., Wenderoth, M. P., Cunningham, M., & Dirks, C. (2010). Teaching the Process of Science: Faculty Perceptions and an Effective Methodology. CBE-Life Sciences Education, 9, 524-535.
6. González, C. I., Ortiz, S., & Solari Colombian, A. (2010). Trypanosoma Cruzi Major Genotypes Circulating in Patients: Minicircle Homologies by Cross-Hybridization Analysis. International Journal for Parasitology, 40, 1685-1692.

7. Gryson N. (2010). Effect of Food Processing on plant DNA Degradation and PCR-Based GMO Analysis: A Review. Analytical and Bioanalytical Chemistry, 396, 2003-2022.
8. Hearn, R. P., & Arblaster, K. E. (2010). DNA Extraction Techniques for Use in Education. Biochemistry and Molecular Biology Education, 3, 161-166.
9. Marchese, A., Esposito, S., Coppo, E., Rossi, G. A., Tozzi, A., Romano, M., Da Dalt, L., Schito, G. C., & Principi, N. (2011). Detection of Streptococcus pneumoniae and Identification of Pneumococcal Serotypes by Real-Time Polymerase Chain Reaction Using Blood Samples from Italian Children ≤5 Years of Age with Community-Acquired Pneumonia. Microbial Drug Resistance, 20, 1-6.
10. Mello, F. C., Souto, F. J., Nabuco, L. C., Villela-Nogueira, C. A., Coelho, H. S., Franz, H. C., Saraiva, J. C., Virgolino, H. A., Motta-Castro, A. R., Melo, M. M., Martins, R. M., & Gomes, S. A. (2007). Hepatitis B Virus Genotypes Circulating in Brazil: Molecular Characterization of Genotype F Isolates. BMC Microbiology, 23, 103.
11. Miller, S. A., Dykes, D. D., & Polesky, H. F. (1988). A Simple Salting out Procedure for Extracting DNA from Human Nucleated Cells. Nucleic Acids Research, 16, 1215.http://dx.doi.org/10.1093/nar/16.3.1215
12. Miller, M. B. (1994). Practical DNA Technology in School. Journal of Biological Education, 28, 203-211. http://dx.doi.org/10.1080/00219266.1994.9655393
13. Mortarino, M., Garagiola, I., Lotta, L. A., Siboni, S. M., Semprini, A. E., & Peyvandi, F. (2011). Non-Invasive Tool for Foetal Sex Determination in Early Gestational Age. Haemophilia, 17, 952-956. http://dx.doi.org/10.1111/j.1365-2516.2011.02537.x
14. Nomura, M., Shigematsu, H., Li, L., Suzuki, M., Takahashi, T. et al. (2007). Polymorphisms, Mutations, and Amplification of the EGFR Gene in Non-Small Cell Lung Cancers. PLoS Medicine, 4, e125. http://dx.doi.org/10.1371/journal.pmed.0040125
15. Poh, J. J., & Gan, S. K. (2014). Comparison of Customized Spin-Column and Salt-Precipitation Finger-Prick Blood DNA Extraction. Bioscience Reports, 34, 629-634.http://dx.doi.org/10.1042/BSR20140105
16. Rogers, N. L., Cole, S. A., Lan, H. C., Crossa, A., & Demerath, E. W. (2007). New Saliva DNA Collection Method Compared to Buccal Cell Collection Techniques for Epidemiological Studies. American Journal of Human Biology,19, 319- 326.http://dx.doi.org/10.1002/ajhb.20586
17. Salazar, L. A., Hirata, M. H., Cavalli, S. A., Machado, M. O. et al. (1998).

Optimized Procedure for DNA Isolation from Fresh and Cryopreserved Clotted Human Blood Useful in Clinical Molecular Testing. Clinical Chemistry, 44, 1748-1450.

18. Sambrook, J., Fritsch, E. R., & Maniatis, T. (1989). Molecular Cloning: A Laboratory Manual (2nd ed.). Cold Spring Harbor, NY: Cold Spring Harbor Laboratory Press.

19. Schijman, A. G., Bisio, M., Orellana, L., Sued, M., Duffy, T., Mejia Jaramillo, A. M. et al. (2011). International Study to Evaluate PCR Methods for Detection of Trypanosoma cruzi DNA in Blood Samples from Chagas Disease Patients. PLoS Neglected Tropical Diseases, 5, e931. http://dx.doi.org/10.1371/journal.pntd.0000931

20. Slonczewski, J. L., & Marusak, R. (2004). A Response to BIO 2010: Transforming Undergraduate Education for Future Research Biologists, from the Perspective of the Biochemistry and Molecular Biology Major Program at Kenyon College. Biochemistry and Molecular Biology Education, 32, 151-155.http://dx.doi.org/10.1002/bmb.2004.494032030342

21. Sitaraman, R. (2012). From Bedside to Blackboard: The Benefits of Teaching Molecular Biology within a Medical Context. Perspectives in Biology and Medicine, 55, 461-466.http://dx.doi.org/10.1353/pbm.2012.0030

22. Takayama, T., Yamada, S., Watanabe, Y., Hirata, K., Nagai, A., Nakamura, I., Bunai, Y., & Ohya, I. (2001). Origin of DNA in Human Serum and Usefulness of Serum as a Material for DNA Typing. Legal Medicine, 3, 109-113. http://dx.doi.org/10.1016/S1344-6223(01)00018-9

23. Tsui, N. B. Y., Kadir, R. A., Chan, K. C. A., Chi, C., Mellars, G., Tuddenham, E. G., Leung, T. Y., Lau, T. K., Chiu, R. W. K., & Dennis Lo, Y. M. (2011). PCR Analysis of Maternal Plasma DNA Noninvasive Prenatal Diagnosis of Hemophilia by Microfluidics Digital. Blood, 117, 3684-3691. http://dx.doi.org/10.1182/blood-2010-10-310789

24. Tibell, L. A., & Rundgren, C. J. (2010). Educational Challenges of Molecular Life Science: Characteristics and Implications for Education and Research. CBE Life Sciences Education, 9, 25-33.

25. Voet, J. G., Bell, E., Boyer, R., Boyle, J., O'Leary, M., & Zimmerman, J. K. (2003). Recommended Curriculum for a Program in Biochemistry and Molecular Biology. Biochemistry and Molecular Biology Education, 31, 161-162.http://dx.doi.org/10.1002/bmb.2003.494031030223

Chapter 2

MOLECULAR CLONING AND EXPRESSION OF A FAMILY 6 CELLOBIOHYDROLASE GENE CBHII FROM PENICILLIUM FUNICULOSUM NCL1

Vanitha Chinnathambi[1], Meera Balasubramanium[2], Ramani Gurusamy[2], and Gunasekaran Paramasamy[2]

[1]Center for Marine Bioprospecting, AMET University Kanathur, Chennai, India

[2]Department of Genetics, Center for Excellence in Genomic Sciences, School of Biological Sciences, Madurai Kamaraj University, Madurai, India

ABSTRACT

Aim: Lignocelluloytic enzymes are the largest class of hydrolase enzyme which utilizes the plant biomass to produce renewable sources. Hence practices for larger production of these enzymes at lower cost received much attention for industrial use. Hence this paper deals with expression and purification of cellobiohydrolase gene from Penicillium funiculosum NCL1. Methods & Results: A cellobiohydrolase gene, cbhII of Penicillium funiculosum NCL1 was cloned and expressed in Pichia pastoris X33. Two exons of the cbhII gene were amplified separately and fused by overlap extension PCR. The fused product was cloned in yeast expression vector pPICZαA and expressed in P. pastoris under the control of the AOX1 promoter. P. pastoris transformants expressing recombinant cellobiohydrolase were selected on CMC agar plate and their ability to produce the cellobiohydrolase was evaluated in flask cultures. P. pastoris X33 (pPICbh6) efficiently secreted the recombinant cellobiohydrolase into the medium and produced the cellobiohydrolase activity (5 U/ml) after 96 h of growth. The recombinant cellobiohydrolase produced by P. pastoris (pPICBH6) showed maximum activity at pH 4.0 and temperature 50°C and higher specificity in hydrolysis of filter-paper.

INTRODUCTION

Lignocellulosic biomass is an important source of the renewable energy for production of bioethanol. Cellulose and hemicelluloses can be hydrolyzed into glucose. Cellulase is an enzyme complex capable of hydrolyzing cellulose into glucose molecules. The complete degradation of cellulose to glucose requires the action of at least three types of enzymes: Endo-β-1, 4-glucanase, Exo-β-1, 4-glucanase (cellobiohydrolase) and β-glucosidase [1]. Cellobiohydrolase is the essential component of the cellulase system to hydrolyze cellulose, consisting of both crystalline and amorphous cellulose. However, fungi are the most studied organisms because of their higher yields and capacities to produce complete cellulase complex. Recently, research has been focused on microbial cellulases due to the large scale production in the industries and their use in biomass degradation biomass for several industrial applications including biofuel production [2] [3]. The most abundant enzymes are two cellobiohydrolase, Cel7A and Cel6A, also called CBHI and CBHII, respectively. These are also the most efficient enzymes on highly crystalline cellulose [4]. Cellulases have proved to be commercially useful in the textile industry, substituting conventional stonewash methods (biostoning).The fungus Penicillium funiculosum is a filamentous fungus an efficient producer of cellulase [5] [6]. The P. funiculosum have been reported to produce high amount cellulases such as CMCase (13 - 15 U/ml), pNPGase (10 - 12 U/ml) and cellobiase (7.5 U/ml). Genes coding cellobiohydrolase GH6 have been cloned and characterized from a variety of fungal sources, including Trichoderma reesei [7], Chaetomium thermophilum [8], Irpex lacteus MC-2 [9], Chaetomium thermophilum [10]. This is the first report on the expression of cbhII from Penicillium funiculosum in pichia pastoris. In the present study, we report the cloning and expression of cbhII gene in pichia expression system.

MATERIALS AND METHODS

Strains, Reagents and Culture Media

Penicillium funiculosum NCL1 was obtained from National Chemical Laboratory, Pune, India. Stock cultures were kept on potato dextrose agar and subcultured monthly. P. funiculosumNCL1 spores were inoculated in Reese Basal medium as described by [11] at a final concentration of 10^8 spores/ml. Flasks were incubated in an orbital shaker (220 rpm) at 30°C for 96 h. The mycelia were recovered by filtration on a nylonfilter (30 μm spore) washed with 0.9% (w/v) NaCl and dried by pressing between two filter papers. Escherichia coli DH5α was used as a host for plasmid propagation. P. pastoris

X33 (Invitrogen, USA) was used as a host for expression of the recombinant cbhII.

PCR

PCRs were performed in a PTC-200 programmable thermal cycler (MJ Research, Massachusetts, USA) with one cycle of 94°C for 5 min followed by 35 cycles of denaturation (60 s at 94°C), annealing (60 s at 50°C - 60°C) and extension (60 s at 72°C), with a final extension of 72°C for 10 min. For analysis, 10 µl of reaction product was electrophoresed on a 1% agarose gel and stained with ethidium bromide (5 µg/ml). Primers used in this study are listed in Table 1.

Table 1. Oligonucleotide primers used in this study.

Primer name	Sequence (5'-3')	Length (bp)
ORFCBH6F	CCTAGCTAGCATGTTGC-GATATCTTTCCATCGTTG	35
ORFCBH6R	TCTTAGGCCG-GCCCTAGAC-CAAAGCTGGGTTGGCA	35
CBH6F	CCNGAYCGYGAYTGYG-CYGC	20
CBH6R	TKRTARTTGCYRTCRAAN-AGCA	22
Cbh6e12R	TACCAGGGATACATT-GTGCGTAGTAAGGGTT-TAGAGTGCTGCA	43
Cbh6e345F	ACGTGCAGCACTCTA-AACCCTTACTACGCA-CAATGTATCC	40
pPICBH6F	CCTAGCTAGCATGTTGC-GATATCTTTCCATCGTTG	35
pPICBH6R	CATGCGGCCGCTACAAA-CATTGAGAGTAGTAAGGGT	36

Cloning of CBHII Cellobiohydrolase

Genomic DNA of P. funiculosum NCL1 was isolated from mycelia by the method of Murray and Thompson. DNA from agarose gel was also extracted using Prefect prep gel extraction column (Eppendorf, Germany) according to the manufacturer's instructions. Exon I of cbhII gene was amplified with overlapping primer sets [Cbh6e345F-ORFCBH6R] and [ORFCBH6F-Cbh6E12R]. Both amplified exon I and exon II were purified and mixed in 1:1 molar ratio. These fragments were joined by the second round of PCR with no primers. Final round of PCR was made with ORF specific primers (ORFCBH6F and ORFCBH6R) specific for cbh6 gene. On the basis of the nucleotide sequence of cbh6 primers were designed to facilitate further cloning into pPICZαA vector. The fusion product cbh6 with its own signal sequence was digested with EcoRI and NotI and cloned into pPICZαA, the resulting plasmid pPICbh6 which was linearized by digestion with SalI to facilitate integration via homologous recombination at the AOX1 locus in P. pastoris X33 strain (Invitrogen, California, USA).

Transformation of P. pastoris

P. pastoris X33 was grown overnight (30°C at 250 rpm) in YPD broth (10 g/l yeast extract, 20 g/l peptone and 20 g/l dextrose) and transformed with the recombinant construct, according to the manufacturer's instructions (Invitrogen, California, USA). P. pastoris X33 (mut+) was transformed with linearized construct (pPICbh6) by electroporation [BTX (ECM399), Germany]. Transformed cells were selected on YPD agar medium Supplemented with 2 M sorbitol and zeocin (100 μg/ml).

Expression of Cellobiohydrolase in Recombinant P. pastoris

The yeast culture medium (BMG medium) consisted of (g/l) 13.4 g of YNB, 4×10^{-4} g of Biotin, 5 ml of glycerol and 100 mM phosphate buffer pH 6.0. P. pastoris was grown in BMG medium in an orbital shaker (250 rpm) at 30°C to an OD_{600} nm of 1.3 to 1.6. The cells were harvested by centrifugation 1500 × g for 10 min and the pellet was resuspended in 25 ml of BMMY medium (BMG medium in which glycerol was replaced by methanol (5 ml/l) and further supplemented with yeast extract 10 g/l and peptone 20 g/l in 250 ml of Erlenmeyer flasks kept at the similar conditions. Methanol (0.5%) was fed to the culture every 24 h for induction and samples were withdrawn at intervals. The cells were removed by centrifugation (10,000 × g for 10 min) and the supernatant was assayed for cellobiohydrolase activity.

Characterization of Cellobiohydrolase

To determine the specificity of the recombinant cellobiohydrolase, the enzyme was assayed with different substrates (Sigma, St Louis, MO, USA) such as avicel, CMC, pNPG and xylan. Reducing sugars were measured by the 3, 5-dinitrosalicylic acid method with glucose and as the standard. The activities were compared with the highest one obtained for any of these substrates. The effect of pH on the cellobiohydrolase activity was determined by measuring the relative activity using sodium acetate (pH 5.0 - 5.5), sodium phosphate (pH 6.0 - 8.5) and sodium carbonate (pH 9.0 - 9.5) buffers. The maximum activity was considered as 100%, and used as reference in determining relative activities at different pH values. The effect of temperature on the reaction rate was determined by performing the standard reaction at different temperatures in the range of 30°C - 90°C.

The stability of the cellobiohydrolase as a function of pH was determined by measuring the residual cellobiohydrolase activity after incubation of the enzyme for 1 h at different pH at 30 °C. The relative activity was expressed considering the activity before incubation as 100%. Thermostability of cellobiohydrolase was deter- mined by incubating the enzyme extract at different temperatures (30°C - 90°C) for 30 min. Protein was esti- mated according to the method of [12] using bovine serum albumin as the standard.

Sequence Analysis

Nucleotide and deduced amino acid sequences were analyzed with the sequence analysis tools. Signal peptide sequence was analyzed by SignalP 3.0 server (http://www.cbs.dtu.dk/services/SignalP). Related sequences were obtained from the databases using the software BLAST. Phylogenetic analyses were performed in MEGA 2.1, using the minimum evolution (ME) approach. GENSCAN online tool (www.genes.mit.edu/GENSCAN.html) was used for identification of gene features such as exons and splice sites in genomic DNA. Bio Edit (version 7.0.4.1) was used for sequence editing and analysis.

RESULTS

Majority of cellobiohydrolase belong to the two families, GH6 and GH7 glycosyl hydrolases. To clone cbhII cellobiohydrolase, degenerate primers were designed (Table 1) based on conserved catalytic domain regions of known GH6 cellobiohydrolases of Penicillium species. Using P. funiculosum NCL1 genomic DNA as template, PCR was performed with degenerate primers Cbh6F & Cbh6R and to amplify cbhII cellobiohydrolase gene. The resulted amplicon of length 900 bp was cloned and sequenced. The partial cellobiohydrolase gene sequences were BLAST with the sequences of other fungal cellobiohydrolase in the GenBank and EMBL databases sequences. From the retrieved sequences, the ORF for family GH6 (cbhII) cellobiohydrolases were predicted and amplified. Sequencing of cellobiohydrolase cbhII gene from P. funiculosum NCL1 showed 100% identity towards sequences of the A. cellulolyticus Y-94

Acc2 gene. Based on comparison of this sequence with A. cellulolyticus Y-94 Acc2 gene cbh6 of P. funiculosum was found to have 5 introns.

Further, to amplify the full length cDNA of cbhII gene sequence, RT-PCR was done. First strand cDNA was synthesized using total RNA isolated from avicel grown culture. Using this first strand cDNA as template, PCR was performed with primers (ORFCBH6F & ORFCBH6R) specific for cbh6 gene. An expected amplicon of 1.4 kb cDNA was obtained. The product was cloned and sequenced. The cDNA sequence showed identity towards cellobiohydrolase (Acc2) of A. cellulolyticus Y-94. The alignment of cDNA and genomic DNA sequence of P. funiculosum cbh6 showed that the second intron was not spliced. As the cbh6 cDNA of P. funiculosum has one intron, attempt was made to fuse the exons of cbh6 gene. Primers were designed to amplify the two exons separately. Fusion primers were designed to fuse the exons. First round of PCR were performed with overlapping primer sets [Cbh6e345F-ORFCBH6R] and [ORFCBH6F-Cbh6E12R] which resulted in amplicons of 1.2 kb and 200 bp respectively. The second round of overlap extentsion PCR was performed without exon-primers. Final round of PCR was made with ORF specific primers (ORFCBH6F and ORFCBH6R) specific for cbh6 gene which resulted in an expected amplicon of 1.4 kb. The cbhII gene sequence was submitted to GenBank accession number (FJ000002).

Nucleotide Sequence Analysis of the cbhII Gene

The cbhII gene consisted of an ORF 1700 bp. Five introns with an average length of 55 bp were identified in the cbhII gene based on alignment with other cellobiohydrolase genes belonging to family GH6 and also with predicted cDNA obtained using GENSCAN software. The size of introns resembled those of other filamentous fungal introns and varied between 55 and 76 bp. Conserved domain search (RPSBLAST) analysis confirmed the presence of catalytic domain of GH6 cellobiohydrolase. Deduced aminoacid sequence of cbhII was compared to other characterized GH6 cellobiohydrolases. The sequence of GH6 exhibited maximum identity with A. cellulolytics Y-94, 100% and followed by A. terreus 46%, N. fischeri NRRL 18.156%, A. fumigatus Af293 cellobiohydrolase 51%, S. sclerotiorum −33%, A. niger −39%, T. ressei (cbhII) 43%, T. viride cellobiohydrolase (cbhII) −43%, Hyocrea koningi −38%. The protein comprises of 529 residues including the signal peptide (Figure 1). The putative signal peptide was predicted with SignalP 3.0 Server and the most likely cleavage site is between position 20th and 21st aminoacid. The mature protein comprises of 422 residues, with a calculated molecular mass of 47.7 kDa and a theoretical pI of 4.62. Phylogenetic analysis revealed that cbhII from P. funiculosum formed a separate cluster and is closely related A.

cellulolyticus Y-94, P. marneffei ATCC 18224, Talaromyces stipitatus ATCC 10500 (Figure 2).

Expression of cbhII of P. funiculosum NCL1 in P. pastoris

The cbhII gene without signal sequence was amplified using primers pPICbh6F and pPICbh6R from the genomic DNA of P. funiculosum NCL and cloned at EcoRI and NotI site of expression vector pPICZαA in order to make in frame fusion of the cbhII gene with the α-factor secretion signal in the vector. Thus generated recombinant plasmid pPICbh6 was confirmed with PCR and restriction analyses. The linearized recombinant plasmid pPICbh6 was transferred into P. pastoris X33 for integration by single crossover recombination. A total of 25 transformants were randomly picked and grown on YPD agar containing 1% avicel and zeocin (100 mg/ml). After a 72 h of growth, three recombinant colonies which produced large zone of hydrolysis were chosen for further study.

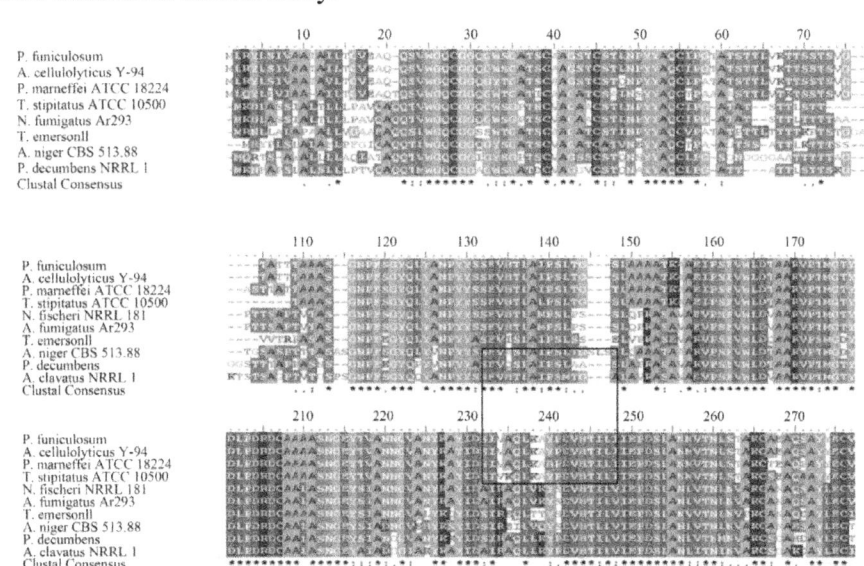

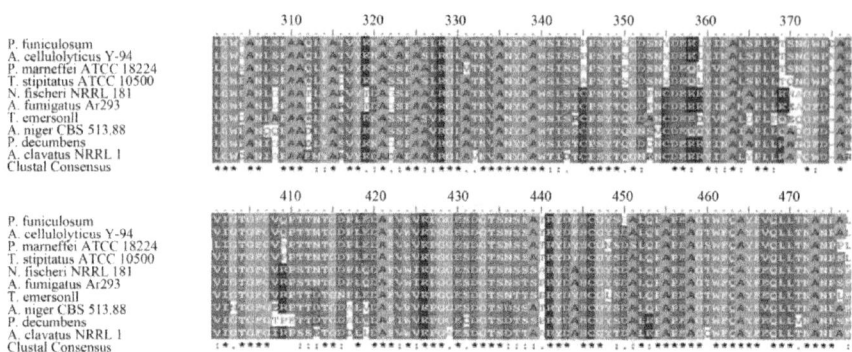

Figure 1. Comparison of GH family 6 cellobiohydrolase with cbh6 of P.funiculosum. Multiple alignments were obtained using ClustalX in Phylip format by the neighbor-joining method. The amino acid residues of are numbered; conserved regions are boxed; "*"—invariant residues; ":"—similar amino acids; "."—less similar amino acids. A. cellulolytics Y-94, A. terreus, N. fischeri NRRL 181, A. fumigatus Af293 cellobiohydrolase, S. sclerotiorum, A. niger, T. ressei (cbhII), T. viride cellobiohydrolase (cbhII), Hyocrea koningi.

The selected clones were grown at 30°C and induced with methanol in BMMY medium. The cellobiohydrolase production was determined every 24 h. The recombinant yeast strains produced 2 - 6 U/ml of cellobiohydrolase in the culture medium while one of the recombinant strains, PIC2 produced maximum cellobiohydrolase activity of 5.5 U/ml (Figure 3). The kinetics of cellobiohydrolase production by the transformant is shown in (Figure 4). The specific activity of the recombinant cellobiohydrolase was 0.8 U/mg. SDS PAGE analysis of culture supernatant showed expected protein band of 47.7 kDa (Figure 5).

Characterization of Recombinant Cellobiohydrolase

The activity profile of cellobiohydrolase in terms of pH was determined in different buffers (pH3.0 - 9.0).

The pH optimum for P. pastoris (pPICBH6) cellobiohydrolase was 5.6 and it showed 85% - 95% of relative activity at pH 3.0 to 5.2. When preincubated for 30 min in different buffers pH 3.0 to 5.0 at 30°C. CBHII exhibited maximum activity at pH 5.0 (Figure 6(a)). When preincubated in different buffers for 1 h at 30°C, cellobiohydro- lase was 100% stable but the enzyme lost its activity at pH 8. Thus, the optimal pH for both cellobiohydrolases P. pastoris (pPICbh6) was 3.0 - 6.0 without much loss of activity. The cbhII higher specificity in hydrolysis of filter-paper.

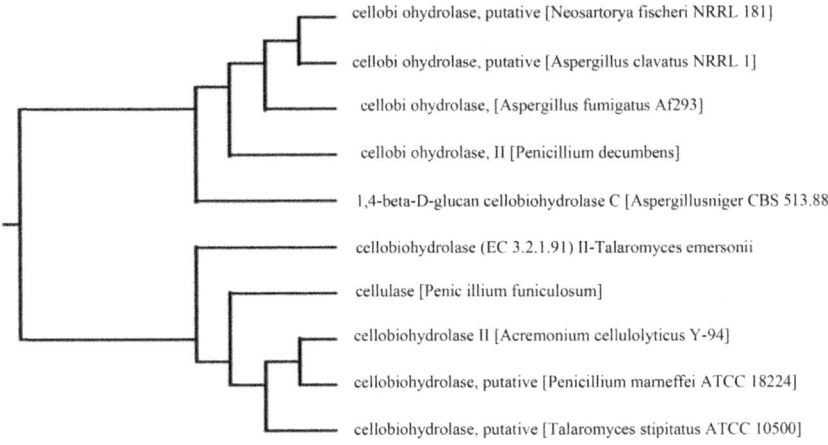

Figure 2. Phylogenetic analysis of cbh6 from P. funiculosum NCL1. The sequences of GH6 cellobiohydrolases were used to access their phylogenetic relationship. The phylogram was generated with ClustalW in FASTA format using the neighbor-joining method and displayed in tree view.

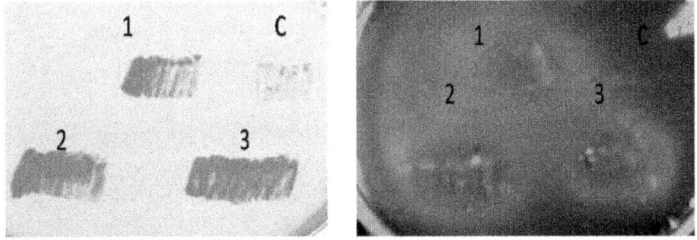

Figure 3. Expression of cellobiohydrolase in P. pastoris (pPICbh6). P. pastoris transformants were patched on minimal methanol agar medium with 0.5% Avicel. Methanol 100 ml was added to the lid of the inverted plate every 24 h. After 72 h, the plate was overlaid with 1% Congo red. Then destained with 1 M NaCl after 20 min and the zone of clearance was observed (C—control, 1—pPIC1, 2—pPIC2, 3—pPIC3).

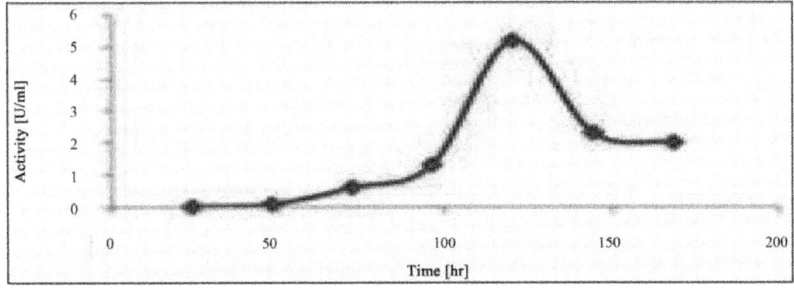

Figure 4. Kinetics of cellobiohydrolase production by recombinant P. pastoris

(pPICbh6). P. pastoris (pPICbh6) was grown in minimal medium with 0.5% methanol. Samples were taken at 24 h intervals and assayed for cellobiohydrolase activity in the culture supernatant.

DISCUSSION

Partial cellobiohydrolase genes of P. funiculosum NCL1 900 bp for GH6 gene were amplified by PCR using degenerate primers. The partial cellobiohydrolase gene sequences were BLAST with the sequences of other fun- gal cellobiohydrolase in the GenBank and EMBL databases sequences. From the retrieved sequences, the ORF for family GH6 (cbhII) cellobiohydrolases were predicted and amplified. Sequenced cellobiohydrolase (cbhII) gene from P. funiculosum NCL1 showed 100% identity towards sequences of the A. cellulolyticus Y-94

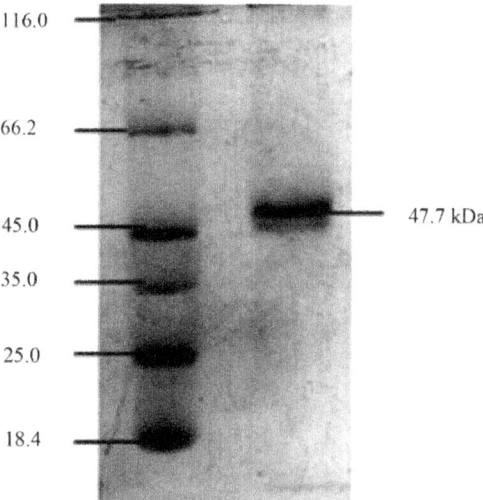

Figure 5. SDS-PAGE of the purified cellobiohydrolase of P. pastoris (pPICbh6). Purified cellobiohydrolase was resolved on a 12% SDS-PAGE and stained with coomassie brilliant blue R-250. Lane 1—purified cellobiohydrolase (coomassie blue stained), Lane M—Molecular weight marker (116.6 - 18.4 kDa).

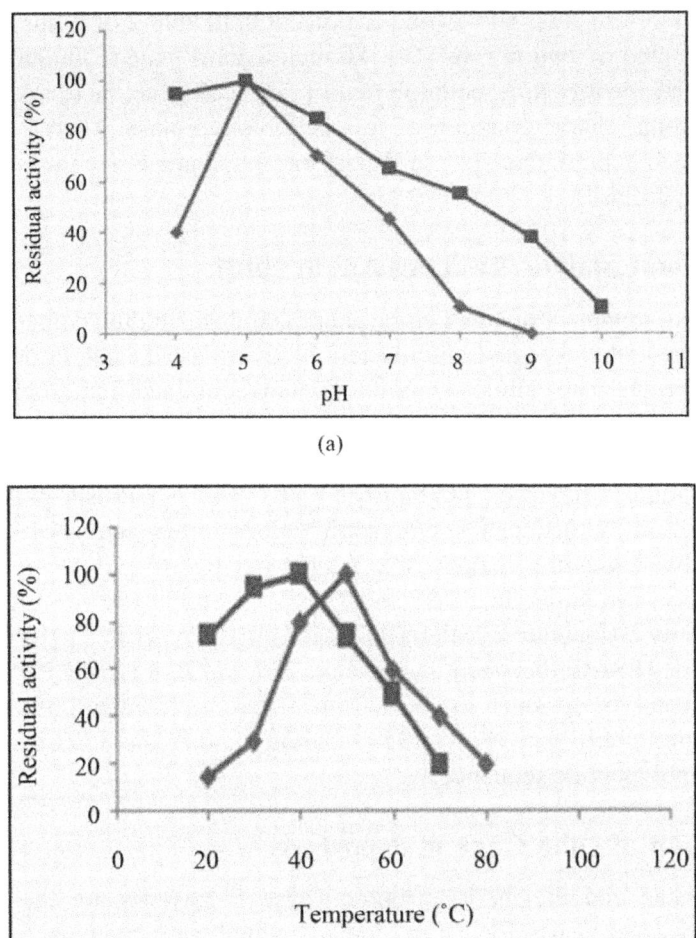

Figure 6. Effect of pH and temperature on the activity of purified cellobiohydrolase from recombinant P. pastoris (pPICbh6). (a) Cellobiohydrolase activity was determined in different buffers (pH 3.0 - 10.0) at 50°C. The stability of the enzyme at different pH levels was determined by incubating the enzyme in different buffers for 1 h at 50°C and the residual activity was measured at pH 4.8, 50°C; (b) The cellobiohydrolase activity was measured at different temperatures (30°C - 80°C) at pH 4.8 the purified enzyme was preincubated at different temperatures for 30 min and the residual activity was measured at pH 4.8 and 50°C. Relative activity was calculated by considering the maximum activity as 100%.

Acc2 gene. The cbh6 gene consists of 1700 bp including five introns 71, 59, 71, 76 and 56 bp with consensus 5'and 3' intron splice sites. Toda et al. reported that cbhII from Irpex lacteus MC-2 is interrupted by eight small introns

53 - 59 bp in size [13]. The coding regions of both cbhI.2 of P. chrysosporium are interrupted by two introns [14]. Though a cbhII gene (FJ000002) showed about 100% identity to A. cellulolyticus Y-94 Acc2 gene, there is no study on the functional characterization of this cellobiohydrolase. Therefore, attempt was made to express this gene in P. pastoris and characterize the recombinant cellobiohydrolase.

Conserved Catalytic Residues Are in cbhII

CbhII multi domain structure composed of cellulose binding domain (CBD), a Ser/Thr/Pro-rich linker, and a family cellobiohydrolase II catalytic domain from the N-terminus. The cellulose binding domain contained four aromatic amino acid residues (Trp25, Phe33, Tyr51, and Tyr52) and four cysteines (Cys28, Cys39, Cys45, Cys55), which are highly conserved among carbohydrate binding modules classified into family 1 type CBM [14]. Prosite pattern search performed on the deduced T. emersonii Cel6A protein sequence indicated a GH family 6 signature 1 pattern V-x-Y-x (2)-P-x-R-D-C-[GSAF]-x (2)-[GSA] (2)-x-G between amino acid 181 and 197 (VVYDLPDRDCAAAASNG) and a GH family 6 signature 2 pattern [LIVM YA]-[LIVA]-[LIVT]-[LIV]- E-P-D-[SAL]-[LI]-[PSAG] between amino acid 229 and 238 (ILVIEPDSLA). The putative catalytic residues in cbhII from T. reesei [15] are conserved in the T. emersonii cbhII gene at AA 235 (aspartate) and AA 413 (aspartate) in the deduced polypeptide sequence.

Expression of cbh6 Gene in P. pastoris

S. cerevisiae and the methylotrophic yeast, P. pastoris are the two most frequently used yeasts for expression of recombinant proteins [16] [17]. A great advantage of P. pastoris is that a low number of endogenous proteins that are secreted into the culture media. However, in most cases, greater success has been obtained with other signal peptides, such as the secretion signal sequence from the S. cerevisiae α-factor prepro peptide [18]. Family GH6 cellobiohydrolase genes from various fungi Irpex lacteus MC 2 [19], Trichoderma parceramosum [20] were cloned and expressed in P. pastoris. To express the cbh6 gene of Penicillium funiculosum NCL 1 in Pichia pastoris, the exons were fused using PCR based method. The fused product was successfully expressed extracellularly with its own signal sequence. In similar manner a family GH6 cellobiohydrolase from Irpex lacteus MC-2 was expressed extracellular with P. pastoris system but with a leader peptide of the a-mating factor from Saccharomyces cerevisiae fused to the N-terminus of the translated product makes it possible to secrete recombinant proteins into the medium. Cellobiohydrolase gene (cbhII) from was successfully expressed in the Saccharomyces cerevisiae α-factor secretion signal [21]. cbhI and cbhII were isolated from Trichoderma viride AS3.3711 and T. viride CICC 13038 were successfully expressed in Saccharomyces cerevisiae H158 and secreted in yeast S. cerevisiae and they were secreted in an active form. The recombinant cellobiohydrolase produced by P. pastoris is active at pH 4.0 like other family GH6 cellobiohydrolases [22]. The recombinant enzyme exhibited optimum catalytic activity at pH 4.0 and 50 degrees C respectively. It was thermostable at 50°C and retained 50% of its original activity after 30 min at 70°C. The stability of the cbhII was observed till pH 6. The cbhII showed maximum activity at 50°C and stable up to 60°C. As the family GH6 cellobiohydrolases are true cellobiohydrolase, the recombinant cellobiohydrolase also showed highest specificity towards filter paper and not on other substrates, CMC and pNPG.

The importance of lignocellulose biotechnology and the many potential applications of lignocellulose enzymes in various industries such as chemicals, fuel, food, brewery and wine, animal-feed, pulp-and-paper, textile and agriculture are well documented [23] -[25]. Since the recombinant cbhII produced by P. pastoris found greater stability through wide range of pH. This property makes it useful for bios toning in the textile and detergent industries.

CONCLUSION

A gene encoding CBHII was successfully isolated from Penicillium funiculosum using RT-PCR. This gene contains the conserved catalytic amino acid residues that have been reported in other fungal cellulases. However, it does not contain a CBD. The enzyme was successfully expressed and secreted in yeast P. pastoris and it was secreted in an active form. Enzymatic properties of CBHII were also determined. Due to its temperature stability and pH stabilities, our cellobiohydrolase might be useful for textiles, pulp and paper industries.

REFERENCES

1. Beguin, P. and Lemaire, M. (1996) The Cellulosome: An Exocellular, Multiprotein Complex Specialized in Cellulose Degradation. Critical Reviews in Biochemistry and Molecular Biology, 31, 201-236. http://dx.doi.org/10.3109/10409239609106584
2. Beg, Q.K., Kapoor, M., Mahajan, L. and Hoondal, G.S. (2001) Microbial Xylanases and Their Industrial Applications: A Review. Applied Microbiology and Biotechnology, 56, 326-338. http://dx.doi.org/10.1007/s002530100704

3. Collins, T., Gerday, C. and Feller, G. (2005) Xylanases, Xylanase Families and Extremophilic Xylanases. FEMS Microbiology Reviews, 29, 3-23. http://dx.doi.org/10.1016/j.femsre.2004.06.005

4. Koivula, A.L., Ruohonen, G., Wohlfahrt, T., Reinikainen, T.T., Teeri, K., Piens, M., Claeyssens, M., Weber, A., Vasella, D., Becker, M.L., Sinnott, J.Y., Zou, G.J. and Kleywegt, M. (2002) Cel6A from Trichoderma reesei: The Roles of Aspartic Acids D221 and D175. Journal of the American Chemical Society, 124, 10015-10024. http://dx.doi.org/10.1021/ja012659q

5. Saloheimo, M., Lehtovaara, P., Penttilä, M., Teeri, T.T. and Stahlberg, J. (1988) EGIII, a New Endoglucanase from Trichoderma reesei: and the Characterization of Both Gene and Enzyme. Gene, 63, 11-21. http://dx.doi.org/10.1016/0378-1119(88)90541-0

6. Mandel, M. and Reese, E.T. (1960) Sophorose as an Inducer of Cellulase in Trichoderma viride. Journal of Bacteriology, 79, 816.

7. Godbole, S., Decker, S.R., Nieves, R.A., Adney, W.S., Vinzant, T.B., Baker, J.O., Thomas, S.R. and Himmel, M.E. (1999) Cloning and Expression of Trichoderma reesei Cellobiohydrolase I in Pichia pastoris. Biotechnology Progress, 15, 828-833. http://dx.doi.org/10.1021/bp9901116

8. Li, Y.L., Li, H., Li, A.N. and Li, D.C. (2009) Cloning of a Gene Encoding Thermostable Cellobiohydrolase from the Thermophilic Fungus Chaetomium thermophilum and Its Expression in Pichia pastoris. Journal of Applied Microbiology, 106, 1867-1875. http://dx.doi.org/10.1111/j.1365-2672.2009.04171.x

9. Toda, H., Nagahata, N. and Amano, Y. (2008) Gene Cloning of Cellobiohydrolase II from the White Rot Fungus Irpex Lacteus MC-2 and Its Expression in Pichia pastoris. Bioscience, Biotechnology, and Biochemistry, 72, 3142-3147. http://dx.doi.org/10.1271/bbb.80316

10. Liu, S.A., Li, D.C., E, S.J. and Zhang, Y. (2005) Cloning and Expressing of Cellulase Gene (cbh2) from the Thermophilic Fungal Chaetomium thermophilum CT2. Chinese Journal of Biotechnology, 21, 892-899. (In Chinese)

11. Medve, J., Stahlberg, J. and Tjerneld, F. (1994) Adsorption and Synergism of Cellobiohydrolase I and II of Trichoderma reesei during Hydrolysis of Microcrystalline Cellulose. Biotechnology and Bioengineering, 44, 1064-1073. http://dx.doi.org/10.1002/bit.260440907

12. Lowry, O.H., Rosebrough, N.J., Farr, A.L. and Randall, R.J. (1951) Protein Measurement with the Folin Phenol Reagent. Journal of

Biological Chemistry, 193, 265-275.

13. Toda, H., Nagahata, N., Amano, Y., Nozaki, K., Kanda, T., Okazaki, M. and Shimosaka, M. (2008) Gene Cloning of Cellobiohydrolase II from the White Rot Fungus Irpex lacteus MC-2 and Its Expression in Pichia pastoris. Bioscience, Biotechnology and Biochemistry, 72, 3142-3147. http://dx.doi.org/10.1271/bbb.80316

14. Sims, P.F.G., Soares-Felipe, M.S., Wang, Q., Gent, M.E., Tempelaars, C.M. and Broda, P. (1994) Differential Expression of Multiple Exocellobiohydrolase I-Like Genes in the Lignin-Degrading Fungus Phanerochaete chrysosporium. Molecular Microbiology, 12, 209-216. http://dx.doi.org/10.1111/j.1365-2958.1994.tb01010.x

15. Rouvinen, J., Bergfors, T., Teeri, T., Knowles, J.K.C. and Jones, T.A. (1990) Three-Dimensional Structure of Cellobiohydrolase II from Trichoderma reesei. Science, 249, 380-386. http://dx.doi.org/10.1126/science.2377893

16. Cereghino, J.L. and Cregg, J.M. (2000) Heterologous Protein Expression in the Methylotrophic Yeast Pichia pastoris. FEMS Microbiology Reviews, 24, 45-66. http://dx.doi.org/10.1111/j.1574-6976.2000.tb00532.x

17. Cregg, J. and Tolstorukov, I. (2012) P. Pastoris ADH Promoter and Use there of to Direct Expression of Proteins. US Patent No. 8222386.

18. Kallas, A. (2006) Heterologous Expression, Characterization and Applications of Carbohydrate Active Enzymes and Binding Modules. Royal Institute of Technology, School of Biotechnology, Dept. Wood Biotechnol., Stockholm.

19. Zahri, S., Zamani, M.R., Motallebi, M. and Sadeghi, M. (2005) Cloning and Characterization of cbhII Gene from Trichoderma parceramosum and Its Expression in Pichia pastoris. Iranian Journal of Biotechnology, 3, 204-215.

20. Liu, Y., Yoshida, M., Kurakata, Y., Miyazaki, T., Igarashi, K., Samejima, M., Fukuda, K., Nishikawa, A. and Tonozuka, T. (2010) Crystal Structure of a Glycoside Hydrolase Family 6 Enzyme, CcCel6C, a Cellulase Constitutively Produced by Coprinopsis cinerea. FEBS Journal, 277, 1532-1542. http://dx.doi.org/10.1111/j.1742-4658.2010.07582.x

21. Gavrilescu, M. and Chisti, Y. (2005) Biotechnology—A Sustainable Alternative for Chemical Industry. Biotechnology Advances, 23, 471-499. http://dx.doi.org/10.1016/j.biotechadv.2005.03.004

22. Howard, R.L., Abotsi, E., Jansen van Rensburg, E.L. and Howard, S. (2003) Lignocellulose Biotechnology: Issues of Bioconversion and Enzyme Production. African Journal of Biotechnology, 2, 602-619.

http://dx.doi.org/10.5897/AJB2003.000-1115

23. Bhat, M.K. (2000) Research Review Paper. Cellulases and Related Enzymes in Biotechnology. Biotechnology Advances, 18, 355-383. http://dx.doi.org/10.1016/S0734-9750(00)00041-0

24. Sun, J. and Cheng, J. (2002) Hydrolysis of Lignocellulose Materials for Ethanol Production: A Review. Bioresource Technology, 83, 1-11. http://dx.doi.org/10.1016/S0960-8524(01)00212-7

25. Hong, J., Tamaki, H., Yamamoto, K. and Kumagai, H. (2003) Cloning of a Gene Encoding Thermostable Cellobiohydrolase from Thermoascus Aurantiacus and Its Expression in Yeast. Applied Microbiology and Biotechnology, 63, 42-50. http://dx.doi.org/10.1007/s00253-003-1379-3

Chapter 3

MOLECULAR CLONING OF A CHITINASE GENE FROM THE OVOTESTIS OF KURODA'S SEA HARE APLYSIA KURODAI

Gaku Matsunaga, Syuuji Karasuda, Ryo Nishino, Hideto Fukushima, and Masahiro Matsumiya

Department of Marine Science and Resources, College of Bioresource Sciences, Nihon University, Kanagawa, Japan

ABSTRACT

In this study, we report that we successfully cloned and sequenced a chitinase gene from the ovotestis of Kuroda's sea hare Aplysia kurodai. By using reverse transcription-polymerase chain reaction (RT-PCR) and a system for the 5' and 3' rapid amplification of cDNA ends, we obtained a 1352 bp chitinase gene (AkChi) from the ovotestis of A. kurodai. AkChi contains a 1263 bp open reading frame that encodes 421 amino acids. The domain structure predicted from the deduced amino acid sequence was an N-terminal signal peptide and a catalytic domain of glycoside hydrolase (GH) family 18 chitinase. A comparative analysis of the deduced amino acid sequences of AkChi with those of the acidic mammalian chitinase of the California sea hare Aplysia californica revealed the highest homology at 83%. The purified chitinase from the ovotestis was digested by trypsin, and 119 residues of digested peptides were consistent with the deduced amino acid sequence of AkChi. We used RT-PCR to evaluate the expression of AkChi in various tissues of A. kurodai, and we observed that AkChi was expressed only in the ovotestis. A phylogenetic tree analysis, performed using the amino acid sequences of AkChi and known GH family 18 chitinases, showed that AkChi was separated from the molluscan chitinases with a chitin binding domain. To our knowledge, this is the first study demonstrating the cDNA cloning of an ovotestis chitinase from a sea hare.

INTRODUCTION

Chitin, a major molecular constituent of the exoskeleton of insects and crustaceans, is a straight-chain homopolymer of β-1,4-linked N-acetyl-D-glucosamine units [1]-[3]. Chitinases (EC 3.2.1.14) are enzymes that randomly hydrolyze the β-1,4 glycosidic bonds of chitin [4]. They have been found in various organisms, and they play important physiological roles in functions such as attack, defense, morphological changes, and digestion [5] [6].

The characterization and cDNA cloning of chitinases from several fishes have been reported [7]-[9]. The stomach chitinases of fish have been identified and are classified into two groups, acidic fish chitinase-1 (AFCase-1) and acidic fish chitinase-2 (AFCase-2) based on the differences in their primary structure and the activity toward short substrates [8]. Chitinases from molluscs play important physiological roles in the digestion of food [10] [11], attacking crustaceans [12], and shell formation [13] [14]. However, reports on the distribution, characterization, and cDNA cloning of molluscan chitinases are limited [10]-[16]. In this study, we were using the Kuroda's sea hare, Aplysia kurodai. A. kurodai is a kind of herbivorous gastropoda seen in the vicinity of the coast from April to June. In addition, this creature was allowed to degenerate shells despite the shellfish. In a previous study, we detected chitinase activity in the ovotestis and egg of A. kurodai [16], whereas lysozyme activity (antibacterial enzyme activity) was not detected in all of the organs [16]. We also reported the purification and properties of a chitinase from the ovotestis of A. kurodai [16]. Together the results indicated that the physiological role of this chitinase was as a defense against nematodes and fungus which had chitin in the body wall as a structural component [16].

In the present study, we cloned the cDNA encoding chitinase from the ovotestis of A. kurodai and determined the primary structure of the chitinase.

MATERIALS AND METHODS

Materials

Kuroda's sea hare Aplysia kurodai and laid egg were captured from the tide pools of Shimoda Bay (Shizuoka, Japan) in June.

Cloning of the Chitinase cDNA from A. Kurodai

The sequences of all primers are presented in Table 1. Total RNA was extracted from the ovotestis of A. kurodai using ISOGEN II reagent (Nippon Gene, Tokyo) according to the manufacturer's instructions. First-strand cDNA was synthesized using 500 ng of total RNA and oligo dT primers with Prime Script Reverse Transcriptase (Takara Bio, Shiga, Japan) according to the manufacturer's instructions. Six degenerate primers were designed for the reverse transcriptase-polymerase chain reaction (RT-PCR) from conserved sequences of molluscan chitinase, including those from California sea hare (Aplysia californica; GenBank: XM_005112601), triangle sail mussel (Hyriopsis cumingii; GenBank: JN582038), Pacific oyster (Crassostrea gigas; GenBank: AJ971239), Hawaiian bobtail squid (Euprymna scolopes; GenBank: KF015222), and golden cuttlefish (Sepia esculenta; GenBank: AB986212).

The first PCR was performed using A. kurodai cDNA as a template and P1 and P2 as primers (Figure 1). The PCR parameters were as follows: 94°C for 2 min, followed by 30 cycles of 94°C for 30 s, 55°C for 30 s, and 72°C for 30 s. Nested PCR was performed using the products of the first PCR as templates and P3, P4, P5, and P6 as primers, with the same PCR parameters as described above. The nucleotide sequence analysis of the RT-PCR amplified chitinase cDNA fragments from the ovotestis of A. kurodai detected one nucleotide sequence (AkChi).

For the 3' rapid amplification of cDNA ends (RACE), we designed primers specific to AkChi (i.e., P7, P8, and P9, respectively; Table 1) based on the detected sequences. We amplified cDNA fragments encoding the 3' region of AkChi using A. kurodai cDNA as the template and the primer pairs P7 and 3R, P8 and 3R, and P9 and 3R (Figure 1). The PCR parameters were as follows: 94°C for 2 min, followed by 30 cycles of 94°C for 30 s, 56°C for 30 s, and 72°C for 30 s. For 5' RACE, specific primers (P10, P11, and P12 for AkChi; Table 1) were designed based on the nucleotide sequences obtained from RT-PCR. cDNA fragments encoding the 5' regions of AkChi were amplified using PCR. The first PCR was performed using the newly synthesized first-strand cDNA as a template and the primer pairs P10 and P11 for AkChi. Nested PCR was performed using the first PCR products as templates and the primer pairs P10 and P12 for AkChi. The PCR parameters were as follows: 94°C for 1 min, followed by 30 cycles of 94°C for 30 s, 49°C for 30 s, and 72°C for 30 s.

Table 1. Primers used for PCR, RACE, and tissue-specific expression.

Primer	Sequence (5' → 3')	Purpose
P1*	TNGCNGCNTTYGARTGGAAYGA	Primary PCR
P2*	CATNCCNSWRAARTCRTCRTTRTC	Primary PCR
P3*	GGNGGNTGGAAYATGGG	Primary PCR
P4*	ACCCAYTGRTTNCCNARNACNA	Primary PCR

P5*	GNAAYTTYGAYGGNYTNGA	Primary PCR
P6*	TTDATCATYTCRCANACYT-CRTARTA	Primary PCR
P7	GCCGGATACGAAGTGGAC	3' RACE
P8	GGAACTTAACGAGTACTT	3' RACE
P9	GACAGACGAGAGCGACTCTGGTCG	3' RACE
3R	CTGTGAATGCTGCGACTACGAT	3' RACE
P10	CACAATGACGTTGCAAG	5' RACE, Full-length PCR
P11	ATGGCCTGGGCTCATTTT	5' RACE
P12	TTATCCTCTGGAGGGCT	5' RACE
P13	CACGTTATGATTGCGAC	Full-length PCR
P14	TCTGCTGCTGTGAGTGCTGGCAAGG	tissue-specific expression
P15	GCATTTCGCACACCTCGTAGTAAGA	tissue-specific expression
β-actin-a*	GAYAAYGGNWSNGGNATGTG	tissue-specific expression
β-actin-b*	TCRAACATDATYTGNGTCAT	tissue-specific expression

Note: *Degenerate primers.

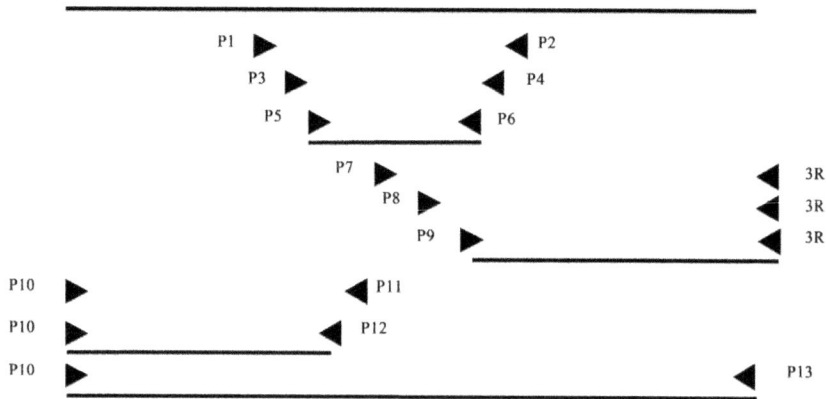

Figure 1. Schematic representation of the cDNA structure of AkChi and location of the primers. Arrowheads indicate the primers, and lines between the arrowheads indicate the amplified cDNA fragments.

The nucleotide sequences of cDNA fragments containing a full-length open reading frame (ORF) were confirmed by PCR using specific primers (P10 and P13 for AkChi; Table 1) and Platinum Pfx DNA Polymerase (Invitrogen, Carlsbad, CA).

Nucleotide Sequence Analysis

The RT-PCR, 3' RACE, and 5' RACE amplification products, and the full-length amplification products were subcloned into pGEM-T Easy Vector (Promega, Madison, WI), according to the manufacturer's instructions. Sequences were determined on an ABI PRISM 3130 genetic analyzer (Applied Biosystems, Foster City, CA) using a Big Dye Terminator v3.1 cycle sequencing kit (Applied Biosystems).

Amino Acid Sequence of the Peptide of the Purified Chitinase from the Ovotestis of A. kurodai

A chitinase from the ovotestis of A. kurodai was purified as described [16]. The purified chitinase was subjected to sodium dodecyl sulfate-polyacrylamide gel electrophoresis (SDS-PAGE) and stained with AE-1360 EzStain Silver (ATTO, Tokyo). A gel slice was cut into small pieces and destained by destaining solution (15 mM $K_3[Fe(CN)_6]$, 50 mM $Na_2S_2O_3$). Destained gel pieces were trypsinized as described in the manual of In-Gel Tryptic Digestion Kit manual (Thermo Scientific, Waltham, MA). The peptide mixtures thus obtained were subjected to a nano-scale liquid chromatography-electrospray ionization-tandem mass spectrometry (nanoLC- ESI-MS/MS) analysis using a Q Exactive mass spectrometer (Thermo Scientific) equipped with a captive spray ionization source (Michrom Bioresources, Auburn, CA) and an Advance UHPLC System (Michrom Bioresources).

Tissue-Specific Expression of AkChi

Total RNA was prepared from the ovotestis, egg, skin, gill, crop, anterior gizzard, and posterior gizzard as described in the cloning methods section (2.2) above. First-strand cDNA was pre-cloned from the RNA isolated from each tissue and egg as described in the RT-PCR section (2.2) above. For tissue-specific expression, we designed primers specific to AkChi (P14 and P15, respectively; Table 1) based on the detected sequences. AkChi was amplified using the first-strand cDNA as template and the primer pairs P14 and P15 (Table 1). The PCR parameters were as follows: 94°C for 1 min, followed by 35 cycles of 94°C for 30 s, 62°C for 30 s, and 72°C for 30 s. To determine the amount of total RNA in each tissue, we amplified β-actin mRNA fragments using specific primer pairs (Table 1).

Phylogenetic Tree Analysis of AkChi

In order to classify the chitinase from the ovotestis of A. kurodai among the GH family 18 chitinases, we constructed a phylogenetic tree based on the enzyme precursor sequences by the neighbor-joining method, using the ClustalW program (http://www.genome.jp/tools/clustalw/). A bacterial chitinase (GenBank: X03657) was used as the out group.

RESULTS AND DISCUSSION

Cloning of A. kurodai Chitinase cDNA

The structure of AkChi and the location of primer sequences are schematically represented in Figure 1. The internal sequence of the cDNA of A. kurodai ovotestis chitinase was amplified by RT-PCR using degenerate primers (from P1 to P6, respectively; Table 1); an amplified product of approx. 400 bp was obtained. The product was sequenced, and 86% homology with the acidic mammalian chitinase of A. californica was confirmed (accession no. XM_005112601). Because the sequence was part of ovotestis chitinase cDNA from A. kurodai, we used it to design gene-specific primers for 3' and 5' RACE (from P7 to P12; Table 1). An amplified product of approx. 430 bp was obtained by 3' RACE, and its sequence contained a stop codon. An amplified product of approx. 520 bp was also obtained by 5' RACE; its sequence contained a start codon. Based on these results, we designed full-length primers (P10 and P13; Table 1) to incorporate these start and stop codons. cDNA was amplified using the primers and the amplified product was sequenced.

The full-length cDNA of A. kurodai ovotestis chitinase (AkChi) was 1352 bp in length and contained an ORF of 1263 bp encoding 421 amino acids (Figure 2). The size of ORF of AkChi was smaller than it from H. cumingii [14], 1962 bp encoding 653 amino acids. A poly-A sequence in eukaryotes was detected at the 3' end of AkChi. AkChi, which encodes A. kurodai ovotestis chitinase, has been registered in the database of the DNA Data Bank of Japan (DDBJ) (accession no. LC085435). We compared the deduced amino acid sequence of AkChi with that of other organisms using BLAST, and the highest homology, 83%, was confirmed with the acidic mammalian chitinase of A. californica (accession no. XM_005112601).

Figure 2. cDNA and deduced amino acid sequences of AkChi. Underlined sequences show matching with the peptide fragments of the purified and tripsinized enzyme (coverage: 35.39%, 119 residues).

Figure 3 compares amino acid sequences from AkChi and some other known molluscan chitinases (A. californica, H. cumingii, C. gigas, E. scolopes, and S. esculenta). The deduced amino acid sequence of AkChi was shown to have a structure of the GH family 18 chitinase, with an N-terminal signal peptide and a GH 18 catalytic domain. The catalytic domain also contained an active site that is a conserved sequence of GH family 18 chitinases (Figure 3). Though the chitinase of H. cumingii [14] and E. scolopes [15] had two chitin binding domains (CBDs) and the chitinase of S. esculenta had one CBD, AkChi lacked a CBD. It was reported that fish chitinases have one CBD [8]. This result suggests that the structure of molluscan chitinase is diverse compared to the fish chitinases.

Amino Acid Sequence of the Chitinase

We analyzed the sequences of the peptide fragments obtained by the tryptic treatment of the purified chitinase from the ovotestis of A. kurodai [16] were analyzed and compared them to the deduced amino acid sequence of AkChi. The obtained sequences from peptide fragments were consistent with the deduced amino acid sequence of AkChi (coverage: 35.39%, 119 residues) (Figure 2). This result suggests that AkChi is a gene coding the purified enzyme. In addition, trypsin is cut the C-terminal side of lysine and arginine. In this result, it was confirmed that the trypsin is working properly in the all of cleavage site.

Tissue-Specific Expression of AkChi

We investigated the tissue-specific expression of AkChi in A. kurodai by RT-PCR using the housekeeping β-ac- tin gene as a control (Figure 4). It is reported that fish express chitinase to the digestive organs for digestion of chitin from food [17]. The expression profile results indicated that AkChi was present only in the ovotestis. We previously detected chitinase activity in the ovotestis and egg from A. kurodai [16], whereas lysozyme activity (antibacterial enzyme activity) was not detected in any of the organs [16]. A. kurodai has to prey on seaweed.

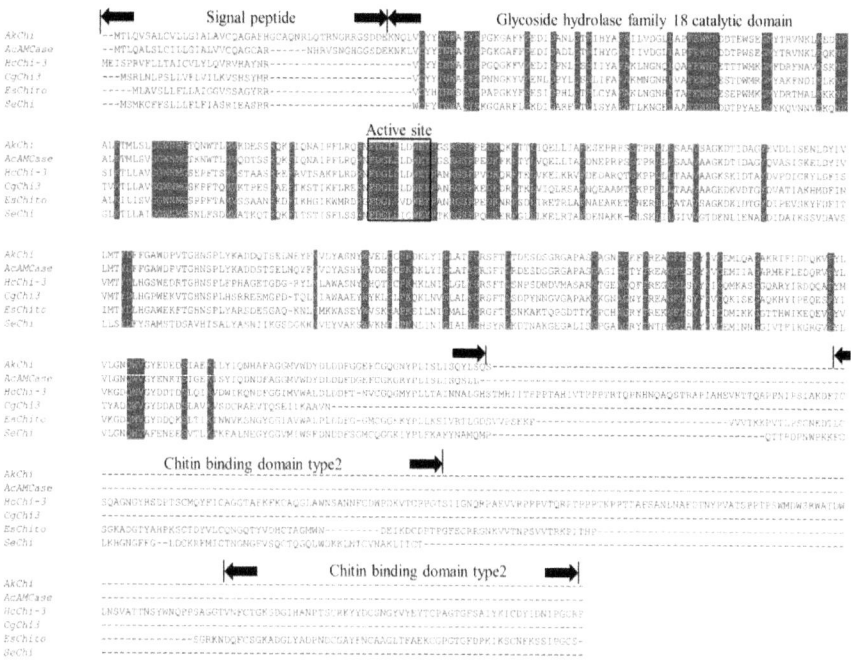

Figure 3. Multiple alignment of duduced amino acid sequences of A. kurodai chitinase (AkChi) with Aplysia californica acidic mammalian chitinase (AcAMCase), Hy-

riopsis cumingii chitinase-3 (HcChi-3), Crassostrea gigas Chit3 protein A (CgChi3), Euprymna scolopes chitotriosidase (EsChito), and Sepia esculenta chitinase (SeChi). GenBank accession nos.: AcAMCase, XM_005112601; HcChi-3, JN582038; CgChi3, AJ971239; EsChito, KF015222; SeChi, AB986212. Matched sequences are shown in black.

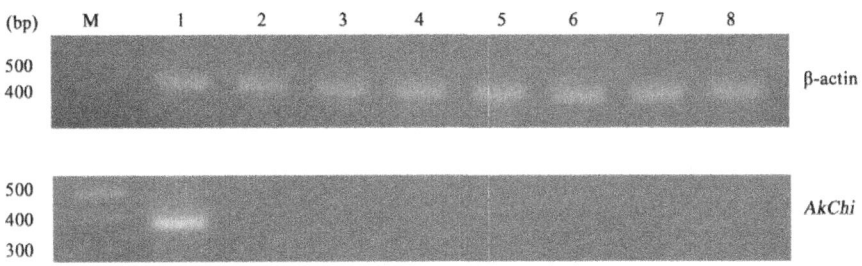

Figure 4. Expression profiles of AkChi and β-actin mRNA in tissue using RT-PCR. M, markers; 1, ovotestis; 2, egg; 3, skin; 4, gill; 5, buccal mass; 6, crop; 7, anterior gizzard; 8, posterior gizzard.

Thus, A. kurodai is not necessary chitinase in digestion and attack of food as squid [10] [11] and octopus [12], respectively. In addition, there is not necessary to shell formation because it does not even have shells. These results suggest that the role of this chitinase is as a defense against nematodes and fungus which have chitin in the body wall as a structural component.

Phylogenetic Tree Analysis of AkChi

We performed a phylogenetic tree analysis of GH family 18 chitinases and AkChi (Figure 5). Acidic mammalian chitinases (AMCases) have been found in the stomach of mammals.

44 Molecular Biotechnology: Principles and Practices

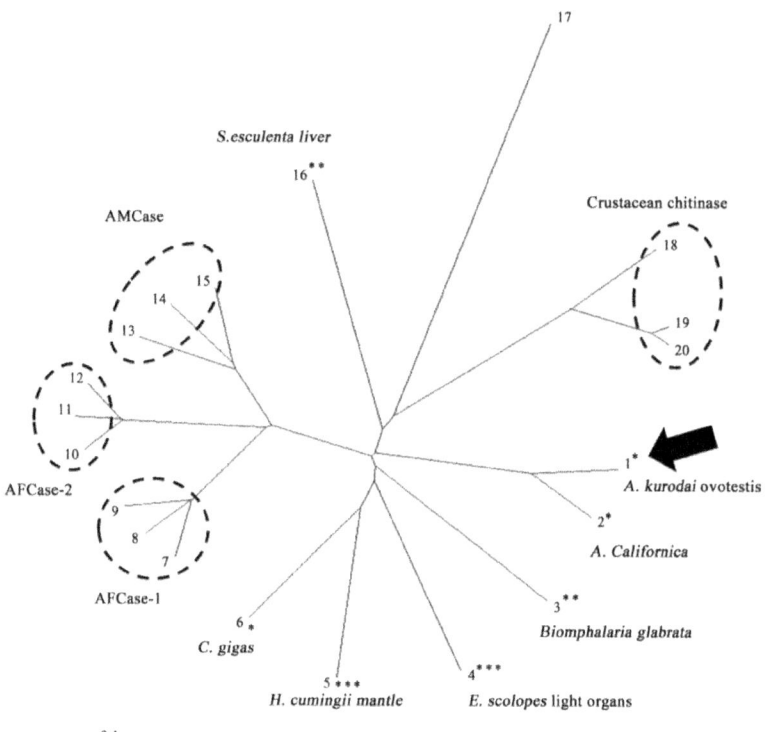

No.	Species	Genbank accession number
1	Aplysia kurodai (chitinase)	LC085435
2	Aplysia californica (acidic mammalian chitinase)	XM_005112601
3	Biomphalaria glabrata (chitinase-3-like protein 1)	XP_013090777
4	Euprymna scolopes (chitotriosidase)	KF015222
5	Hyriopsis cumingii (chitinase-3)	JN582038
6	Crassostrea gigas (Chit3 protein)	AJ971239
7	Epinephelus coioides (chitinase1)	AB686658
8	Sebastiscus marmoratus (chitinase1)	FJ169895
9	Parapristipoma trilineatum (chitinase1)	AB642677
10	Epinephelus coioides (chitinase2)	FJ169894
11	Parapristipoma trilineatum (chitinase2)	AB642678
12	Sebastiscus marmoratus (chitinase2)	AB686659
13	Bos Taurus (chitin binding protein b04)	AB051629

14	Mus musculus (acidic chitinase)	EF094027
15	Homo sapiens (acidic mammalian chitinase)	AF290004
16	Sepia esculenta (chitinase)	AB986212
17	Serratia marcescens (chiA protein precursor)	X03657
18	Portunus trituberculatus (chitinase1)	AB874469
19	Portunus trituberculatus (chitinase2)	AB890123
20	Scylla serrata (chitinase)	EU402970

Figure 5. Phylogenetic tree analysis of chitinase amino acid sequence by the neighbor-joining method of the program Clustal W. A bacterial chitinase, Serratia marcescens chitinase, was used as the out group. The scale bar indicates the substitution rate per residue. The arrow shows AkChi obtained in the present study. * Molluscan chitinase without a CBD; ** Molluscan chitinase with one CBD; *** Molluscan chitinase with two CBDs.

Two chitinase groups with different structures and activity toward short substrates, AFCase-1 and AFCase-2, have been found in the stomach of fish [8]. Crustacean showed a chitinase group [18]. In contrast, molluscan chitinases did not show clear chitinase groups. The reason for this might be the differences in the chitinase domain structure that are due to the presence or absence of a CBD and the number of CBDs. We previously detected chitinase activity in the ovotestis and oviduct from the Walking sea hare Aplysia juliana [16]. If the success in cloning the chitinase from A. juliana, it will be conceivable to form a group of sea hare chitinase.

CONCLUSION

The cDNA of the ovotestis chitinase obtained from A. kurodai contained a 1263 bp open reading frame with a coding potential for 421 amino acid peptides. AkChi had the structural motifs of GH family 18 chitinase, but it did not have chitin binding domain. This study is the first report of the cloning of chitinase from the ovotestis of a sea hare.

ACKNOWLEDGEMENTS

This work was supported in part by a Grant-in-Aid for Scientific Research (C) (no. 25450309) and a College of Bioresource Science, Nihon-University Grant (2015).

REFERENCES

1. Khandeparker, L., Gaonkar, C.C. and Desai, D.V. (2013) Degradation of Barnacle Nauplii: Implications to Chitin Regulation in the Marine Environment. Biologia, 68, 696-706. http://dx.doi.org/10.2478/s11756-013-0202-6

2. Arbia, W., Arbia, L., Adour, L. and Amrane, A. (2013) Chitin Extraction from Crustacean Shells Using Biological Methods—A Review. Food Technology and Biotechnology, 51, 12-25.

3. Kramer, K.J. and Koga, D. (1986) Insect Chitin: Physical State, Synthesis, Degradation and Metabolic Regulation. Insect Biochemistry, 16, 851-877. http://dx.doi.org/10.1016/0020-1790(86)90059-4

4. Umemoto, N., Ohnuma, T., Mizuhara, M., Sato, H., Skriver, K. and Fukamizo, T. (2013) Introduction of a Tryptophan Side Chain into Subsite +1 Enhances Transglycosylation Activity of a GH-18 Chitinase from Arabidopsis thaliana, AtChiC. Glycobiology, 23, 81-90. http://dx.doi.org/10.1093/glycob/cws125

5. Gooday, G.W. (1999) Aggressive and Defensive Roles for Chitinases, Chitin and Chitinases. Cellular and Molecular Life Sciences, 87, 157-169.

6. Henrissat, B. (1991) A Classification of Glycosyl Hydrolases Based on Amino Acid Sequence Similarities. Biochemical Journal, 280, 309-316. http://dx.doi.org/10.1042/bj2800309

7. Ikeda, M., Miyauchi, K. and Matsumiya M. (2012) Purification and Characterization of a 56 kDa Chitinase Isozyme (PaChiB) from the Stomach of the Silver Croaker, Pennahia argentatus. Bioscience, Biotechnology, and Biochemistry, 76, 971-979. http://dx.doi.org/10.1271/bbb.110989

8. Ikeda, M., Kondo, Y. and Matsumiya, M. (2013) Purification, Characterization, and Molecular Cloning of Chitinases from the Stomach of the Threeline Grunt Parapristipoma trilineatum. Process Biochemistry, 48, 1324-1334. http://dx.doi.org/10.1016/j.procbio.2013.06.016

9. Laribi-Habchi, H., Dziril, M., Badis, A., Mouhoub, S. and Mameri, N. (2012) Purification and Characterization of a Highly Thermostable Chitinase from the Stomach of the Red Scorpionfish Scorpaena scrofa with Bioinsecticidal Activity toward Cowpea Weevil Callosobruchus maculates (Coleoptera: bruchidae). Bioscience, Biotechnology, and Biochemistry, 76, 1733-1740. http://dx.doi.org/10.1271/bbb.120344

10. Nishino, R., Suyama, A., Ikeda, M., Kakizaki, H. and Matsumiya, M.

(2014) Purification and Characterization of a Liver Chitinase from Golden Cuttlefish, Sepia esculenta. Journal of Chitin and Chitosan Science, 2, 238-243. http://dx.doi.org/10.1166/jcc.2014.1065

11. Matsumiya, M., Miyauchi, K. and Mochizuki, A. (2002) Characterization of 38 kDa and 42 kDa Chitinase Isozymes from the Liver of Japanese Common Squid Todarodes pacificus. Fisheries Science, 68, 603-609. http://dx.doi.org/10.1046/j.1444-2906.2002.00467.x

12. Ogino, T., Tabata, T., Ikeda, M., Kakizaki, H. and Matsumiya, M. (2014) Purification of a Chitinase from the Posterior Salivary Gland of Common Octopus Octopus vulgaris and Its Properties. Journal of Chitin and Chitosan Science, 2, 135-142. http://dx.doi.org/10.1166/jcc.2014.1049

13. Zhang, G., Fang, X., Guo, X., Li, L., Luo, R., Xu, F., Yang, P., Zhang, L., Wang, X., Qi, H., Xiong, Z., Que, H., Xie, Y., Holland, P.W.H., Paps, J., Zhu, Y., Wu, F., Chen, Y., Wang, J., Peng, C., Meng, J., Yang, L., Liu, J., Wen, B., Zhang, N., Huang, Z., Zhu, Q., Feng, Y., Mount, A., Hedgecock, D., Xu, Z., Liu, Y., Domazet-Loso, T., Du, Y., Sun, X., Zhang, S., Liu, B., Cheng, P., Jiang, X., Li, J., Fan, D., Wang, W., Fu, W., Wang, T., Wang, B., Zhang, J., Peng, Z., Li, Y., Li, N., Wang, J., Chen, M., He, Y., Tan, F., Song, X., Zheng, Q., Huang, R., Yang, H., Du, X., Chen, L., Yang, M., Gaffney, P.M., Wang, S., Luo, L., She, Z., Ming, Y., Huang, W., Zhang, S., Huang, B., Zhang, Y., Qu, T., Ni, P., Miao, G., Wang, J., Wang, Q., Steinberg, C.E.W., Wang, H., Li, N., Qian, L., Zhang, G., Li, Y., Yang, H., Liu, X., Wang, J., Yin, Y. and Wang, J. (2012) The Oyster Genome Reveals Stress Adaptation and Complexity of Shell Formation. Nature, 490, 49-54. http://dx.doi.org/10.1038/nature11413

14. Wang, G.-L., Xu, B., Bai, Z.-Y. and Li, J.-L. (2012) Two Chitin Metabolic Enzyme Genes from Hyriopsis cumingii: Cloning, Characterization, and Potential Functions. Genetics and Molecular Research, 11, 4539-4551. http://dx.doi.org/10.4238/2012.October.15.4

15. Kremer, N., Philipp, E.E.R., Carpentier, MC., Brennan, C.A., Kraemer, L., Altura, M.A., Augustin, R., H&aumlsler, R., Heath-Heckman, E.A.C., Peyer, S.M., Schwartzman, J., Rader, B., Ruby, E.G., Rosenstiel, P. and McFall-Ngai, M.J. (2013) Initial Symbiont Contact Orchestrates Host-Organ-Wide Transcriptional Changes that Prime Tissue Colonization. Cell Host & Microbe, 14, 183-194. http://dx.doi.org/10.1016/j.chom.2013.07.006

16. Karasuda, S., Ikeda, M., Miyauchi, K. and Matsumiya, M. (2011) Existence and Physiological Role of Chitinase in the Gonad of Two Species of Sea Hare, Kuroda's Sea Hare Aplysia kurodai and Walking Sea Hare Aplysia Juliana. Proceedings of the 9th Asia-Pacific Chitin and

Chitosan Symposium, Vietnam, 3-6 August 2011, 169-172.

17. Kakizaki, H., Ikeda, M., Fukushima, H. and Masahiro, M. (2015) Distribution of Chitinolytic Enzymes in the Organs and cDNA Cloning of Chitinase Isozymes from the Stomach of Two Species of Fish, Chub Mackerel (Scomber japonicus) and Silver Croaker (Pennahia argentata). Open Journal of Marine Sciences, 5, 398-411. http://dx.doi.org/10.4236/ojms.2015.54032

18. Fujitani, N., Hasegawa, H., Kakizaki, H., Ikeda, M. and Masahiro, M. (2014) Molecular Cloning of Multiple Chitinase Genes in Swimming Crab Portunus trituberculatus. Journal of Chitin and Chitosan Science, 2, 149-156. http://dx.doi.org/10.1166/jcc.2014.1046

Chapter 4

WHOLE-GENOME MOLECULAR HAPLOTYPING OF SINGLE CELLS

H Christina Fan[1], Jianbin Wang[1], Anastasia Potanina[2], and Stephen R Quake[1,3]

[1]Department of Bioengineering, Stanford University, Stanford, California, USA
[2]Howard Hughes Medical Institute, Stanford University, Stanford, California, USA
[3]Department of Applied Physics, Stanford University, Stanford, California, USA

ABSTRACT

Conventional experimental methods of studying the human genome are limited by the inability to independently study the combination of alleles, or haplotype, on each of the homologous copies of the chromosomes. We developed a microfluidic device capable of separating and amplifying homologous copies of each chromosome from a single human metaphase cell. Single-nucleotide polymorphism (SNP) array analysis of amplified DNA enabled us to achieve completely deterministic, whole-genome, personal haplotypes of four individuals, including a HapMap trio with European ancestry (CEU) and an unrelated European individual. The phases of alleles were determined at ~99.8% accuracy for up to ~96% of all assayed SNPs. We demonstrate several practical applications, including direct observation of recombination events in a family trio, deterministic phasing of deletions in individuals and direct measurement of the human leukocyte antigen haplotypes of an individual. Our approach has potential applications in personal genomics, single-cell genomics and statistical genetics.

INTRODUCTION

The sequencing of the human reference genome and the development of high-throughput short-read sequencing technologies have enabled partial decoding of an increasing number of individual human genomes[1, 2, 3, 4, 5, 6, 7]. However, all of these 'personal genomes' are incomplete, and should essentially be regarded as rough draft genomes. Although they all suffer from imperfections, such as gaps, miscalled bases and difficulties in determining large-scale structural variation, they are missing fundamental information of the unique haploid structure of homologous chromosomes. Haplotypes, the combinations of alleles at multiple loci along a single chromosome, are difficult to measure with current technologies but are an essential feature of the genome. A simple example of how the lack of this information limits the interpretation of existing genomes is to consider an individual having two mutations in a certain gene. If both mutations are on the same allele, then this individual would have one normal (that is, putatively functional) version of the gene and one mutated version. If the mutations are on different alleles, this individual would have two mutated versions and no normal version of the protein. In the absence of haplotype information, it is impossible to distinguish between these two cases.

Knowledge of complete haplotypes of individuals (personal haplotypes) would therefore be useful in personalized medicine. Notably, several studies have linked specific haplotypes to drug response and to resistance or susceptibility to diseases. A well-known example is the association of human leukocyte antigen (HLA) haplotypes with autoimmune diseases and clinical outcomes in transplantations[8, 9, 10]. Haplotypes within the apolipoprotein gene cluster may influence plasma triglyceride concentrations and the risk toward atherosclerosis[11]. Research suggests that a specific β-globin locus haplotype is associated with better prognosis of sickle cell disease[12], and other studies have linked haplotypes in the matrix metalloproteinase gene cluster with cancer development[13]. Haplotypes are also important in pharmacogenomics, an example being the association of β-2 adrenergic receptor to responses to drug treatment of asthma[14]. Deterministic haplotyping may greatly increase the power of genome-wide association studies in finding candidate genes associated with common but complex traits. It will also contribute to the understanding of population genetics and historical human migrations and the study of *cis*-acting regulation in gene expression.

Direct experimental determination of the haplotypes of an individual is challenging. The International HapMap Consortium has performed extensive SNP genotyping on different human populations, and by using family trios and statistical methods, has been able to catalog commonly occurring haplotype blocks in the human populations. However, in the best cases, when members of a family trio are analyzed, this approach leads to errors in resolving haplotype at approximately every ~3–8 megabases, and in the most general case, when an individual genome analyzed in the absence of family information, errors occur every 300 kilobases[15, 16]. In the context of personalized genomics and medicine, the approaches used in the HapMap project have limited applicability, as materials from family members are not always available and computational approaches using statistical models have inherent statistical uncertainty and are limited to regions with strong linkage disequilibrium. Mate-pair shotgun genome sequencing has been demonstrated to achieve partial haplotype reconstruction of an individual but the haplotype blocks have limited sizes[5, 17]. Other techniques have been demonstrated, including PCR in various forms[18, 19, 20, 21, 22], atomic force microscopy with carbon nanotubes[23], fosmid/cosmid cloning[24] and hybridization of probes to single DNA molecules[25]. Weaknesses of these methods include the inability to phase SNPs (that is, determine their relative arrangement on homologous chromosomes) more than tens of kilobases apart and/or a limitation in the number of markers that could be phased in a single assay. Whole-genome haplotyping can in principle be achieved by chromosome microdissection[26] or by the construction of somatic cell hybrids[27]. Yet the former is time-consuming and expensive, and the latter requires specialized and expensive equipment. So far, direct whole-genome haplotyping has not been accomplished for any individual. Here we address these issues using microfluidics.

RESULTS

Single-Cell Chromosome Separation and Amplification

We developed an approach termed direct deterministic phasing (DDP) in which the intact chromosomes from a single cell are dispersed and amplified on a microfluidic device (Fig. 1). The device consists of a cell-sorting region, where a single metaphase cell is identified microscopically and captured from a cell suspension; a chromosome release region, where metaphase chromosomes are released by protease digestion of the cytoplasm; a chromosome partitioning region, where the chromosome suspension is randomly separated into 48 partitions of a long narrow channel; an amplification region, where isolated chromosomes are individually amplified by multiple strand displacement amplification; and a product retrieval region, where amplified products are collected. The products are recovered independently, thus allowing direct genetic interrogation and genome-wide determination of haplotypes without the need for family information or statistical inference.

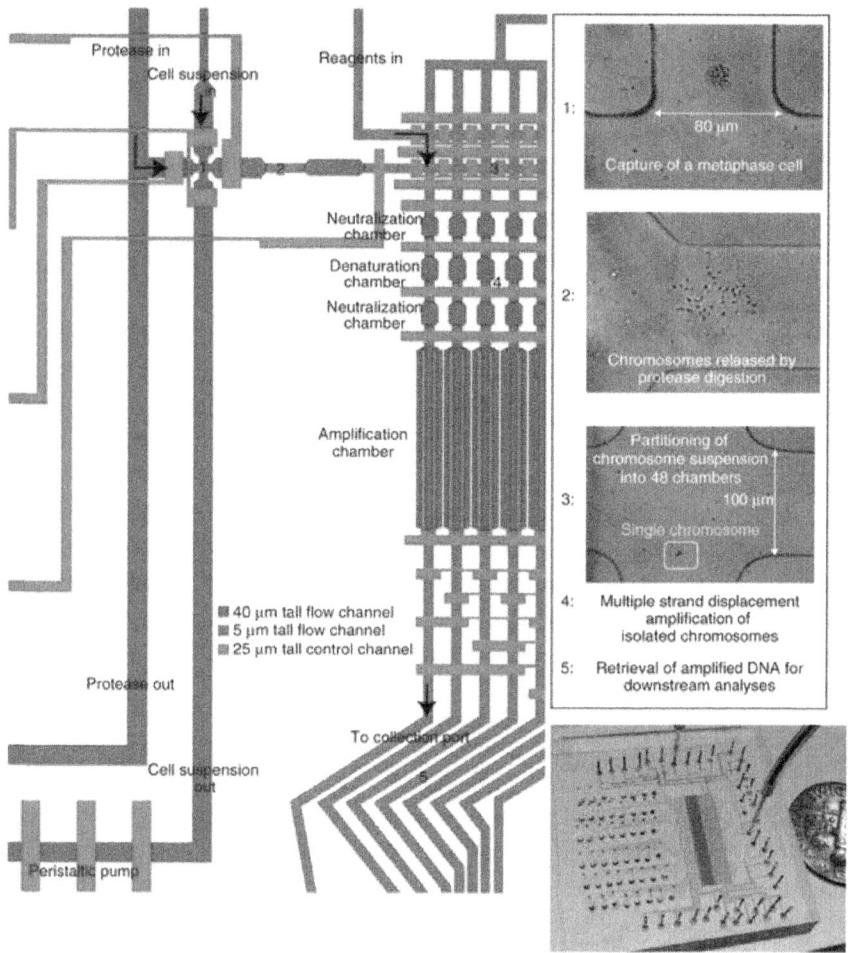

Figure 1: Microfluidic device designed for the amplification of metaphase chromosomes from a single cell. A single metaphase cell is recognized microscopically and captured in region 1. Protease (pepsin at low pH) is introduced to generate chromosome suspension in region 2. Chromosome suspension is partitioned into 48 units (region 3). Content in each partition is individually amplified (region 4). Specifically, chromosomes at low pH are first neutralized and treated with trypsin to digest chromosomal proteins. Chromosomes are denatured with alkali and subsequently neutralized for multiple strand displacement amplification to take place. As reagents are introduced sequentially into each air-filled chamber, enabled by the gas permeability of the device's material, chromosomes are pushed into one chamber after the next and finally arrive in the amplification chamber. Amplified materials are retrieved at the collection ports (region 5). In the overview image of the device, control channels are filled with green dye. Flow channels in the cell-sorting region and amplification region are filled with red and blue dyes, respectively.

Whole-Genome Haplotyping of Members in a HapMap CEU Trio

We first verified DDP with three lymphoblastoid cell lines, GM12891, GM12892 and GM12878, representing a father-mother-daughter trio in the CEPH European (CEU) 1463 family. These cell lines have been extensively genotyped by the HapMap project.

For each single-cell experiment, the chromosomal origins of the contents of each microfluidic chamber were established by a 46-loci Taqman genotyping PCR. In this stage of metaphase, the chromosomes have duplicated but sister chromatids are still bound together at the centromere; therefore each metaphase cell has 46 separable chromosomes and no more than two chambers should contain templates for a given PCR genotyping assay. As expected, for assays that yielded PCR signals in two chambers, the alleles for both chambers matched that of the genomic DNA if the individual was homozygous for the tested locus, and the alleles of the two chambers were different if the individual was heterozygous for the tested locus (Fig. 2). There was no obvious bias in the distribution and no chromosome pairs were particularly difficult to separate. Because the chromosomes are randomly dispersed into chambers, it is possible that both homologous copies of a chromosome will co-locate in the same chamber. This probability can be made arbitrarily small by increasing the number of chambers, and in practice when co-location occurs we simply repeat the experiment with another cell.

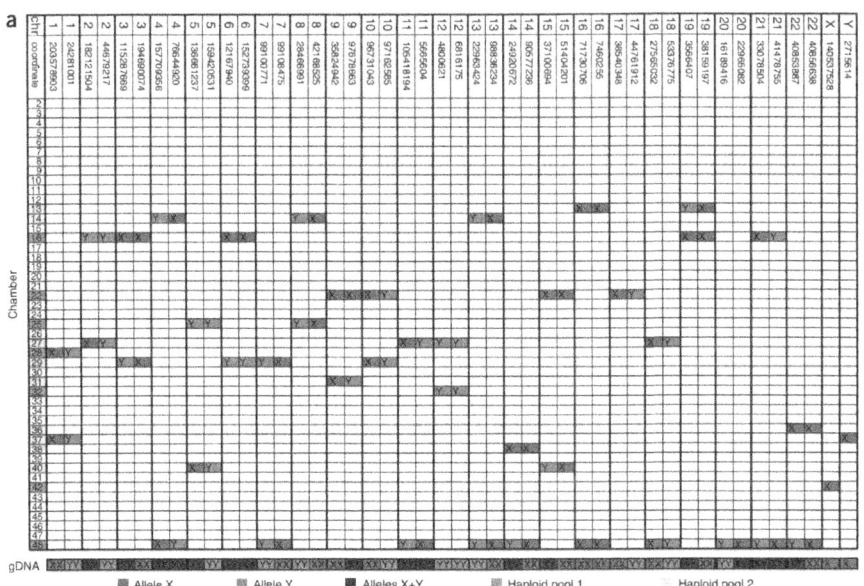

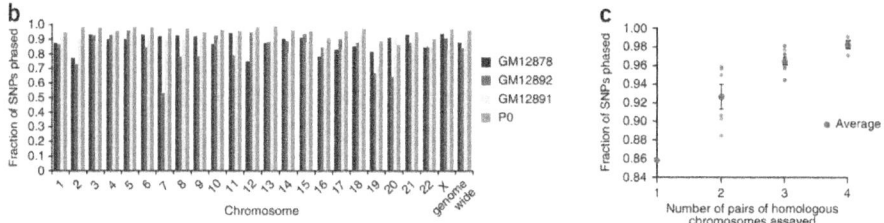

Figure 2: Whole-genome haplotyping. (a) Determining the chromosomal origin of amplification products in a microfluidic device using 46-loci PCR. This table represents results from an experiment using a single metaphase cell of P0's cultured whole blood. A row represents the content inside a chamber on the microfluidic device, and a column represents a locus, with specified chromosome and coordinate (NCBI Build 36.1). Each locus, except those on chromosomes 17 and 20, was found in two chambers. The two alleles of a SNP are highlighted in red and green. Heterozygous loci are labeled in blue. Chamber numbers labeled yellow were pooled together and genotyped on one HumanOmni1-Quad array, and chamber numbers labeled orange were pooled together and genotyped on another array. Genomic DNA extracted from cultured whole blood was also tested with the same 46-loci PCR. (b) Statistics of whole-genome haplotyping. The fraction of SNPs present on the array phased for each chromosome of each individual (GM12891, GM12892, GM12878 and a European individual 'P0') is shown as a colored bar. (c) Fraction of SNPs phased as a function of the number of pairs of homologous chromosomes assayed. This is based on the results from four single-cell experiments of P0. Each point represents the coverage of an autosome. The error bars represent s.e.m.

Products from multiple chambers were pooled together into two mixtures such that each mixture contained one of the two homologous copies of most chromosomes. The two 'haploid' mixtures were separately genotyped on whole-genome genotyping arrays (Illumina's HumanOmni1-Quad BeadChip). For each individual, three to four single-cell experiments were performed, and each homologous chromosome had, on average, ~2 to 3 biological replicates. Phases were established for ~87.9%, ~89.9% and ~83.8% of ~970,000 refSNPs present on the array for GM12878, GM12891 and GM12892, respectively (Fig. 2 and Supplementary Data Sets 1,2,3). By counting the number of inconsistent allele calls among biological replicates of each chromosome homolog, we estimated the error originating from amplification and genotyping for a single phase measurement to be 0.2–0.4%. The actual phasing error per SNP was much smaller because the final phases of most SNPs were determined by the consensus among replicates (Supplementary Fig. 1) and can be made as small as desired by increasing the number of replicates.

We compared our experimental phasing data of the child (GM12878) with haplotype data available from the HapMap project. In the HapMap project, haplotypes in the CEU population were obtained by studying the genotypes of family trios. About 80% of the heterozygous SNPs of the child can be unambiguously phased given that one parent is homozygous for the SNP. The remaining ~20% of heterozygous SNPs in the child are ambiguous and require statistical phasing because both parents are heterozygous. Comparison of DDP and HapMap data on unambiguous SNPs provides an estimate of the accuracy of DDP. The concordance rate between the two data sets was 99.8%. The small number of inconsistencies arose from either error in DDP genotyping or error in genotyping in HapMap data (Fig. 3a). When considering ambiguous SNPs alone, the incongruence rate between the two data sets was 5.7%. The majority of these inconsistencies (96.0%) came from incorrect statistical phasing in the HapMap project, as we could confirm the phases of these ambiguous SNPs in the child from the experimentally determined phases of the two parents (Fig. 3). These data agree with previous evaluations of the accuracies of statistical phasing in CEU trios[15, 28] and highlight the utility of direct experimental phasing even when family data are available.

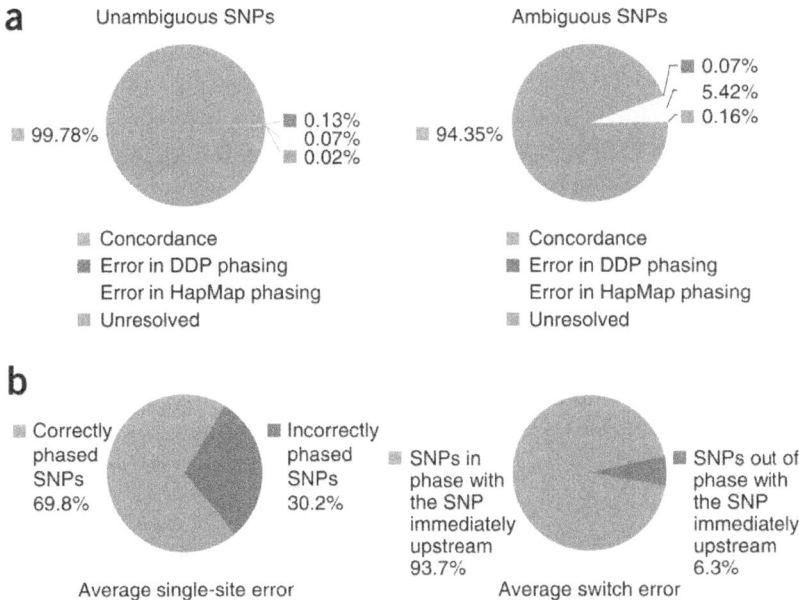

Figure 3: Comparison of statistically determined phases with experimentally determined phases. (a) Comparison of experimentally determined phases of ~160,000 heterozygous SNPs of GM12878 (child of the trio) and those determined by phase III of the HapMap project. Unambiguous SNPs refer to those that are homozygous for at least one parent and are deterministically phased using family data in HapMap. This

comparison shows the accuracy of DDP. Ambiguous SNPs refer to those that are heterozygous for all members of the trio and statistical phasing is used in HapMap. This comparison provides an evaluation of statistical phasing. (b) Comparison of experimentally determined phases of P0 and those determined by PHASE. Seventy-six regions on the autosomal chromosomes were randomly selected and statistically phased three times. Each region carried 100 heterozygous SNPs and spanned an average of ~2 Mb. Switch error rate was calculated as the proportion of heterozygous SNPs with different phases relative to the SNP immediately upstream. Single-site error rate was calculated as the proportion of heterozygous SNPs with incorrect phase. A SNP was considered correctly phased if it had the dominant phase. For each region, the average values from the three runs were reported. Presented here are the average switch error and single-site error per region. The deterministic phases measured by DDP are taken as the ground truth.

Whole-Genome Haplotyping of a European Individual

Having validated the DDP approach on well-characterized HapMap samples, we applied it to determine the haplotypes of an individual, labeled 'P0', whose genome has been sequenced[6] and clinically annotated[29]. As only a few cells are required for DDP, we collected a blood sample from a finger prick. Whereas some of the early microfluidic devices used for experiments with the family trio contained defects leading to the failure to retrieve products from some chambers, refinement in device fabrication yielded fully functional devices and thus improved the number of SNPs phased per single-cell experiment for P0. The average number of pairs of autosomal chromosomes separated per single cell of P0 was 17.5. We obtained ~96.1% coverage of the ~1.2 million SNPs present on the HumanOmni1S array using four single cells (Fig. 2 and Supplementary Data Set 4). An additional ~861,000 SNPs were phased using materials from three single cells and the HumanOmni1-Quad array (Supplementary Data Set 5). For homologous chromosomes that were separated in all four single-cell replicates (that is, four biological replicates of each homologous copy), up to 99.2% of all SNPs assayed on a chromosome were phased (Fig. 2). We noticed that the SNPs that were not phased tended to cluster together and closer inspection revealed that they were usually located in regions with higher GC content (Supplementary Fig. 2). Stronger molecular associations between DNA strands at regions with higher GC content might have led to more difficult amplification, and such phenomena associated with phi29 have been previously reported[30].

Phasing of SNPs was also achieved by direct sequencing. We lightly sequenced amplified material from three single copies of P0's chromosome 6, at an average read depth of $3.5\times$ to $7.7\times$ per copy (Supplementary Table 1).

About 46,000 heterozygous SNPs on chromosome 6 determined by previous genome sequencing were phased, including several of the medically relevant rare variants that were identified in the clinical annotation of the genome[29]. For alleles called by threefold or greater coverage, the concordance rate of phasing by sequencing and phasing by genotyping arrays was 99.8% (Supplementary Fig. 3a). This indicates that allele calling with haploid materials can be achieved accurately with relatively low coverage, an advantage over conventional genotyping by sequencing, which requires much higher fold coverage to guarantee accuracy of heterozygous SNPs. The amplification of minute amount of materials using the polymerase phi29 has been known to cause amplification bias and formation of nonspecific products that would undermine sequencing performance.

Our group previously demonstrated improved performance of whole-genome amplification of single bacteria by reducing amplification volumes by ~1,000-fold using microfluidic devices similar to the one in this study[31, 32]. The present sequencing experiments show that nonspecific products constituted a very small amount of the amplified materials and provide a characterization of the amplification bias for human chromosome–sized, single-molecule templates (Supplementary Table 1 and Supplementary Fig. 3b–d). The coverage across the chromosome was non-uniform, yet distribution of reads over most of the chromosome in all sequenced copies was within two orders of magnitude (Supplementary Fig. 3c).

Comparison of Experimental and Statistical Phasing

Statistical inference has been commonly used to estimate haplotypes in unrelated individuals, yet the lack of true haplotypes means that few studies have been conducted to evaluate the accuracy of these computational approaches. We used the statistical inference software PHASE[33, 34, 35] to infer haplotypes for P0 using CEU haplotypes, determined by family trios in the HapMap Project, as the background and compared the inferred haplotypes to the P0 haplotypes determined by DDP. Evaluation of a total of 76 ~2-Mb regions, each defined by 100 heterozygous SNPs, revealed an average of 6.3 block switches per region and an average block size of ~260 kb. An average of 30.2% of heterozygous SNPs were incorrectly phased using the statistical method (Fig. 3b and Supplementary Fig. 4). These results agree with two previous studies that compared statistical haplotype inference with real phases obtained from somatic cell hybrids and complete hydatidform moles[36, 37], and illustrate the importance of direct experimental phasing especially when family data are not available.

Direct Measurement of Recombination Events within a Family Trio

The availability of parental haplotypes allowed us to directly measure the products of recombination events that led to an individual's unique genome, which could previously only be inferred using three-generation families[38] or two-generation families with large sibships[39]. We aligned each homologous chromosome of the child to the pair of chromosomes of the parent from whom the chromosome was inherited. Figure 4 illustrates the crossover events resulting from the paternal and maternal meioses. We detected 26 and 38 events in the male meiosis and female meiosis, respectively, with a median resolution of ~43 to 44 kb (Supplementary Table 2). The number of detected recombination events matched those in previous reports and supports the notion that the number of recombination events in females is generally higher than that in males[38, 40]. At least 60% of these regions had recombination rates above the median sex average according to the deCODE genetic map[39].

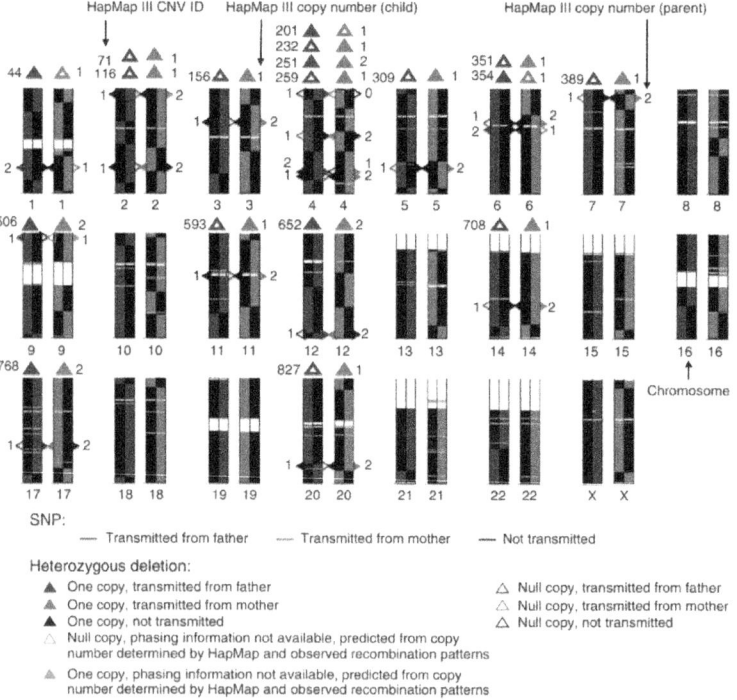

Figure 4: Direct observation of recombination events and deterministic phasing of heterozygous deletions in the family trio. Each allele with DDP data available for the child and the parent is represented by a colored line (blue, alleles transmitted to the child from the father; red, alleles transmitted to the child from the mother; black,

untransmitted alleles). Centromeres and regions of heterochromatin are not assayed by genotyping arrays and are thus in white. Heterozygous deletions in the parents are represented as triangles along each homologous chromosome. A solid triangle represents one copy and a hollow triangle represents a null copy. The phases of deletions are determined for each parent independently. The triangles are color coded according to the state of transmittance as determined by the location of the deletion relative to spots of recombination. The phases of the deletions in the child are determined independent of the parents and are shown on top of the parental chromosomes. The integers on the left are the IDs of each region given by HapMap phase III. The numbers on the right are the copy number of a region in the child as determined by HapMap. Chromosomes are plotted with the same length.

In addition to the switchover of large blocks of homologous chromosomes as a result of recombination, we observed switchover at single sites, constituting ~0.4% of the total number of SNPs in each parent-child comparison; these are presumably products of gene conversion or cell culture–induced mutations, as well as DDP error.

Phasing of Heterozygous Deletions

Although copy number variants (CNVs) can be phased using statistical methods similar to those used to phase SNPs[41, 42, 43], direct experimental phasing of structural variation such as copy number polymorphisms has largely been unexplored. We experimentally phased the heterozygous deletions accessible with genotyping arrays, as determined by the HapMap Project, of the three individuals in the family trio. We phased 12 and 6 heterozygous deletions present within the family trio using genotyping array data (Supplementary Table 3a) and real-time PCR (Supplementary Table 3b), respectively. All of the phased heterozygous deletions within the trio agreed with the inheritance pattern (Fig. 4). We also phased all eight heterozygous deletions that had been detected by genome sequencing of P0 (ref. 6) using data from genotyping arrays and real-time PCR. Results from all platforms among all single-cell replicates were consistent (Supplementary Table 4). Phasing of other types of CNVs with the current approach of genotyping array and PCR is challenging, but we envision that deep sequencing of amplified materials would eventually allow each chromosome to be assembled and thus enable phasing of all CNVs.

Direct Determination of the HLA Haplotypes of an Individual

An important application of DDP is the determination of the HLA haplotypes within an individual. The HLA loci are highly polymorphic and are distributed over ~4 Mb on chromosome 6. The ability to haplotype the HLA genes within the region is clinically important because this region is associated with autoimmune and infectious diseases[44], and the compatibility of HLA haplotypes between donor and recipient can influence the outcomes of transplantation[8]. Yet molecular techniques to measure HLA haplotypes in individuals are still limited[45].

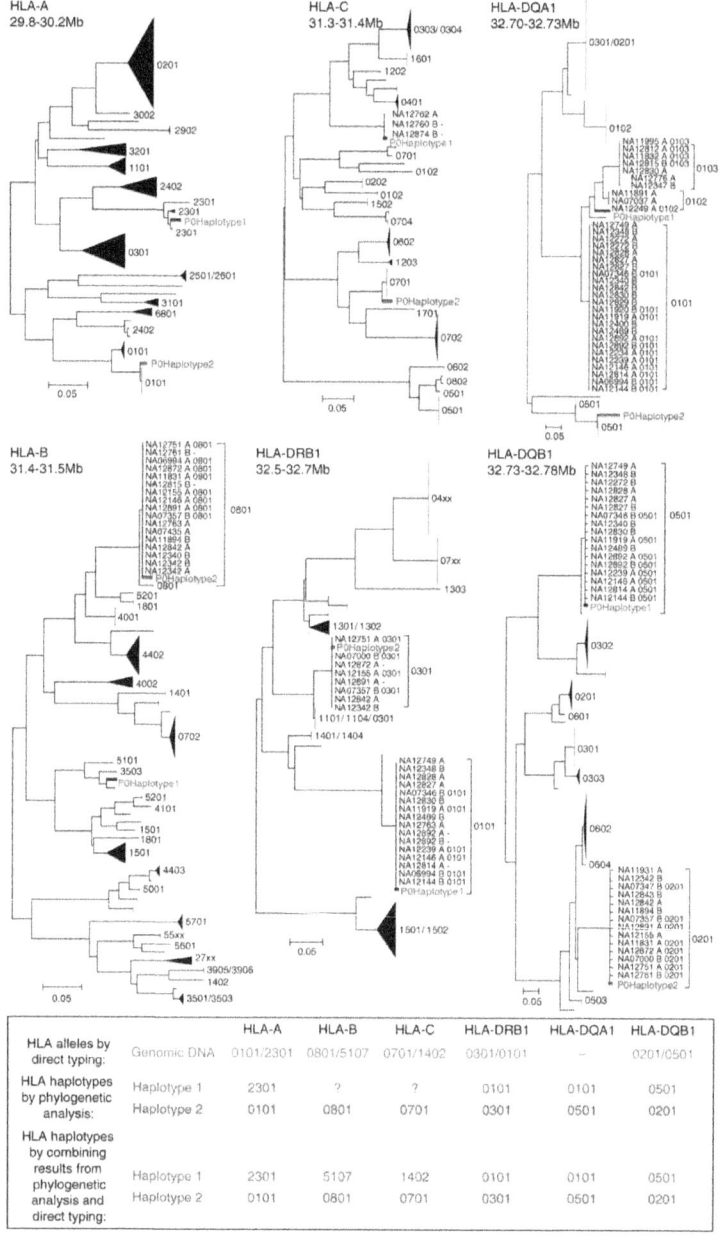

Figure 5: HLA haplotypes of P0 determined using DDP. At each of the six classi-

cal HLA loci, the experimentally phased SNP haplotypes of P0 and 176 phased SNP haplotypes of CEU trios available from HapMap phase III were placed on a neighbor-joining tree. The two haplotypes of P0 are labeled in red and blue. For haplotypes in the CEU panel with HLA typing data, the four-digit HLA allele is presented next to the sample label. Most of each tree is compressed. Each compressed subtree is labeled with the HLA allele associated with members inside the subtree, if HLA allele information is available. The allelic identities of HLA-B and HLA-C on haplotype 1 were not determined with DDP because CEU individuals with similar SNP haplotypes as P0's SNP haplotypes did not have HLA typing data at these loci but could be inferred from the results of direct HLA typing of genomic DNA (first row of table). HLA-DQA1 was not directly typed.

To determine the HLA haplotypes, we first had to determine the HLA allele at each locus, which is usually achieved by direct sequencing. Here, we sought a simpler approach to determine the allele at each HLA locus by taking advantage of the experimentally determined SNP haplotypes of P0. Briefly, we used phylogenetic analyses to compare the SNP haplotypes of P0 within each HLA gene to those of CEU individuals whose HLA genes were typed previously (Fig. 5). The combination of the alleles at each HLA locus determined by phylogenetic analyses agreed with direct HLA typing of genomic DNA. Combining the results from all loci yielded the two HLA haplotypes of P0. One of the HLA haplotypes is the 8.1 ancestral haplotype, which is one of the most frequently observed haplotypes in Caucasians[46] and is associated with elevated risks of immunopathological diseases[47].

DISCUSSION

The DDP approach is scalable. Multiple cells can be processed simultaneously by modifying device design. Currently, the most labor-intensive procedure is the manual identification of metaphase cells. We anticipate that automation of this with a relatively simple engineering solution, such as the combination of computer vision and fluorescent labeling of mitotic cells, will dramatically increase throughput. The majority of the cost of the project went to the genotyping arrays, and as sequencing costs continue to drop, it may become more cost effective to sequence rather than to genotype.

DDP requires the presence of metaphase chromosomes because during metaphase chromosomes are most condensed and can be physically separated. DDP therefore requires sources of cells that can undergo mitosis, such as blood samples and cell lines. Yet DDP requires as little as a single cell, and thus may also have important applications in single-cell genomics, in fields such as preimplantation genetic diagnosis, prenatal diagnosis, aging, and cancer diagnosis and research.

To our knowledge, the work described here represents the first demonstration of a molecular-based, whole-genome haplotyping technique amenable for personal genomics. Whereas the bulk of the experiments described here focus on direct deterministic phasing of ~1 million variants accessible by genotyping arrays, DDP can be used to phase all variants in the genome. DDP of tagSNPs present on the genotyping arrays inherently provides phasing information for common variants that are in strong linkage disequilibrium with the tagSNPs. In addition, we showed that amplified materials from separated chromosome homologs could be directly sequenced, yielding phasing information for variants, including the rare and private ones, which are absent on standard genotyping arrays. Combining DDP SNP analysis with shotgun genome sequencing could allow the determination of the complete personal haplotype of an individual, even in the absence of family information.

METHODS

Microfluidic Device Design, Fabrication and Operation.

The microfluidic device was made of polydimethylsiloxane (PDMS) and was fabricated using soft lithography by the Stanford Microfluidic Foundry. The two-layered device had rectangular 25-µm tall control channels at the bottom and rounded flow channels at the top. The device was bonded to a glass slide coated with a thin layer of PDMS. In the cell-sorting region of the device, flow channels were 40 µm high and 200 µm wide. In the amplification region of the device, flow channels were 5 µm and 100 µm wide and reaction chambers were 40 µm tall. A membrane valve was formed when a control channel crossed over with a flow channel and was actuated when the control channel was pressurized at 20–25 p.s.i. The area of each valve was 200 µm × 200 µm for the 40-µm flow channels, and 100 µm × 100 µm for the 5-µm flow channels. Membrane valves were controlled by external pneumatic solenoid valves that were driven by custom electronics connected to the USB port of a computer. A Matlab program was written to interface with the valves.

Cell Culture

The Epstein-Barr virus–transformed lymphoblastoid cell lines GM12891, GM12892 and GM12878, belonging to the pedigree CEU 1463 (Coriell Cell Repositories), were cultured in RPMI 1640, supplemented with 15% FBS. Each culture was treated with 2 mM thymidine (Sigma) for 24 h at 37 °C to enrich the population of mitotic cells. Followed by multiple washings, cells were cultured in normal medium for 3 h and treated with 200 ng/ml nocodazole (Sigma) for 2 h at 37 °C to arrest cells at metaphase.

Whole blood (~250 μl) obtained from a finger-prick of Patient Zero (‹P0›) was treated with sodium heparin and cultured in PB-Max medium (Invitrogen) for 4 days. The culture was treated with 50 ng/ml colcemid (Invitrogen) for 6 h. The culture was layered on top of Accuspin System-Histopaque-1077 (Sigma) and centrifuged for 8 min at 590g. Nucleated cells at the interface was removed and washed once with HBSS.

Metaphase arrested cells incubated with 75 mM KCl at 25 °C for 10 to 15 min. Acetic acid was added to the cell suspension at a final concentration of 2% to fix the cells. After fixation on ice for 30 min, cells were washed multiple times and finally suspended in 75 mM KCl-1mM EDTA-1% Triton X-100. Cells were treated with 0.2 mg/ml RNaseA (Qiagen) before loading onto the microfluidic device.

Cell sorting, Chromosome Release and Multiple Strand Displacement Amplification

Before the loading of the cell suspension, the cell-sorting channel of the device was treated with Pluronic F127 (0.2% in PBS). Cell suspension was introduced into the device using an on-chip peristaltic pump and an off-chip pressure source. Metaphase cells could be distinguished from interphase cells microscopically by morphological differences. Once a single metaphase cell was recognized at the capture chamber, surrounding valves were actuated to isolate it from the remaining cell suspension. Pepsin solution (0.01% in 75 mM KCl, 1% Triton X-100, 2% acetic acid) was introduced to digest the cytoplasm and release the chromosomes. The chromosome suspension was pushed into a long narrow channel and partitioned into 48 180-pl compartments by actuating a series of valves along the channel. Trypsin (0.25%) in 150 mM Tris-HCl (pH 8.0) (1.2 nl) was introduced to neutralize the solution and to digest chromosomal proteins. Ten minutes later, denaturation buffer (Qiagen's Repli-G Midi kit's buffer DLB supplemented with 0.8% Tween-20; 1.4 nl) was introduced. The device was placed on a flat-topped thermal cycler set at 40 °C for 10 min. This was followed by the introduction of neutralization solution (Repli-G kit›s stop solution; 1.4 nl) and incubation at 25 °C for 10 min. A mixture of reaction buffer (Qiagen›s Repli-G Midi Kit), phi29 polymerase (Qiagen›s Repli-G Midi Kit), 1× protease inhibitor cocktail (Roche) and 0.5% Tween-20 (16 nl) was fed in. The total volume per reaction was 20 nl and the device was placed on the flat-topped thermal cycler set at 32 °C for about 16 h. Amplification products from each chamber was retrieved from its corresponding outlet by flushing the chamber with TE buffer (pH 8.0) supplemented with 0.2% Tween-20. About 3–5 μl of products were collected from each chamber. Products were incubated at 65 °C for 3 min to inactivate the phi29 enzyme.

Initial Genotyping with 46 Loci

To determine the identity of chromosomes in each chamber, we performed Taqman PCR using a set of 46 genotyping assays (two assays per autosome and one assay per sex chromosome) on the products of each chamber on the 48.48 Dynamic Array (Fluidigm). The assays used are listed in Supplementary Table 5.

Whole-Genome Phasing Using Genotyping Arrays

To generate sufficient materials for genotyping array experiments, we amplified DNA products from the microfluidic device a second time in 10 μl volume using the Repli-G Midi Kit›s protocol for amplifying purified genomic DNA. Products from multiple chambers were pooled together into two mixtures such that each mixture contained one of the homologous copies of each chromosome. Each mixture, containing roughly one haploid genome of a cell, was genotyped on the HumanOmni1-Quad or HumanOmni1S BeadChips (Illumina). For GM12891, GM12878 and P0, four single cells were haplotyped. For GM12892, three single cells were haplotyped. Haplotyping data for cell lines were obtained from HumanOmni1-Quad array. Haplotyping data for P0 were obtained from both HumanOmni1-Quad and HumanOmni1S arrays. Genomic DNA extracted from each cell line was also genotyped on the HumanOmni1-Quad array. Genomic DNA of P0 was genotyped on the HumanOmni1S array.

For each chromosome homolog, the allelic identity of a SNP was determined from the consensus among the biological replicates. If equal numbers of both alleles were observed at the site, no consensus was drawn. We estimated the error of a single genotyping measurement by counting the number of inconsistent allele calls at sites typed more than once. For SNPs of which only one of the alleles was observed, the identity of the other allele was determined using the genotypes of genomic DNA. For the trio, the genotypes of genomic DNA were measured on the HumanOmni1-Quad BeadChip. The concordance rate of these genotypes with HapMap data was ~99.1% for each cell line. For P0, genotypes of genomic DNA were measured on the HumanOmni1S BeadChip. The combination of the consensus alleles from the two homologs at each SNP site should, in principle, agree with the genotype call of the genomic DNA control. SNPs that did not follow this rule (~0.3% for cell lines and ~0.4% for P0) were eliminated from downstream analyses.

Data files containing the phased haplotypes of the members of the trio and P0 are available as Supplementary Data Set 1 (GM12891), Supplementary Data Set 2 (GM12892), Supplementary Data Set 3 (GM12878), Supplementary

Data Set 4 (P0 Omni1S) and Supplementary Data Set 5(P0 Omni1Quad). Each file contains whole-genome haplotypes of each individual of the CEU trio (GM12891, GM12892, GM12878) and P0. For the trio, refSNPs present on the Omni1-Quad array (Illumina) that were phased by DDP are included in these files. For P0, refSNPs present on the Omni1-Quad and SNPs present on the Omni1S arrays phased by DDP are included. For P0›s data on Omni1-Quad arrays, only SNPs with both alleles directly observed were included. Alleles are presented relative to the forward strand, and SNPs with A/T and G/C alleles are not included. Column 1: SNP name; column 2: chromosome (chromosome X designated as ‹23›); column 3: position on chromosome (hg18); column 4: allele on homologous copy 1; column 5: allele on homologous copy 2.

Phasing of Chromosome 6 Using High-Throughput Sequencing

Three chambers containing amplified materials from a single copy of chromosome 6 were selected from the four single-cell experiments of P0 for paired-end sequencing on Illumina's Genome Analyzer II. Two chambers contained materials from chromosome 6 only, whereas the third chamber contained materials from a homolog of chromosomes 6, 16 and 18. Second-round amplified materials from these chambers were fragmented through a 30-min 37 °C incubation with 4 µl dsDNA Fragmentase (New England Biolabs) in a 20-µl reaction. Fragmented DNA was end-repaired, tailed with a single ‹A› base, and ligated with adaptors. A 12-cycle PCR was carried out and PCR products with sizes of 300–500 bp were selected using gel extraction. Sequencing libraries were quantified with digital PCR[48]. Thirty-six base pairs were sequenced on each end.

Image analysis, base calling and alignment were performed using Illumina's GA Pipeline version 1.5.1. The first 32 bases on each read were aligned to the human genome (Build 36.1). SNP calling was carried out using Illumina's CASAVA version 1.6.0. Positions covered at least three times according to the 'sort.count' intermediate files were used in downstream analyses. A list of heterozygous SNPs was obtained from the sequenced genome of P0, using quality score >2.8 and heterozygous score of 20 (ref. 6). The phases of heterozygous SNPs were determined either from the direct observation of both alleles in the different homologs, or by inferring the identity of the unobserved allele if only one allele was detected.

Data Sources

Genotypes, CNVs and phasing data of the three lymphoblastoid cell lines were downloaded from the website of the International HapMap Project (http://hapmap.ncbi.nlm.nih.gov/). Genotypes of the merged phase I+II and III data were used. Phasing information from phase III was used. CNV data from phase III was used.

Comparison between Experimentally Determined Phases of the Family Trio with HapMap Data

We compared the experimentally determined phases of the heterozygous SNPs of the child (GM12878) to those determined by phase III of the HapMap Project. SNPs with A/T and G/C alleles were excluded from comparison. To determine the accuracy of experimental phasing and to locate spots of crossovers, the phases of heterozygous SNPs of each parent (GM12891, GM12892) were compared to those in the child (GM12878) inherited from that parent.

Phasing of Heterozygous Deletions

For the trio, a list of heterozygous deletions was obtained from phase III of the HapMap project. For P0, heterozygous deletions that were detected by previous genome sequencing and subsequently verified by digital PCR were studied[6]. For P0, the assays were the same as those used in a previous study[6]. For the trio, the sequences of primers and probes are listed in Supplementary Table 6. The assumption was that one of the chromosome homologs should give no calls for SNP markers or no PCR amplification within a region of heterozygous deletions. Digital PCR (Fluidigm's 48.770 digital array) was also performed using genomic DNA of each member of the trio to verify copy numbers.

Statistical Phasing of P0 using PHASE.

We evaluated the accuracy of statistical phasing by comparing statistically phased haplotypes and experimentally determined haplotypes of P0. We inferred the haplotypes of P0 using PHASE 2.1 (a Bayesian method–based program for haplotype reconstruction)[33, 34, 35]. Due to computational capacity, we randomly chose four regions on each autosomal chromosome (except chromosomes 4, 20, 21), each having 100 bi-allelic SNPs that were heterozygous in P0. We only selected SNPs with both alleles directly haplotyped and with perfect concordance with genotype determined by whole-genome sequencing. Each region covered a range of ~0.7 to ~3.3 Mb, with an average SNP to SNP distance of ~20 kb. We used the 176 phased CEU haplotypes in phase III of the HapMap project as known haplotypes for the inference. For each region, we ran the reconstruction three times with the same default settings but different random seeds and compared the results with the experimentally determined haplotypes. Switch error rate was calculated as the proportion of heterozygous

SNPs with different phases relative to the SNP immediately upstream. Single-site error rate was calculated as the proportion of heterozygous SNPs with incorrect phase. A SNP was considered correctly phased if it had the dominant phase. For each region, the average values from the three runs were reported.

Determination of HLA Haplotypes of P0.

A total of 176 phased CEU haplotypes obtained from phase III of the HapMap project, together with experimentally phased haplotypes of P0, were used to construct neighbor-joining trees at the six classical HLA loci on chromosome 6. The coordinate boundaries of which haplotyped SNPs were used for each locus are presented in Figure 5. Only SNPs with both alleles directly observed were used. The number of SNPs used for HLA-A, HLA-B, HLA-C, HLA-DRB, HLA-DQA, and HLA-DQB were 420, 139, 89, 59, 14 and 34, respectively. Allele sharing distances were computed for each pair of haplotypes as $\frac{1}{n}\sum_{i=1}^{n} d_i$, where n is the number of loci and d_i equals 0 for matched alleles and 1 for unmatched alleles at the ith SNP locus. Trees were constructed using MEGA 4.1 (ref. 49). A list of HLA alleles of individuals in the CEU panel typed in a previous study[9] was downloaded from http://www.inflammgen.org/. The allelic identity of each homologous chromosome of P0 at each HLA locus was determined by the allelic identities of its nearest neighbors in the tree.

ACKNOWLEDGMENTS

We thank J. Melin and the Stanford Microfluidic Foundry for fabrication of microfluidic devices. We thank N. Neff and G. Mantalas for performing sequencing experiments. We thank D. Pushkarev for providing data and primers from the genome sequencing project of P0. We thank M. Anderson and D. Tyan at Stanford Histocompatibility Laboratory for providing HLA typing results. We thank Y. Marcy, P. Blainey, J. Jiang and A. Wu for helpful discussions. The project was supported by the US National Institutes of Health (NIH) Pioneer Award and an NIH U54 award. H.C.F. was supported by a scholarship from the Siebel Foundation. J.W. was supported by a scholarship from the China Scholarship Council.

Contributions

H.C.F. and S.R.Q. conceived the experiments. H.C.F. designed the microfluidic device. A.P. developed protocols for device fabrication. H.C.F. and J.W. performed the experiments. H.C.F., J.W. and S.R.Q. analyzed the data and wrote the manuscript.

Competing Financial Interests

S.R.Q. is a founder, consultant and shareholder of Fluidigm Corporation and Helicos Biosciences Corporation, and a consultant and shareholder of Artemis Health. H.C.F. was previously employed at Fluidigm Corporation. All other authors declare no conflict of interest.

SUPPLEMENTARY INFORMATION

Supplementary Tables

Supplementary Table 1. Statistics of high-throughput sequencing of the homologous copies of P0's chromosome 6. Reads were mapped to NCBI Build 36.

Single Cell Experiment ID / Library	1		2		3	
Content as determined by Initial 46-loci Genotyping	Chr 6		Chr 6, 16, 18		Chr 6	
Which Homolog of Chr6	1		1		2	
Paired-End Read	Read 1	Read 2	Read 1	Read 2	Read 1	Read 2
% Passed Filters	94.25	94.25	94.60	94.60	93.78	93.78
Total Number of Reads (1E6)	27.77	27.77	24.92	24.92	17.58	17.58
% No Match	6.06	6.29	3.57	4.55	5.91	6.47
% Failed QC	0.01	0.19	0.02	0.21	0.02	0.23
% Mapped to Multiple Locations in the Genome	15.54	15.61	16.63	16.65	14.74	14.74
% Mapped to Unique Location on Chr6	77.70	77.17	39.69	39.04	78.95	78.13
% Mapped to Unique Location on Chr16	-	-	16.74	16.46	-	-
% Mapped to Unique Location on Chr18	-	-	23.01	22.65	-	-
% Mapped to Unique Location on Other Chr	0.69	0.74	0.35	0.44	0.38	0.42
% Paired Chr6 Reads	94.06		93.03		94.06	
Median Insert Size (chr6)	75		74		74	
% Unique within Chr6 Reads	42.92		70.25		54.91	
% Chr6 Covered (Sequenced Bases in hg18)	37.75		40.34		50.11	
% Chr6 Covered for Each Homolog (Sequenced Bases in hg18)	59.74				50.11	
% Chr6 Covered for Both Homologs (Sequenced Bases in hg18)	72.37					

Supplementary Table 2. Cross-over regions in paternal (GM12891) and maternal (GM12892) genomes.

female meiosis (GM12892)

chr	lower bound marker	lower bound position	upper bound marker	upper bound position	distance between markers (bp)
1	rs1441521	34027288	rs10787373	34030058	2770
1	rs2133356	161155882	rs4500290	161156665	783
1	rs10802219	243765708	rs9919234	243770613	4905
2	rs11680550	139213779	rs4322786	139222178	8399
2	rs12612937	192583070	rs2356971	192809355	226285
3	rs4611855	31822364	rs6773772	31826730	4366
3	rs9837853	120581739	rs2077599	120584940	3201
4	rs12502642	2683821	rs17834108	2880500	196679
4	rs6820169	36313842	rs4833153	36355721	41879
4	rs17325744	77952318	rs2172496	77953963	1645
4	rs10034059	115082601	rs11098213	115087327	4726
5	rs10071667	125813137	rs13162396	125819039	5902
5	rs322358	172121053	rs322353	172125473	4420
6	rs7764439	42797733	rs9296399	42886445	88712
7	rs2077279	7335129	rs10265974	7631221	296092
7	rs2107550	30884888	rs11771444	30990698	105810
8	rs1503386	68597780	rs7017897	68604153	6373
9	rs2053852	93722239	rs278565	94237796	515557
9	rs9407180	39263076	rs7013627	70223358	309602B2
9	rs12378919	104963406	rs390249	105122135	158729
10	rs7072496	6648319	rs1556003	6674626	26307
10	rs10458640	75450844	rs7086415	76129675	678831
10	rs2984132	105647882	rs813944	105795487	147605
11	rs360140	9733143	rs11601212	9745400	12257
11	rs10765774	95292966	rs17806593	95370551	77585
12	rs7299315	50062761	rs7300192	50080310	17549
12	rs4620786	93625533	rs892494	93641011	15478
13	rs1537787	19701063	rs4769989	19772910	71847
13	rs746208	100192727	rs9518148	100199991	7264
14	rs4900134	92036873	rs7142236	92081630	44757
15	rs9R6040	91795443	rs7182163	91869966	74023
16	rs8050499	66985827	rs1299487	66992384	6557
17	rs8079495	67829583	rs9905851	67944104	114521
18	rs4798490	6745963	rs3810046	6931662	185699
20	rs6085806	6912082	rs4815965	6959958	47876
20	rs9974012	54966696	rs6064496	55129245	162549
X	rs5935714	14073189	rs4333771	14077601	4412
X	rs3896120	138327957	rs2805901	138636668	308711

male meiosis (GM12891)

chr	lower bound marker	lower bound position	upper bound marker	upper bound position	distance between markers (bp)
1	rs3005911	56567553	rs3005916	56640111	72558
1	rs2992650	219214335	rs2808220	219218429	4094
2	rs17013760	77237223	rs1837426	77248658	11435
2	rs3843310	129157811	rs7594747	129204976	47165
2	rs2385852	220635693	rs6726031	220668618	32925
3	rs336550	162595561	rs1450525	162634135	38574
4	rs1503979	21244020	rs10938878	21790095	546075
4	rs12511321	115528209	rs4834401	115534577	6368
4	rs7601348	168544633	rs2319242	168601213	56580
5	rs395	9670945	rs13188988	9671077	132
5	rs11749892	126624509	rs10478767	126651535	27026
5	rs10070923	171559996	rs10069019	171650318	90322
8	rs11996752	3563798	rs2623749	3571831	8033
9	rs2519803	135795610	rs10761407	135951119	155509
10	rs2942366	17070860	rs12780809	17111972	41112
10	rs10502025	103449835	rs3612181	103465197	15562
12	rs1240259	68520201	rs7314810	68591587	71386
12	rs10847323	126351529	rs2656828	126368391	16862
14	rs2749527	93896821	rs11624472	93969665	72844
16	rs288580	61115666	rs1424026	61135369	19703
17	rs2920351	72568803	rs480518	72626322	57519
18	rs1785061	4591573	rs16947309	4599374	7801
19	rs9676093	72211572	rs1893282	72484246	272674
20	rs2256454	51066483	rs1936966	51125186	58703
21	rs8132121	41298033	rs17000509	41348612	50579
22	rs2223271	33250388	rs2103554	33318577	68189

Supplementary Table 3. Phasing of heterozygous deletions in the CEU family trio. Supplementary

Table 3A. Phasing of heterozygous deletions using data from SNP arrays.

HapMap III CNP ID	chr	start	end	size	HapMap III copy number			12878		12891		12892	
					12878	12891	12892	Paternally Inherited	Maternally Inherited	Homolog 1	Homolog 2	Homolog 1	Homolog 2
HM3_CNP_44	1	188038628	188043864	5236	1	2	1	8/8[#]	1/8	8/8	8/8	0/7	7/7
HM3_CNP_156	3	65163493	65189725	26232	1	1	2	0/5	5/5	0/4	4/4	5/5	3/5
HM3_CNP_232	4	87195118	87198968	3850	1	1	2	0/2	2/2	2/2	0/2	2/2	2/2
HM3_CNP_259	4	161271979	161291569	19590	1	1	2	0/4	4/4	0/4	4/4	4/4	3/4
HM3_CNP_251	4	153010030	153012241	2211	2	2	1	1/3	3/3	3/3	3/3	3/3	0/3
HM3_CNP_351	6	55934096	55954486	20390	1	1	2	0/9	9/9	3/3	0/3	9/9	4/9
HM3_CNP_354	6	67065532	67105350	39818	1	2	1	10/10	1/10	9/9	8/9	1/9	9/9
HM3_CNP_389	7	12988634	12994810	6176	1	1	2	1/4	4/4	3/3	0/3	3/3	0/3*
HM3_CNP_506	9	6691130	6698357	7227	2	1	1	2/2	1/2	0/2	2/2	0/0*	0/0*
HM3_CNP_652	12	127796462	127798498	2036	1	2	2	4/4	1/4	6/6	0/6	6/6	5/6
HM3_CNP_768	17	51515464	51527822	12358	2	1	2	10/10	10/10	10/10	0/10	9/10	10/10
HM3_CNP_827	20	52081215	52092058	10843	1	1	2	0/6	6/6	0/6	6/6	0/6*	6/6

[#]Number of typed markers / number of markers typed in at least one homolog within the region.

*At least one homolog did not contain any typed markers.

In Figure 4, 'Homolog 1' is the plotted on the right, and 'Homolog 2' is plotted on the left.

The homolog carrying a copy of the region is shaded.

Supplementary Table 3B. Verification of phases of heterozygous deletions using real-time PCR

HapMap III CNP ID	chr	start	end	size	HapMap III copy number			Digital PCR copy number			12878		12891		12892	
					12878	12891	12892	12878	12891	12892	Paternally Inherited	Maternally Inherited	Homolog 1	Homolog 2	Homolog 1	Homolog 2
HM3_CNP_71	2	14622400	14627543	5143	1	1	2	1	1	2	0/2[$]	1/1	0/3	3/3	1/1	1/1
HM3_CNP_116	2	183793698	183798067	4369	1	1	2	1	1	2	0/2	1/1	0/3	3/3	1/1	1/1
HM3_CNP_201	4	9819751	9844366	24615	1	1	0	1	1	0	2/4	0/4	2/2	0/2	0/3	0/3
HM3_CNP_309	5	137835860	137844856	8996	1	1	2	1	1	2	0/4	1/4	3/3	0/3	3/3	3/3
HM3_CNP_371[#]	6	132751050	132754736	3686	1	1	2	1	1	2	1/3	3/3	3/3	1/3	2/2	2/2
HM3_CNP_593	11	55124465	55209585	85120	1	1	2	1	1	2	0/2	3/3	0/3	3/3	1/1	1/1
HM3_CNP_708	14	73311441	73315360	3919	1	1	2	1	1	2	0/2	3/3	1/1	0/1	2/2	2/3

[$]Number of homologs giving positive PCR amplification / number of homologs tested.

[#]Both homologs gave positive PCR amplification, although the copy number of this CNV was 1 for the two individuals.

The homolog carrying a copy of the region is shaded.

Supplementary Table 4. Phasing of heterozygous deletions of P0.

ID*	Chr	Start Coordinate`	End Coordinate`	A heterozygous refSNP that defines the two homologous copies of a chromosome	Allele of Chosen refSNP in Homolog 1	Allele of Chosen refSNP in Homolog 2	Method	Single Cell ID^	Homolog 1	Homolog 2
5-109	5	1093870	109408	rs1003973	C	T	SNP	C	17/17#	0/17
							PCR$	1	+	-
								2	+	-
								3	+	-
6-10	6	1038440	103869	rs6596815	A	C	SNP	C	-	-
							PCR	1	+	-
								2	-	-
								3	+	-
6-64	6	6478100	647960	rs6596815	A	C	SNP	C	0/3	3/3
							PCR	1	-	+
								2	-	+
								3	-	+
15-89	15	8994800	899920	rs3929082	C	T	SNP	C	0/16	16/16
							PCR	1	No data^%	No data^%
								2	-	+
								3	-	+
13-63	13	6312300	631350	rs2801749	A	G	SNP	C	1/1	0/1
							PCR	1	+	-
								2	+	-
								3	+	-
1-15	1	1508220	150853	rs3131972	A	G	SNP	C	0/0	0/0
							PCR	1	-	+
								2	-	+
								3	-	+
8-25	8	2502900	250460	rs2003497	A	G	SNP	C	0/0	0/0
							PCR	1	-	+
								2	-	+
								3	-	+
19-56	19	5682500	568400	rs2312724	C	T	SNP	C	0/0	0/0
							PCR	1	-	+
								2	No data^%	No data^%
								3	-	+

The chromosome homolog that contains a copy of the region is shaded.

*Same labeling as in Supplementary Table 4 of Pushkarev et al (Nature Biotechnology, 2009).

`HG18.

#Number of typed markers / number of markers typed in at least one homolog within the region.

^PCR experiments were done on amplified materials from separated chromosome homologs obtained from 3 single cells; 'C' refers to combined genotyping results from the same homolog in 3 whole-genome haplotyped cells

$Positive or negative PCR signal.

%Homologous copies were not separated.

Supplementary Table 5. Information of Taqman genotyping assays used.

Chromosome	position (build 36.3)	ABI Taqman Genotyping Assay ID	rs number
1	24281001	C___234618_10	rs4313343
1	203578903	C___9114654_10	rs7549293
2	44679217	C__11166500_10	rs17032420
2	182121504	C___1276208_10	rs12997453
3	115287669	C__11907549_1_	rs1872575
3	194690074	C__25749280_10	rs6444724
4	76644920	C___1880371_10	rs13134862
4	157709356	C___7428940_10	rs1554472
5	136661237	C___2556113_10	rs13182883
5	159420531	C___1995608_10	rs7704770
6	12167940	C___9371416_10	rs13218440
6	152739399	C___2515223_10	rs214955
7	99100771	C__30203950_10	rs10264272
7	99108475	C__26201809_30	rs776746
8	28466991	C___2049946_10	rs10092491
8	42168525	C___2533651_10	rs732612
9	35824942	C___3005619_10	rs2236291
9	97678663	C___665481_10	rs689697
10	96731043	C__27104892_10	rs1057910
11	105418194	C___1636106_10	rs6591147
12	4800621	C___488643_10	rs12423234
12	6816175	C___2184724_1_	rs2269355
13	22963424	C___1760747_10	rs4770456
13	98836234	C___1619935_1_	rs1058083
14	24920672	C___2120263_10	rs1454361
14	90577236	C___1310541_10	rs11847557
15	37100694	C__11673733_10	rs1821380
15	51404201	C__29375514_10	rs8037429
16	7460255	C__31419546_10	rs7205345
16	71730706	C__26053735_10	rs6499616
17	38540348	C__11631183_10	rs2175957
17	44761912	C___1121246_10	rs747039
18	27565032	C___7459903_10	rs985492
18	53376775	C___3285337_1_	rs1736442
19	3566407	C___2576338_10	rs768963
19	38159197	C___8582892_1	rs2304102
20	16189416	C___1274218_10	rs12480506
20	22965082	C___3206279_1_	rs2567608
21	33078504	C__11354606_10	rs12627142
21	41478755	C__31891795	rs8130833

22	40853887	C__27102425_10	rs16947
22	40856638	C__11484460_40	rs1065852
X	140537528	C__30576569_20	rs6636416
Y		Custom (non-genotyping)*	SRY gene

Forward primer: CGCTTAACATAGCAGAAGCA
Reverse primer: AGTTTCGAACTCTGGCACCT
Probe: TGTCGCACTCTCCTTGTTTTGACA

Supplementary Table 6. Sequences of primers and probes used for the phasing of heterozygous deletions within the CEU family trio.

HapMap III CNP ID	Chr	Start Coordinate	End Coordinate	Size	Forward primer	Reverse primer	Probe
HM3_CNP_71	2	14622400	14627543	5143	AAGGCTTCCTTAGTGCTCCT	TGAAAGAAACTCATGGCTCA	TTGCTTTCTGATAGTTTCTTGGCCTG
HM3_CNP_116	2	183793698	183798067	4369	GCCTCAGTTTCAGGCATTAT	CTGCTGCCTCATAGAAGTGA	CCACCCCTTACCTTCTCTTTCTGC
HM3_CNP_201	4	9819751	9844366	24615	TCCACTGACCAGATACATGC	ACCATCAGACCAAGACATCC	TCCGTGGTTATAGGCCATGCC
HM3_CNP_309	5	137835860	137844856	8996	ATTCATAATAGCCGGAGTGG	AGGCCTATTTCATTGTTTGG	AATCCAAATGTCTATTCAACCAGTGCA
HM3_CNP_371	6	132751050	132754736	3686	CTGAATGGCTGTGTATGGAA	AATTCAACCTGCAGAAGCAT	TCCCCATGTCCTATGTCTGCAATT
HM3_CNP_593	11	55124465	55209585	85120	TAAAATGGAGCGATGACTGA	GTCCCAACTCTGAAGTTTGC	TCAAGAAGTGTCCTACTTTGCCTGAGA
HM3_CNP_708	14	73311441	73315360	3919	CTGTGGTTGTGAGTCCTGAG	TGTTACGGTTCTGAAGAGCA	TCCACTCTGCTGGGGAGGCT

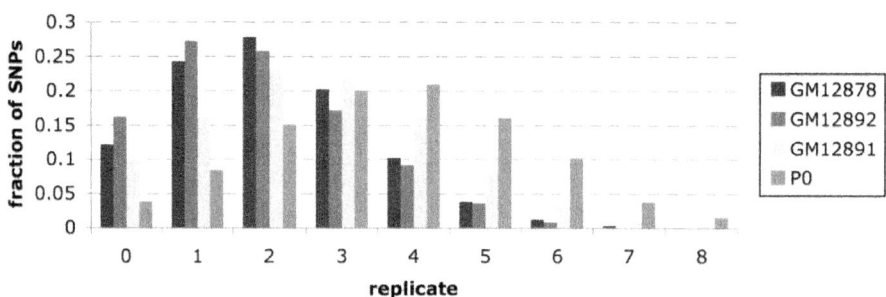

Supplementary Figure 1. Number of replicates phased per SNP for each individual.

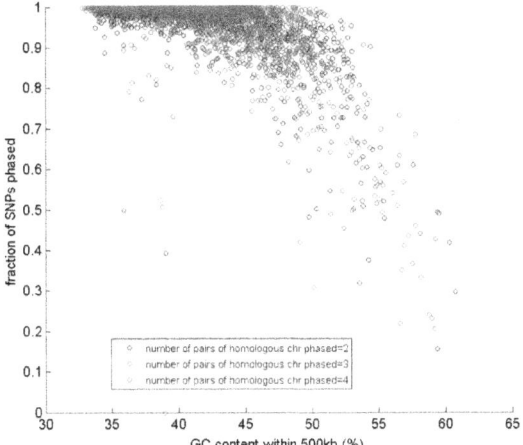

Supplementary Figure 2. SNPs in regions with relatively higher GC content are less accessible by genotyping arrays, potentially resulted from phi29's reduced amplification efficiency in regions with higher GC content. Plotted here is the fraction of SNPs phased by genotyping arrays (based on the ability of the arrays to type the alleles in amplified materials) as a function of GC content of regions where SNPs are located. Shown here are data from whole-genome haplotyping of P0 using Illumina's Omni1S arrays. Fraction of SNPs successfully phased within each 500kb bin was measured and plotted against the GC content of the bin. The 22 autosomes are separated into 3 groups and given three colored labels, depending on how many pairs of homologous copies were assayed (out of the four single cells experimented).

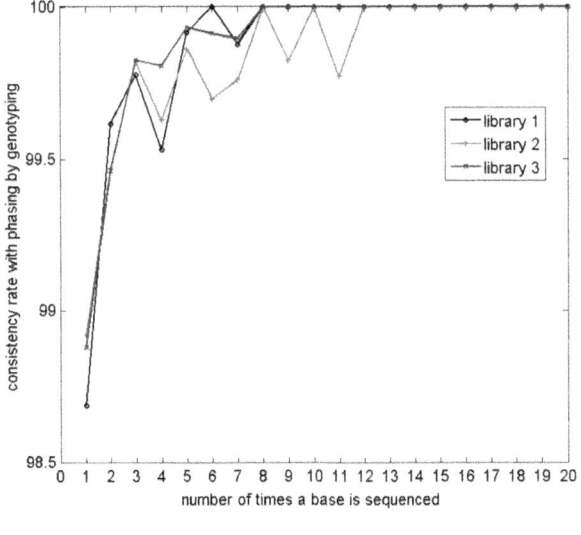

(A)

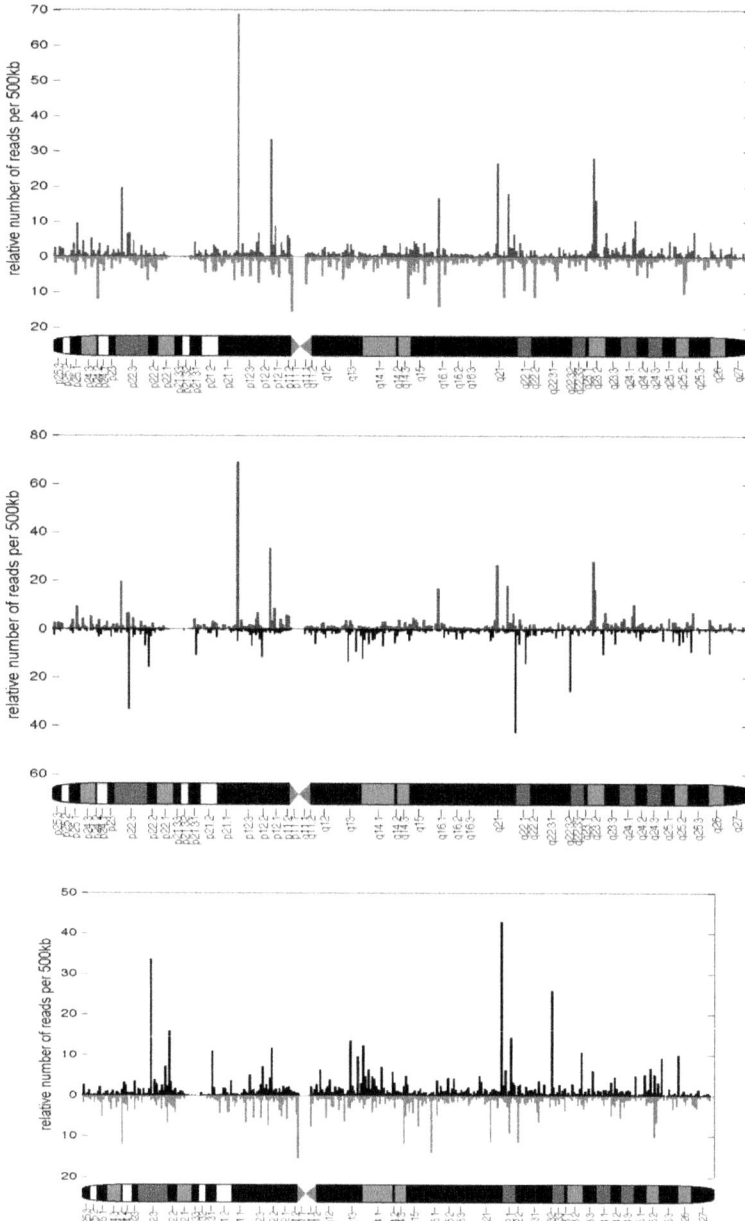

(B)

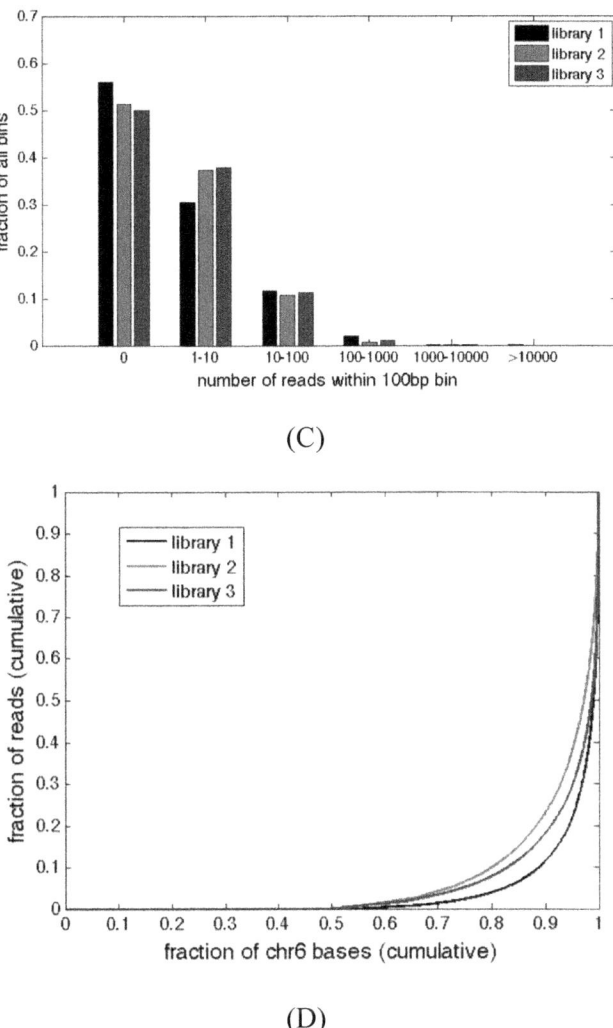

(C)

(D)

Supplementary Figure 3. High-throughput sequencing of the homologous copies of P0's chromosome 6. Three copies, labeled as libraries 1, 2, and 3, were sequenced. (A) Concordance of phasing by sequencing and phasing by genotyping arrays as a function of sequencing coverage. Only SNPs that were phased more than twice with genotyping arrays were compared. (B) Distribution of 32bp reads across chromosome 6 for three different homologous copies of chromosome 6 sequenced (libraries 1, 2, and 3), represented as bars in black, red, and blue. Plotted here is the number of reads per 500kb relative to the sample median. Each plot shows a pair-wise comparison. Sequences within the centromeric and the polymorphic MHC regions could not properly align. The distribution of reads across the chromosome is highly variable among libraries, as a result of amplification bias of a single chromosome template. The regions with excessive number of reads are located at different positions along the chromosome in different libraries. (C) Histogram showing the number of reads per 100bp bin for each of the libraries. For the majority of the chromosome covered with at least one read, the number of reads fall within two orders of magnitude. (D) Fraction of chromosome 6 covered given the fraction of total sequenced reads. This reflects on the fact that a small fraction on the chromosome is excessively over-represented.

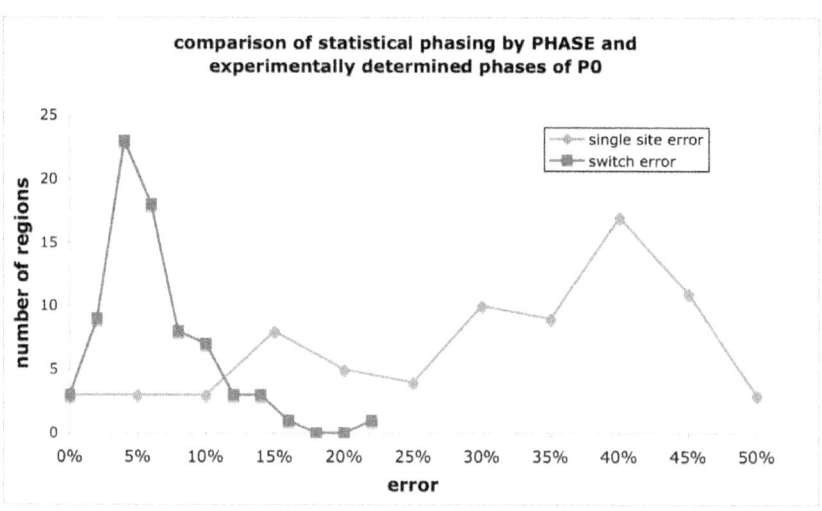

Supplementary Figure 4. Comparison of experimentally determined phases (by DDP) of P0 and those determined by PHASE, a software for statistical haplotype inference. Seventysix regions on the autosomal chromosomes were randomly selected and statistically phased three times. Each region carried 100 heterozygous SNPs and spanned an average of ~2Mb. Presented here are the distributions of the average switch error and average single site error per region.

REFERENCES

1. Wheeler, D.A. *et al.* The complete genome of an individual by massively parallel DNA sequencing. *Nature* **452**, 872–876 (2008).
2. Bentley, D.R. *et al.* Accurate whole human genome sequencing using reversible terminator chemistry. *Nature* **456**, 53–59 (2008).
3. Ahn, S.M. *et al.* The first Korean genome sequence and analysis: full genome sequencing for a socio-ethnic group. *Genome Res.* **19**, 1622–1629 (2009).
4. Kim, J.I. *et al.* A highly annotated whole-genome sequence of a Korean individual. *Nature* **460**, 1011–1015 (2009).
5. Wang, J. *et al.* The diploid genome sequence of an Asian individual.

Nature **456**, 60–65(2008).

6. Pushkarev, D., Neff, N.F. & Quake, S.R. Single-molecule sequencing of an individual human genome. *Nat. Biotechnol.* **27**, 847–850 (2009).
7. Schuster, S.C. *et al.* Complete Khoisan and Bantu genomes from southern Africa. *Nature* **463**, 943–947 (2010).
8. Petersdorf, E.W., Malkki, M., Gooley, T.A., Martin, P.J. & Guo, Z. MHC haplotype matching for unrelated hematopoietic cell transplantation. *PLoS Med.* **4**, e8 (2007).
9. de Bakker, P.I. *et al.* A high-resolution HLA and SNP haplotype map for disease association studies in the extended human MHC. *Nat. Genet.* **38**, 1166–1172 (2006).
10. Stewart, C.A. *et al.* Complete MHC haplotype sequencing for common disease gene mapping. *Genome Res.* **14**, 1176–1187 (2004).
11. Groenendijk, M., Cantor, R.M., de Bruin, T.W. & Dallinga-Thie, G.M. The apoAI-CIII-AIV gene cluster. *Atherosclerosis* **157**, 1–11 (2001).
12. Nagel, R.L. *et al.* The Senegal DNA haplotype is associated with the amelioration of anemia in African-American sickle cell anemia patients. *Blood* **77**, 1371–1375 (1991).
13. Sun, T. *et al.* Haplotypes in matrix metalloproteinase gene cluster on chromosome 11q22 contribute to the risk of lung cancer development and progression. *Clin. Cancer Res.* **12**, 7009–7017 (2006).
14. Drysdale, C.M. *et al.* Complex promoter and coding region beta 2-adrenergic receptor haplotypes alter receptor expression and predict in vivo responsiveness. *Proc. Natl. Acad. Sci. USA* **97**, 10483–10488 (2000).
15. The International HapMap Consortium. A haplotype map of the human genome. *Nature* **437**, 1299–1320 (2005).
16. Frazer, K.A. *et al.* A second generation human haplotype map of over 3.1 million SNPs. *Nature* **449**, 851–861 (2007).
17. Levy, S. *et al.* The diploid genome sequence of an individual human. *PLoS Biol.* **5**, e254(2007).
18. Zhang, K. *et al.* Long-range polony haplotyping of individual human chromosome molecules. *Nat. Genet.* **38**, 382–387 (2006).
19. Mitra, R.D. *et al.* Digital genotyping and haplotyping with polymerase colonies. *Proc. Natl. Acad. Sci. USA* **100**, 5926–5931 (2003).
20. Ding, C. & Cantor, C.R. Direct molecular haplotyping of long-range genomic DNA with M1-PCR. *Proc. Natl. Acad. Sci. USA* **100**, 7449–

7453 (2003).

21. Michalatos-Beloin, S., Tishkoff, S.A., Bentley, K.L., Kidd, K.K. & Ruano, G. Molecular haplotyping of genetic markers 10 kb apart by allele-specific long-range PCR. *Nucleic Acids Res.* **24**, 4841–4843 (1996).

22. Ruano, G., Kidd, K.K. & Stephens, J.C. Haplotype of multiple polymorphisms resolved by enzymatic amplification of single DNA molecules. *Proc. Natl. Acad. Sci. USA* **87**, 6296–6300 (1990).

23. Woolley, A.T., Guillemette, C., Li Cheung, C., Housman, D.E. & Lieber, C.M. Direct haplotyping of kilobase-size DNA using carbon nanotube probes. *Nat. Biotechnol.* **18**, 760–763 (2000).

24. Burgtorf, C. *et al*. Clone-based systematic haplotyping (CSH): a procedure for physical haplotyping of whole genomes. *Genome Res.* **13**, 2717–2724 (2003).

25. Xiao, M. *et al*. Direct determination of haplotypes from single DNA molecules. *Nat. Methods* **6**, 199–201 (2009).

26. Ma, L. *et al*. Direct determination of molecular haplotypes by chromosome microdissection.*Nat. Methods* **7**, 299–301 (2010).

27. Douglas, J.A., Boehnke, M., Gillanders, E., Trent, J.M. & Gruber, S.B. Experimentally-derived haplotypes substantially increase the efficiency of linkage disequilibrium studies.*Nat. Genet.* **28**, 361–364 (2001).

28. Marchini, J. *et al*. A comparison of phasing algorithms for trios and unrelated individuals.*Am. J. Hum. Genet.* **78**, 437–450 (2006).

29. Ashley, E.A. *et al*. Clinical assessment incorporating a personal genome. *Lancet* **375**, 1525–1535 (2010).

30. Bredel, M. *et al*. Amplification of whole tumor genomes and gene-by-gene mapping of genomic aberrations from limited sources of fresh-frozen and paraffin-embedded DNA. *J. Mol. Diagn.* **7**, 171–182 (2005).

31. Marcy, Y. *et al*. Nanoliter reactors improve multiple displacement amplification of genomes from single cells. *PLoS Genet.* **3**, e155 (2007).

32. Marcy, Y. *et al*. Dissecting biological "dark matter" with single-cell genetic analysis of rare and uncultivated TM7 microbes from the human mouth. *Proc. Natl. Acad. Sci. USA* **104**, 11889–11894 (2007).

33. Stephens, M., Smith, N.J. & Donnelly, P. A new statistical method for haplotype reconstruction from population data. *Am. J. Hum. Genet.* **68**, 978–989 (2001).

34. Stephens, M. & Donnelly, P. A comparison of Bayesian methods for haplotype reconstruction from population genotype data. *Am. J. Hum. Genet.* **73**, 1162–1169 (2003).

35. Stephens, M. & Scheet, P. Accounting for decay of linkage disequilibrium in haplotype inference and missing-data imputation. *Am. J. Hum. Genet.* **76**, 449–462 (2005).
36. Kukita, Y. *et al.* Genome-wide definitive haplotypes determined using a collection of complete hydatidiform moles. *Genome Res.* **15**, 1511–1518 (2005).
37. Andres, A.M. *et al.* Understanding the accuracy of statistical haplotype inference with sequence data of known phase. *Genet. Epidemiol.* **31**, 659–671 (2007).
38. Broman, K.W., Murray, J.C., Sheffield, V.C., White, R.L. & Weber, J.L. Comprehensive human genetic maps: individual and sex-specific variation in recombination. *Am. J. Hum. Genet.* **63**, 861–869 (1998).
39. Kong, A. *et al.* A high-resolution recombination map of the human genome. *Nat. Genet.* **31**,241–247 (2002).
40. Frazer, K.A. *et al.* A second generation human haplotype map of over 3.1 million SNPs.*Nature* **449**, 851–861 (2007).
41. Conrad, D.F. *et al.* Origins and functional impact of copy number variation in the human genome. *Nature* **464**, 704–712 (2010).
42. Su, S.Y. *et al.* Inferring combined CNV/SNP haplotypes from genotype data. *Bioinformatics***26**, 1437–1445 (2010).
43. McCarroll, S.A. *et al.* Integrated detection and population-genetic analysis of SNPs and copy number variation. *Nat. Genet.* **40**, 1166–1174 (2008).
44. Shiina, T., Hosomichi, K., Inoko, H. & Kulski, J.K. The HLA genomic loci map: expression, interaction, diversity and disease. *J. Hum. Genet.* **54**, 15–39 (2009).
45. Guo, Z., Hood, L., Malkki, M. & Petersdorf, E.W. Long-range multilocus haplotype phasing of the MHC. *Proc. Natl. Acad. Sci. USA* **103**, 6964–6969 (2006).
46. Maiers, M., Gragert, L. & Klitz, W. High-resolution HLA alleles and haplotypes in the United States population. *Hum. Immunol.* **68**, 779–788 (2007).
47. Price, P. *et al.* The genetic basis for the association of the 8.1 ancestral haplotype (A1, B8, DR3) with multiple immunopathological diseases. *Immunol. Rev.* **167**, 257–274 (1999).
48. White, R.A. III, Blainey, P.C., Fan, H.C. & Quake, S.R. Digital PCR provides sensitive and absolute calibration for high throughput sequencing. *BMC Genomics* **10**, 116 (2009).

49. Tamura, K., Dudley, J., Nei, M. & Kumar, S. MEGA4: molecular evolutionary genetics analysis (MEGA) software version 4.0. *Mol. Biol. Evol.* **24**, 1596–1599 (2007).

Chapter 5

PROJECT-BASED LEARNING TO PROMOTE EFFECTIVE LEARNING IN BIOTECHNOLOGY COURSES

Farahnaz Movahedzadeh, Ryan Patwell, Jenna E. Rieker, and Trinidad Gonzalez

Department of Biological Sciences, Harold Washington College, 30 East Lake Street, Chicago, IL 60601, USA

ABSTRACT

With enrollment in the fields of science, technology, engineering, and mathematics (STEM) shrinking, teachers are faced with the problem of appealing to a new generation of students without sacrificing educational quality. Evidence has shown that this problem can be reduced with the use of a number of pedagogical strategies of which project-based learning (PBL) is one. PBL addresses the fundamental challenge of increasing students' motivation, their mastery of course material, and finding applications for what they have learned to apply in various situations. This study demonstrates the benefits of redesigning a standard lab-based molecular biology course to create a more effective learning environment. Using PBL, students who enrolled in Bio-251 at Harold Washington College in Chicago were given the responsibility of cloning a bacterial gene from one species into a new host species. They were then tasked with the expression and purification of the resulting protein for future research purposes at University of Illinois-Chicago, a leading 4-year research institute. With use of the PBL method, students showed improvement in the areas of self-confidence, lab technical skills, and interest in STEM-related fields and, most of all, the students showed a high level of performance and satisfaction.

INTRODUCTION

In 2003, the National Research Council suggested that opportunities for gaining a greater understanding of science could be realized through project-based laboratory courses [1]. Project-Based Learning (PBL) allows students to work actively with the applied techniques of the laboratory setting, while incorporating critical thinking, collaboration, and problem solving skills in the context of content-based knowledge that influences comprehension and academic self-confidence. This method also benefits students by granting them accountability for laboratory projects, an approach that yields a deeper understanding of how science is practiced by scientists through problem solving and the formulation and testing of a hypothesis-based research [2]. This fundamental understanding is a direct byproduct of the project-based learning method.

In this study, we intend to highlight the outcomes of transforming a standard molecular biology course in a community college into a more effective learning experience using project-based learning. This is simply because project-based learning utilizes both the tactile and visual senses to create a more complete learning experience, one that proves crucial in promoting the growth of STEM program enrollments [3]. Such a system would allow for a greater majority of students already enrolled in STEM courses to promote its success.

With career opportunities in STEM fields expected to increase 22% by the year 2014 [4], these fields are going to need even more highly skilled graduates to keep pace with the demand. Educators must rise to the challenge of identifying and addressing the shortcomings of classic teaching models to create an optimized curriculum and pedagogy that better suits today's students and helps meet the needs of today's societies. Fortunately, this transition is aided by the growing amount of research that has been conducted on how students learn. This research brings new insights on how to better educate the researchers of tomorrow for both academic and professional development.

After all, as Cox et al. [5] argued, conveying facts does not alone produce scientifically literate students. Today's students of science need "a better window on what science is and how it is done, a clear presentation of key concepts that rises above the recitation of details, an articulation of philosophical underpinnings of the scientific discipline at hand, exercises that demand analysis of real data, and an appreciation for the contributions of science to the well-being of humans throughout the world" (p. xxi). We believe that by its nature, PBL can provide the opportunity for students to achieve just that.

METHODS AND MATERIALS

Bio-251 is an introductory molecular biology course and one of the main courses in the biotechnology certificate program at Harold Washington College (HWC) in Chicago. The course had been experiencing low enrollment and retention rates and, in the term of our study, had been scheduled to meet once a week on Fridays from 4:15 to 9 p.m., not a time most students want to spend at school.

Typical introductory molecular biology courses at 2-year colleges often contain a sequence of topic-based lectures and various weekly laboratory assignments. The assignment-based labs are often directed by a manual containing a generic set of assignments tailored to introduce several, often unrelated, techniques. Although students may enjoy learning new techniques, they fall short in understanding how the individual techniques are applied in professional scientific investigations. Furthermore, the connection between the knowledge and the applications is often not clearly made.

Based on realizations such as these, the Bio-251 course was redesigned to offer hands-on experience that is clearly made to be seen and applied by implementing the project-based learning model with a single term-long group laboratory project. Figure 1 shows the concept-map of the relationship between learned concepts, weekly lab assignments, and the single term-long laboratory project. Other redesigned features included a number of quizzes, three exams, homework assignments, a weekly lab report, and a number of discussions/presentations. Lectures were designed with respect to the project in a way that would deliver theoretical knowledge and relevant information that students could immediately apply in laboratory learning and lab experiments. The first time the redesigned Bio-251 course was taught, fourteen students took part in this unique endeavor.

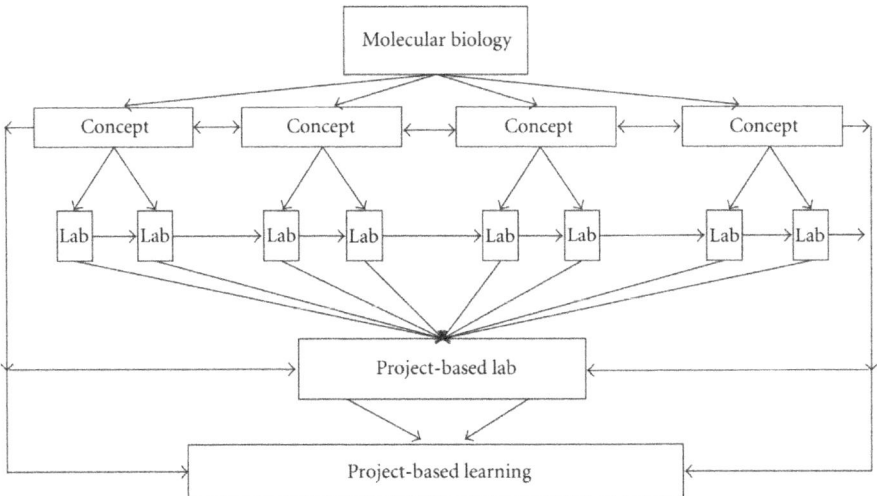

Figure 1: Concept map for project-based learning.

The Project

A unique project for the redesigned Bio-251 class at HWC was developed in collaboration with the nearby College of Pharmacy at the University of Illinois-Chicago (UIC). One of the interests of UIC was to clone a gene and express a protein to be used in medicinal chemistry research. Bringing this project to Bio-251 at Harold Washington College offered a wonderful opportunity for students to do a meaningful project in class through a project-based learning approach and under course supervision.

To truly understand the concept and successfully engage in the project-based learning approach, students need to know that if the first step is not performed properly, they cannot move on to the second one. For example, as seen in Table 1 and Figure 1, students must successfully complete the amplification of the gene before doing the ligation and transformation and so on. The premise was that the instructor's professional research experience in tandem with her teaching experience could provide this critical understanding to the students, while contributing something new to the scientific community.

Table 1: Scheduled labs in Bio-251.

Week	Scheduled labs
1	Basic laboratory safety techniques Basic equipment and bacteriological techniques
2	Basic calculations for preparing solutions of different concentrations
3	Bacterial media prep, inoculation, and growth
4	Isolation of chromosomal DNA and preparation of agarose gel
5	PCR and agarose gel analysis
6	Restriction enzyme digestion, purification, and preparation for cloning
7	Ligation
8	Transformation Making competent cells of E. coli and transforming the recombinant plasmid.
9	Isolation of Plasmid, gel electrophoresis, and screening for clone
10	Preparation for gene expression and transformation to a suitable strain for expression
11	Protein induction and expression

12	Polyacrylamide gel electrophoresis
13	Protein purification
14	Protein purification troubleshooting
15	Bioinformatics (genome analysis, Blast, Primer3 for designing PCR, Web-cutter)
16	Computer analysis of the gene product

Pedagogical Strategy and Rationale

In a typical biology laboratory, students usually follow the direction of a protocol, often with or without really understanding the concepts. Since the lab assignment they are doing is just one of many they have to complete during the semester, personal attachment is rare and boredom quickly follows. The students' focus shifts to getting the lab assignment over quickly and the main objective could easily be lost. In the redesigned Bio-251 course, students were surprised to learn that there was no specific lab manual for the lab project. Instead, the students had access to a number of various lab manuals available in the class/lab, as well as on the Internet, that they could access any time before, during, or after a lab session. Students had to collaborate with each other and rely on each other for support in a problem-solving style, as opposed to reading a cookbook-style lab book with all of the answers neatly laid out. They were required to reflect on their experiences in the form of a lab report each week containing a short introduction, materials and methods, as well as results, and a conclusion. In the discussion section of each lab's conclusion, students had to identify how their results related to the results of previous labs and how these results might help them complete upcoming labs. In addition, students reported all of the problems they encountered in conducting the lab experiments and how they were able to deal with them.

After a few sessions, some of the students would go as far as to predict our next procedure in their conclusions. This was yet another indication that they were developing a deeper understanding of the concepts. Achieving success in each step brought joy and self-confidence to the students and promoted their enthusiasm for the course. Often, students mentioned that they could not wait for the next class meeting.

The Course at Work

Figure 1 shows the concept map of the course and how the various components of this course are related to each other. Concepts learned, weekly labs, project-based labs, and how all of these elements lead to project-based learning.

Topics and Concept Learned

Topics and concepts covered in this class included DNA & RNA structure, DNA replication, bacteria and viruses as model organisms, transcription, translation, gene regulation, recombinant DNA technology and molecular cloning, polymerase chain reaction and its applications, and tools in molecular cloning: restriction enzymes, agarose gel analysis, usage of plasmids, tools for analyzing gene expression, genetically modified organisms, protein expression, protein purification, and an introduction to bioinformatics.

Weekly Lab Assignments

While the weekly lab assignments shown in Table 1 cover a basic molecular biology course, in Bio-251 each lab was designed and conducted with the goal of contributing to the completion of the project-based lab that all of the students were involved in.

Project-Based Lab for Project-Based Learning

On the first day of the class, we introduced project-based learning to the students and explained its meaning, goals, procedures, and why they were going to be actively engaged in PBL instead of conducting the typical labs for a class like this. The first two weeks of the class focused on safety procedures, basic knowledge of bacterial culture, media and solution preparation, and proper usage of equipment (Table 1). At the beginning of each session, we discussed the theory and concepts behind the laboratory procedure of the week. Throughout the semester, students were given an optional bonus question, such as "What is the G/C ratio in the human genome?" At the end of every class session, we devoted time to discuss the relationship between the concept learned and the conducted lab, and identify the potential authentic contribution of the learned elements into the project-based lab.

After a few sessions, students began to take the initiative to start conversations with each other about the learned subjects, protocols, and lab procedures. These open discussions contributed to a higher level of understanding for almost all, as sharing of results allowed students to piece together their individual perspectives into a unified collective understanding through an authentic purpose. This method provided the opportunity for the students to clear up any confusion that could have potentially harmed their confidence. The method also helped students see the relevance of the information presented to them as it related to their own way of thinking, thus allowing them to make the connection between the subject matter, the lab project, and their own experiences.

At the end of each class, we spent 15 minutes on a laboratory briefing where we talked about how the lab experiments went, what types of problems were encountered (technical skills, conceptual knowledge, etc.), and how we solved them, as well as how we could use the outcomes of the lab experiments and the lab techniques in upcoming lab experiments. Students indicated that this was a very useful experience for their cognitive development and helped build and develop their self-confidence.

On a few occasions we had to start certain procedures that had long processing times at the beginning of class in order to finish on time. The students did not receive the corresponding lectures and were just following protocols, very similar to a standard lab course. Unlike a standard lab, the students were exceedingly interested in what they were doing and how it fit into the big picture. They would ask so many questions that discussion had to be reluctantly limited in order to make the deadline. Since the entire course was focused around this project, the students' pronounced efforts to ensure their understanding indicated that the approach was working as intended.

Class Presentation

By mid-semester, students had to choose a topic related to the course and design a presentation using sources from peer-reviewed journals and review articles. Before the presentations were to begin, the students and the instructor discussed the basic elements of an effective presentation, and how important it was to be honest when evaluating one another. From these discussions came a rubric that the students used to grade each other's presentations. This enabled the presenter to receive valuable constructive criticism from multiple perspectives and the students watching had a good reason to pay attention. This approach also allowed for a lighter workload and more time to prepare for the class. The success of this method (which took all of 30 minutes to create) showed that even a routine task, like grading presentations, could be optimized for high-efficiency learning.

Assessment of Student's Academic Performance

Student's academic performances were measured based on a number of quizzes, 3 exams, homework assignments, a weekly lab report, a number of discussions/presentations, and their contribution into the project-based lab.

Technical Outcomes

In our course, the project-based learning was a great success on many levels. Student success indicators chosen included the quality of their performance, satisfaction, and retention. Twelve of the fourteen students enrolled went on to successfully complete the course, ten of whom earned a B or higher. An anonymous survey was given three weeks after the course in which the students reported a high level of understanding, confidence and technical proficiency (Figures 1, 3, and 4). Students also reported that they felt encouraged to take more biology courses (Table 3) which demonstrates the course's ability to promote the growth of STEM courses.

At the pedagogical level, instead of wasting the material by doing unrelated labs and achieving marginal learning outcomes, students learned the subject matter and the lab techniques from an applied approach in which they embraced the project, cloned a gene, and expressed the protein. In the end, the project-based lab was completed with the highest technology at the lowest expense and made a new construct for future research purposes at UIC that was proudly named "pBio-251"! The students' satisfaction is confirmed by the fact that every one of them indicated that they would recommend this course to others (Figure 2).

Did your experience in Bio 251 encourage you to take more boiology courses in the future?

	Response percent	Response count
Yes	91.7%	11
No	8.3%	1

Figure 2: Survey Excerpt: The influence of the course on students' intention in taking more biology courses.

Would you recommend this course to other students?

	Response percent	Response count
Yes	100%	12
No	0%	0

Figure 3: Survey Excerpt: Students' response on whether or not recommending the course to others.

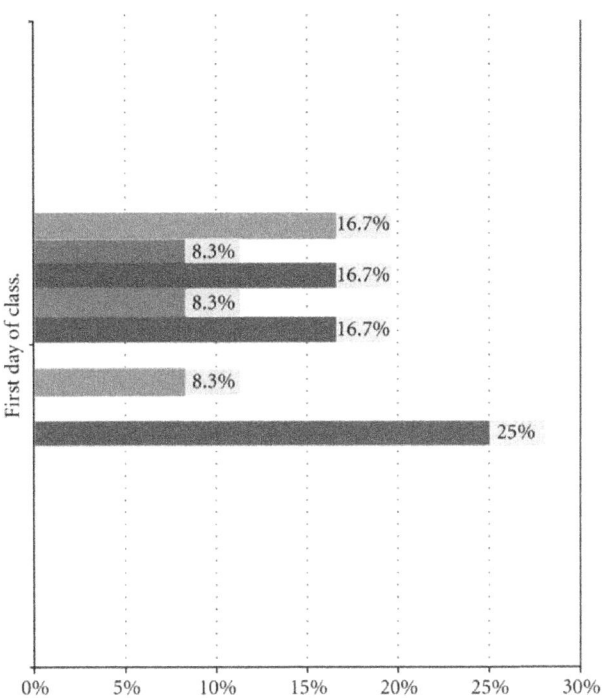

Figure 4: Survey Excerpt: Students' perception of their own comfort with the biotechnology on the first day of the class.

SURVEY ANALYSIS AND EXPLANATION

At the end of semester, a survey was given to the students which contained 10 questions. However, in this paper, we have only selected 6 questions that directly pertained to the topic of this paper. Table 2 shows that when students were asked to rate various statements related to their own learning outcomes, the majority of them, (66.7%), felt confident in their ability to perform the biotechnology lab procedures learned in the class. More than 80% of the students indicated that they understood the real world application of biotechnology lab procedures learned in the class. Furthermore, 75% of the students indicted that they gained an understanding of the fundamental processes involving prokaryotic genome. Finally, more than half of the students (58.3%) indicated that they had gained insight on how medical research is conducted.

Table 2: Survey Excerpt: Students' reflections on their own learning outcomes.

Please rate the following statements about your learning outcomes							
	Strongly disagree	Disagree	Neutral	Agree	Strongly agree	Rating average	Response count
I am confident in my ability to perform the biotechnology procedures I learned from Bio-251	0.0% (0)	8.3% (1)	0.0% (0)	25.0% (3)	66.7% (8)	4.50	12
I understand the real world application of the biotechnology procedures I learned from Bio-251	0.0% (0)	0.0% (0)	8.3% (1)	8.3% (1)	83.3% (10)	4.75	12
I have gained insight on how medical scientific research is conducted	0.0% (0)	0.0% (0)	0.0% (0)	41.7% (5)	58.3% (7)	4.58	12
I have gained an understanding of the fundamental processes involving the prokaryotic genome	0.0% (0)	0.0% (0)	0.0% (0)	25.0% (3)	75.0% (9)	4.75	12

Table 3: Survey Excerpt: Student's reflection on the Bio-251.

Please rate the following statements about the Bio-251 Course							
	Strongly disagree	Disagree	Neutral	Agree	Strongly agree	Rating average	Response count
The course provided a challenging learning environment	0.0% (0)	0.0% (0)	0.0% (0)	8.3% (1)	91.7% (11)	4.92	12
The course content was focused and relevant	0.0% (0)	0.0% (0)	0.0% (0)	16.7% (2)	83.3% (10)	4.83	12
I was encouraged to research topics outside of the classroom	0.0% (0)	0.0% (0)	0.0% (0)	16.7% (2)	83.3% (10)	4.83	12
The course promoted student collaboration	0.0% (0)	0.0% (0)	0.0% (0)	0.0% (0)	100.0% (12)	5.00	12

Table 3 shows that when students were asked to rate various statements related to the course; all of the students (100%) indicated that the course promoted students' collaboration, both in and outside of the class. Furthermore, the majority of the students, (91.7%), indicated that the course challenged them and provided a challenging learning environment that helped them to learn. In addition, more than 80% of the students indicated that the course content was focused and relevant to their learning, and the course and the learning environment encouraged them to research topics outside of the classroom.

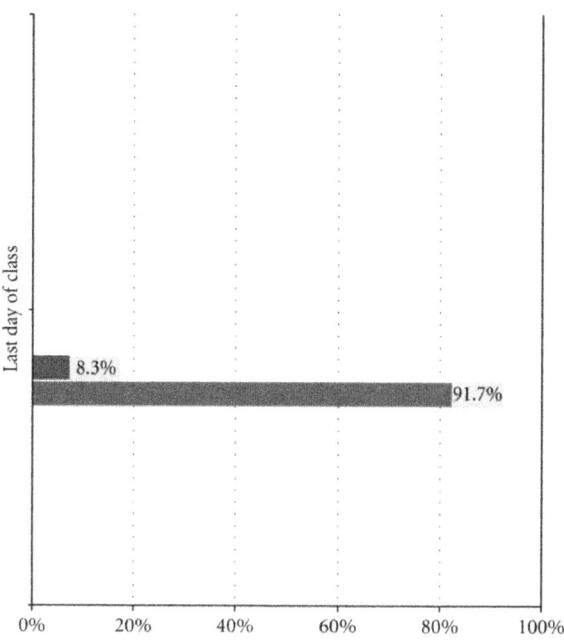

Figure 5: Survey Excerpt: Students' perception of their own comfort with biotechnology on the last day of the class.

Figure 2 shows that the majority of students, (91.7%), indicated that the course encouraged them to take more biology courses.

Furthermore, as shown in Figure 3, all of the students indicated that they felt confident in recommending the course to other students.

Figure 4 shows that only 25% of the students indicated that they were comfortable in performing biotechnology procedures on their own at the beginning of the class.

Figure 5 shows the significant gains the students made in their confidence in describing the methods of biotechnology, and performing biotechnology procedures by the end of the course. While at the beginning of the semester only 25% of the students indicated that they were comfortable in performing biotechnology procedures on their own, by the end of the semester, this level of confidence had increased to 91.7%. This increase is significant because it represents a 66.7% improvement in student confidence.

DISCUSSION AND CONCLUSION

Many educators would agree that the classic methods of teaching science is currently plagued by ineffective artifacts and thus is in dire need of modernization. While the facts delivered by the standard methods seldom change, the audience does. Even though certain students in each class can learn from the current methods, the rest of the students are too often weeded out, regardless of their innate talents. New methods of teaching are needed for the students of today, especially in STEM fields. The objective of promoting growth in the number of students enrolled in STEM programs is shifting from an advantage to a necessity in the modern world.

Project-Based Learning is a method in which students engage in intellectually challenging tasks that drive inquiry questions through gaining content knowledge and academic skills to solve complex problems and informatively defend their solution and outcomes. At HWC, project-based learning enabled the Bio-251 students to practice real world lab methods currently used in biotechnology, to comprehend the process of scientific inquiry that is practiced by working scientists, and to gain knowledge and demonstrate evidence of achievement. Learning research skills through practical experience is one of the main objectives of the project-based learning method, which has been supported by the NSF [6] and the NRC [7]. The individual experience of utilizing the laboratory stimulates increased excitement, knowledge, and confidence in performing applied scientific procedures. Important conceptual connections are made between critical thinking and practical applications through student learning on specifically designed projects. Throughout the semester, the class developed a good rapport that allowed them to work through and troubleshoot labs together, effectively changing the instructor's role from teacher to mentor. The students consistently worked together to overcome the obstacles encountered in class, many of which are issues encountered by professional scientists, as one student stated

The overall mood and environment of this course were unlike anything I had previously experienced. At first, I was wary about taking a 5-hour course on Friday evenings, but when coupled with the casual structure and team emphasis, it felt more like getting together with friends to work on a hobby project than a molecular biology project. The sessions were highly productive and when something did go wrong, we were able to call for a time-out to discuss, we were quick to diagnose, and there was never a shortage of volunteers to help keep things moving.

The team-oriented environment is further supported by the survey (Table 2). It is apparent that very few, if any, of these Bio-251 students will forget this experience. Students were able to understand the concepts presented to them, while at the same time contribute valuable data and insight for research at UIC and for the world of molecular biology. Furthermore, some of the students started seriously thinking about continuing their education beyond the associate degree level.

ACKNOWLEDGMENTS

The authors would like to thank Mr. William Thompson for his technical assistance; Dr. Maris Roze and Dr. Abour Cherif for helpful discussion and critically reading the paper and providing comments, and the all the students of Bio-251 (Section Z, Spring 2011) from Harold Washington College who participated actively in this study. The Biotechnology Program at HWC was funded by the Advanced Technological Education (ATE) of the National Science Foundation (NSF), grant no. 0903067. They also would like to acknowledge SENCER for the subaward from DUE/NSF, grant no. 0717407.

REFERENCES

1. D. J. Treacy, S. M. Sankaran, S. Gordon-Messer et al., "Implementation of a project-based molecular biology laboratory emphasizing protein structure-function relationships in a large introductory biology laboratory course," CBE Life Sciences Education, vol. 10, no. 1, pp. 18–24, 2011.
2. P. Brickman, C. Gormally, N. Armstrong, and B. Hallar, "Effects of inquiry-based learning on students' science literacy skills and confidence," International Journal for the Scholarship of Teaching and Learning, vol.

3, no. 2, pp. 1–22, 2009.
3. W. Ketpichainarong, B. Panijpan, and P. Ruenwongsa, "Enhanced learning of biotechnology students by an inquiry-based cellulase laboratory," International Journal of Environmental and Science Education, vol. 5, no. 2, pp. 169–187, 2010.
4. GAO, Higher Education, Federal Science, Technology, Engineering, and Mathematics Programs and Related Trends: A Report to the Chairman, Committee on Rules, House of Representatives, GAO Highlights, United States Government Accountability Office, 2005, GAO-06-114.
5. M. M. Cox, J. A. Doudna, and M. O'Donnell, Molecular Biology: Principles and Practice, W.H. Freeman and Company, New York, NY, USA, 2012.
6. NSF, Shaping the Future: New Expectations for Undergraduate Education in Science, Mathematics, Engineering, and Technology, National Science Foundation, Arlington, Va, USA, 1996.
7. NRC, Transforming Undergraduate Education in Science, Mathematics, Engineering, and Technology, National Academy of Sciences, Washington, DC, USA, 1999.

Chapter 6

CAROTENOIDS FROM HALOARCHAEA AND THEIR POTENTIAL IN BIOTECHNOLOGY

Montserrat Rodrigo-Baños[1], Inés Garbayo[2], Carlos Vílchez[2], María José Bonete[1], and Rosa María Martínez-Espinosa[1]

[1]Biochemistry and Molecular Biology Division, Agrochemistry and Biochemistry Department, Faculty of Sciences, University of Alicante, Ap. 99, E-03080 Alicante, Spain
[2]Algal Biotechnology Group, University of Huelva and Marine International Campus of Excellence (CEIMAR), CIDERTA and Faculty of Sciences, 21071 Huelva, Spain

ABSTRACT

The production of pigments by halophilic archaea has been analyzed during the last half a century. The main reasons that sustains this research are: (i) many haloarchaeal species possess high carotenoids production availability; (ii) downstream processes related to carotenoid isolation from haloarchaea is relatively quick, easy and cheap; (iii) carotenoids production by haloarchaea can be improved by genetic modification or even by modifying several cultivation aspects such as nutrition, growth pH, temperature, *etc.*; (iv) carotenoids are needed to support plant and animal life and human well-being; and (v) carotenoids are compounds highly demanded by pharmaceutical, cosmetic and food markets. Several studies about carotenoid production by haloarchaea have been reported so far, most of them focused on pigments isolation or carotenoids production under different culture conditions. However, the understanding of carotenoid metabolism, regulation, and roles of carotenoid derivatives in this group of extreme microorganisms remains mostly unrevealed. The uses of those haloarchaeal pigments have also been poorly explored. This work summarizes what has been described so far about carotenoids production by haloarchaea and their potential uses in biotechnology and biomedicine. In particular, new scientific evidence of improved carotenoid production by one of the better known haloarchaeon (*Haloferax mediterranei*) is also discussed.

INTRODUCTION

Carotenoids are pigments that have received considerable attention due to their biotechnological applications and, more importantly, their potential beneficial effects on human health [1,2,3]. These compounds are the second most abundant naturally occurring pigments in nature [4], and they are mainly C_{40} lipophilic isoprenoids ranging from colorless to yellow, orange, and red [5]. The production of such as kind of pigment has been described from plants and some microorganisms such as algae, cyanobacteria, yeast [6] and fungi [7,8].

Plants, algae, yeast, cyanobacteria and fungi have been considered good sources to isolate and even to produce carotenoids at high scale so far [9,10,11,12]. In fact, general characterizations of carotenoids isolated from those organisms are abundant in the literature, in which techniques such as spectrophotometry, thin layer chromatography (TLC), high performance liquid chromatography-mass spectrometry (HPLC-MS) and nuclear magnetic resonance spectroscopy (NMR) are used to define the carotenoids profile from specific species as well as the carotenoids chemical structure [13,14,15,16,17,18]. However, not too much attention has been paid to halophilic microorganisms, and in particular, to haloarchaea as microorganisms with high capability of carotenoids production.

Halophiles comprise a heterogeneous group of microorganisms that require salts for optimal growth. Even high salt concentration up to 4 M is required for some extremophilic species such those belonging the *Haloferacaceae* family, Archaea domain. The pigments produced by these halophilic organisms include phytoene, β-carotene, lycopene, derivatives of bacterioruberin, and salinixanthin [19]. *Dunaliella salina* is one of the better known halophilic microorganisms in terms of carotenoids production [20,21,22,23]. However, apart from that halophilic microalgae, only few studies have been carried out about production of carotenoids by halophiles and in most of the cases, the studies are focused on carotenoids isolation and characterisation by traditional biochemical procedures as those mentioned before [24,25,26].

Within the halophiles there is a family of particular interest in several fields of applications: micro-ecology, biotechnology and extreme metabolic adaptations. This is the case of the *Haloferacaceae* family (previously mentioned) grouping extreme halophilic archaea inhabiting salty environments such as marshes or salty ponds from where NaCl is obtained for human consumption [27,28,29,30]. The first study about carotenoid production by halophilic microorganisms from the *Haloferacaceae* family (previously called *Halobacteriaceae* family*)* were published in the latter half of the 1960s [31,32]. During the last two decades of the last century, several research works demonstrated that some haloarchaeal species not only produce carotenoids but also produce them at high concentration. This fact makes possible to propose haloarchaea as a good natural source for carotenoids production at large scale by means of suitable bioprocess engineering tools, namely specifically designed bioreactors.

This review summarised what it has been described up to now about carotenoids production by haloarchaea and its potential uses in Biotechnology and Biomedicine. The effect of different parameters on carotenogenesis in haloarchaea such as temperature, salt concentration, pH and carbon/nitrogen ration is also discussed.

CAROTENOIDS: STRUCTURE AND FUNCTIONALITY

Carotenoids are hydrophobic compounds which essentially consist of a C_{40} hydrocarbon backbone in the case of carotenes (*i.e.*, they contain 40 carbon atoms in eight isoprene residues), often modified by various oxygen-containing functional groups to produce cyclic or acylicxanthophylls. So, all carotenoids possess a long conjugated chain of double bond and a near bilateral symmetry around the central double bond, as common chemical features [33].This chain may be terminated by cyclic groups (rings) and can be complemented with oxygen-containing functional groups [34].

Carotenoids can be classified into different groups on the basis of the criteria used. Based on the basic chemical structure and the oxygen presence, carotenoids are classified into two types: carotenes or carotenoid hydrocarbons, composed of carbon and hydrogen only; and xanthopylls or oxygenated carotenoids, which are oxygenated and may contain epoxy, carbonyl, hydroxyl, methoxy or carboxylic acid functional groups [35]. Lycopene and β-carotene are examples of carotene carotenoids and lutein, canthaxanthin, zeaxanthin, violaxanthin, capsorubin and astaxanthin are xanthopyll carotenoids [36].

The degree of conjugation and the isomerization state of the backbone polyene chromophore determine the absorption properties of each carotenoid. Due to the numerous conjugated double bonds and cyclic end groups, carotenoids present a variety of stereoisomers with different chemical and physical properties. The most important forms commonly found among carotenoids are geometric (*E-/Z-*). A double bond links the two residual parts of the molecule either in an *E*-configuration with both parts on opposite sides of the plane, or a *Z*-configuration with both parts on the same side of the plane. Geometrical isomers of this type are inter-convertible in solution. This stereoisomerism exerts a marked influence on the physical properties. The conjugation system described imparts carotenoids with excellent light absorbing properties in the blue-green (450–550 nm) range of the visible spectrum. Because of this reason, biochemical techniques such as UV-Vis spectrophotometry or Raman spectroscopy can be used to analyze carotenoids production by plants and microorganisms [37]. When the criterion used to classify carotenoids is related to vitamin A, then the carotenoids can be categorized as follows: (a) vitamin A precursors that do not pigment such as β-carotene; (b) pigments with partial vitamin A activity such as cryptoxanthin, β-apo-8'-carotenoic acid ethyl ester; (c) non-vitamin A precursors that do not pigment or pigment poorly such as violaxanthin and neoxanthin; and (d) non-vitamin A precursors that pigment such as lutein, zeaxanthin and canthaxanthin [38].

Some of the most important carotenoids in terms of biotechnological and biomedical uses explored so far are: Astaxanthin (3,3'-dihydroxy-β,β'-carotene-4,4'-dione) [36,39,40], β-Carotene (β,β-carotene) [38,41,42,43,44], Canthaxanthin (β,β-carotene-4,4'-dione) [45,46,47,48], β-Cryptoxanthin (hydroxy-β-carotene) [38,49,50,51,52,53,54], Fucoxanthin [38,55], Lycopene (ψ,ψ-carotene) [33,56,57], Lutein (β,ε-carotene-3,3'-diol) [42,58,59,60,61], Zeaxanthin (β,β-carotene-3,3'-diol) [38,62,63], and Violaxanthin (5,6:5',6'-diepoxy-5,5',6,6'-tetrahydro-β-carotene-3,3'-diol) [64,65,66,67,68].

CAROTENOIDS IN THE CONTEXT OF LIFE

Carotenoids have received much attention because of their various and important biological roles in all living systems [4,69,70,71]. Although some of those biological roles have been already mentioned in the previous section, this Section includes details about carotenoids biological roles.

In most of the organisms, the most relevant biological functions of carotenoids are linked to their antioxidant properties, which directly emerge from their molecular structure. Xanthophyll carotenoids in particular are free radical scanvengers, potent quenchers of reactive oxygen species (ROS) and nitrogen oxidative species (NOS), and chain-breaking antioxidants. Asthaxanthin and canthaxantin, for example, are better antioxidants and scanvengers of free radicals than β-carotene. In recent years, the understanding of ROS-induced oxidative stress mechanisms and the search for suitable strategies to fight oxidative stress has become one the major goals of medical research efforts [3]. On the other hand, carotenoid pigments are one of these natural products responsible for colours: yellow, orange, red, and purple colours in a wide variety of plants, animals, and microorganisms are due to those compounds [72].

In animals and humans, these compounds are precursors of vitamin A (provitamin A activity) and retinoid compounds required for morphogenesis and embryonic development [35,73]. Vitamin A is well recognized as a factor of great importance for child health and survival, its deficiency causes disturbances in vision and various related lung, trachea and oral cavity pathologies. Animals and humans cannot synthesize carotenoids *de novo*, although are able to convert them into vitamin A. Diet is the only source for these precursors for retinol synthesis, fruits, vegetables and microalgae being the major suppliers of provitamin A active carotenoids [3,35]. Other biological roles and functions of carotenoids in these organisms include: absorbers of light energy, oxygen transporters, scavengers of active oxygen [2,74], antitumor and enhancers of *in vitro* antibody production [75,76]. In birds and fish, carotenoids are an important signal of good nutritional condition and are used in ornamental displays as a sign of fitness and to increase sexual attractiveness [71,77].

In algae and higher plants, carotenoids serve as regulators of plant growth and development, as accessory pigments in photosynthesis and as a photoprotectors. Thus, they contribute to light harvesting, maintaining the structure and function of photosynthetic complexes, quenching chlorophyll triplet states, scavenging ROS, and dissipating excess energy [34]. On the other hand, carotenoids are also precursors for the hormones abscisic acid (ABA) and strigolactones, and as attractants for other organisms, such as pollinating insects and seed-distributing herbivore [35]. In plants, those pigments are involved in various biological processes, such as photosynthesis, hormones synthesis, photomorphogenesis, photoprotection and development [4]. Apart from these important roles, due to their striking and rich color, carotenoids are important floral pigments serving to attract pollinators and seed dispersers.

Finally, microorganisms are a great source of diverse carotenoids. As mentioned before for other organisms, in microorganisms, carotenoids are in charge of light protection, cell color and antioxidative stress mechanisms. It is important to highlight that some carotenoids, such as salinixanthin or thermozeaxanthin, are only produced by some extremophilic microorganisms [78,79]. During the last 30 years, researchers, as well as research and development companies, have paid attention to microorganisms due to the high capability of carotenoid production that some species exhibit. This fact, coupled with new insights on molecular biology techniques and downstream process make those microorganisms good sources for carotenoids production at large scale.

CAROTENOIDS METABOLISM

Carotenoids are derived from the general isoprenoid biosynthetic pathway, along with a variety of other important natural substances such as steroids and gibberellic acid. The starting product required to synthetize all the isoprene derivatives is mevalonic acid which is transformed into a phosphorylated isoprene upon phosphorylation; this isoprene subsequently polymerises. In the course of polymerization, the number and position of the double bonds are fixed.

The synthesis and degradation of carotenes and xanthophylls, the regulation of carotenogenesis, as well as the role of these compounds, have been very well described in plants [4,80] and mammals [81]. Multi-gene engineering approaches have also contributed to better understanding of carotenoid metabolism [82].

The conversion of two molecules of geranylgeranyl pyrophosphate (GGPP) to phytoene, a compound common to all C_{40} carotenogenic organisms, constitutes the first reaction unique to the carotenoid branch of isoprenoid metabolism. From this step, slightly different reactions can be found in different organisms. Anoxygenic photosynthetic bacteria, non photosynthetic bacteria, and fungi desaturate phytoene either three or four times to yield neurosporene or lycopene, respectively. In contrast, oxygenic photosynthetic organisms (cyanobacteria, algae, and higher plants) convert phytoene to lycopene via carotene in two distinct sets of reactions. At the level of neurosporene or lycopene, the carotenoid biosynthesis pathways of different organism's branch to generate the huge diversity of carotenoids found in nature.

In photosynthetic organisms and tissues, the lipophilic carotenoid and bacteriochlorophyll (Bchl) or chlorophyll (Chl) pigment molecules associate non-covalently but specifically with integral membrane proteins. In non photosynthetic organisms and tissues, carotenoids, often protein bound, occur in cytoplasmic or cell wall membranes, oil droplets, crystals, and fibrils.

As mentioned before, animals are not able to synthesize carotenoids *de novo*. They are acquired throughout the diet. In human beings, it has been well demonstrated that most of the ingested carotenoids are absorbed into the gastrointestinal mucosal cells and appear unchanged in the circulation and tissues. In the intestine, the carotenoids are absorbed by passive diffusion after being incorporated into the micelles that are formed by dietary fat and bile acids. The micellar carotenoids are then incorporated into the chylomicrons and released into the lymphatic system [33]. Carotenoids are transported in the plasma exclusively by lipoproteins. Oxygen-functionalized carotenoids are more polar than carotenes. Thus, α-carotene, β-carotene and lycopene tend to predominate in low-density lipoproteins (LDL) in the circulation, whereas high-density lipoproteins (HDL) are major carriers of xanthophylls, such as cryptoxanthins, lutein and zeaxanthin. The delivery of carotenoids to extrahepatic tissues is accomplished through the interaction of lipoprotein particles with receptors and the degradation by lipoprotein lipase [83,84].

Although no less than forty carotenoids are usually ingested in the diet, only six carotenoids and their metabolites have been found in human tissues, suggesting selectivity in the intestinal absorption of carotenoids. In contrast, thirty-four carotenoids and eight metabolites are detected in breast milk and serum of lactating mothers. Recently, facilitated diffusion in addition to simple diffusion has been reported to mediate the intestinal absorption of carotenoids in mammals. The selective absorption of carotenoids may be due to uptake to the intestinal epithelia by means of facilitated diffusion and an unknown mechanism of excretion into the intestinal lumen. It is well known that β-carotene can be metabolised to vitamin A after intestinal absorption of carotenoids, but little is known about the metabolic transformation of non-provitamin A xanthophylls. The enzymatic oxidation of the secondary hydroxyl group leading to keto-carotenoids would occur as a common pathway of xanthophyll metabolism in mammals [38].

PRODUCTION OF CAROTENOIDS BY HALOARCHAEA

Type, Content and Biosynthesis of Haloarchaeal Carotenoids

Halophilic archaea are extreme halophilic microorganisms mainly grouped into the *Haloferacaceae* family, phylum *Euryarchaeota*, *Archaea* domain. They are (mostly) aerobic and generally red-pigmented. They constitute the predominant microbial communities in extreme halophilic environments as it was mentioned before. To be alive under those conditions they have adopted several strategies: (i) amino acidic residues predominate in halophilic proteins surface; (ii) cells accumulate high KCl intracellular concentrations to deal with high ionic strength or some osmolytes such as 2-sulfotrehalose [85]; (ii) cellular bilayers have different composition and structure, *etc*. Due to these adaptations, haloarchaea have become a good and innovative source of different molecules of high interest in biotechnology such as enzymes able to be active at high temperature and high ionic strength [86,87], PHB and PHA, carotenoids, *etc*.

Related to carotenoids, there is little information in the literature about the carotenoid profile of extremophile microorganisms compared with the information available from other organisms, and only few of them are focused on carotenoid production by archaea in general, and by haloarchaea in particular [58]. Figure 1 summarises the number of publications focused on carotenoids. It is important to highlight that despite the huge number of publications on that subject, only 1.3% of them are related to haloarchaeal carotenoids (780 papers about haloarchaeal carotenoids *vs.* 61590 papers about carotenoids in general). What is clearly supported by the literature is that most members of the family*Haloferacaceae* can synthesize C_{50} carotenoids, including bacterioruberin (as the most abundant C_{50} in most of the analysed haloarchaeal species) and its precursors (2-isopentenyl-3,4-dehydrorhodopin (IDR), bisanhydrobacterioruberin (BABR), and monoanhydrobacteriorub erin(MABR)) [32,88]. Several other derivatives have been found in minor amounts, such as 3,4-dehidromonoanhydrobacterioruberin, haloxanthin (which is a derivative of the previous one containing a peroxide end group) and 3,4-epoxymonoanhydrobacterioruberin, identified in *Haloferax volcanii* [26,89]. Other carotenoids such as phytoene, lycopene, and β-carotene are also produced by these species but at lower concentration [7]. Those carotenoids are located in the cell membrane and they are in charge of the colour shown by the red colonies when haloarchaea cells grow on solid media or the red colour shown by salted coastal ponds (mainly in summer). In fact, the content of bacterioruberin pigments in the biomass has been used to monitor the density of halophilic archaeal communities in halophilic environments [90].

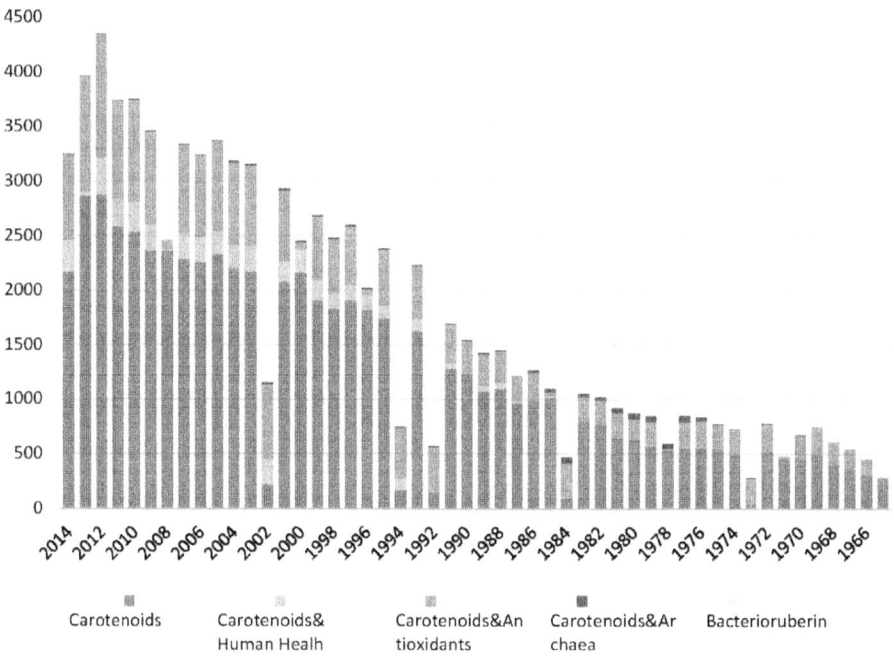

Figure 1. Bars plot summarizing details about the number of publications per year related to carotenoids. The key words used to perform the search were: carotenoids, carotenoids & human health, carotenoids & antioxidants; carotenoids & archaea and bacterioruberin. Pubmed and Scopus were used as databases to do the search.

Other carotenoids have been identified at very low concentrations in halophilic archaea: lycopersene, *cis*- and *trans*-phytoene, *cis*- and *trans*-phytofluene, neo-β-carotene and neo-α-carotene. The low concentrations of these compounds suggest that they may be used as precursors for the synthesis of other carotenoids including lycopene, retinal and the members of the bacterioruberin group. Some species may also produce the ketocarotenoid canthaxanthin in addition to other carotenoids [58]. Although this is the general carotenoid profile exhibited for most of the haloarchaeal species, it is important to note that some of them can produce high amounts of canthanxanthin, β-carotene and *trans*-astaxantin [91].

The presence of characteristic carotenoids (α-bacterioruberin and derivatives) in haloarchaea cells is easy to identify by Raman spectroscopy [30,92,93]. Thanks to this technique, α-bacterioruberin has been identified as the mayor carotenoid in the following haloarchaea: *Halobacterium salinarum* strains NRC-1 and R1, *Halorubrum sodomense, Haloarcula vallismortis* [78] and *Haloarcula japonica* (68.1% of the total carotenoids (mol %) [5]. The last species is also able to produce monoanhydrobacterioruberin (22.5%), bisanhydrobactrioruberin (9.3%), and isopentenyldehidrorhodopin (<0.1%) [5]. The main carotenoids produced by *Halorubrum* sp. TBZ126 were bacterioruberin, lycopene and β-carotene [58,79], while the major carotenoid produced by *Halococcus morrhuae* and *Halobacterium salinarum* was all-*trans*-bacterioruberin, accounting for 69% of the carotenoids, respectively [79].

Ronnekleivand colleagues [26] reported that *Haloferax volcanii* contained the (2S,2'S)-bacterioruberin (82% of total carotenoid), monoanydrobacterioruberin (7%), (2S,2□S)-bisanhydrobacterioruberin (3%), 3,4-dihydromonoanhydrobacterioruberin (2%) and two undecaene $C_{50}H_{74}O_4$ carotenoids (each 2%), the C_{45}-carotenoid (2S)-2-isopentenyl-3,4-dehydrorhodopin (1%) and lycopene (0.3%). The lipid composition of the extremely halophilic archaeon*Haloquadratum walsbyi* was investigated by thin layer chromatography and electrospray ionization-mass spectrometry. The results confirmed the presence of the carotenoids carotene and bacterioruberin, the C_{30}-isoprenoid compound squalene and the menaquinone with eight isoprenoid units vitamin MK-8 [94].

The total carotenoid content in *Haloarcula japonica* was 335 µg·g^{-1} of dry mass, although the contents in *Halobacterium salinarum* and *Halococcus morrhuae* were 89 and 45 µg·g^{-1}, respectively [79].

Although general knowledge about carotenoids biosynthesis and their assimilation in higher plants and human beings is considerable, nutritional functions, as well as metabolic pathways and their regulation, have not been examined in detail in haloarchaea [38]. The biosynthesis of carotenoids in haloarchaea was studied for the first time in the later 1970s. At that time, it was stated that the biosynthetic pathway for the formation of C_{40} carotenes in *Halobacterium* proceeds as follows: isopentenyl pyrophosphate leads to *trans*-phytoene, leads to *trans*-phytofluene, leads to ζ-carotene, leads to neurosporene, leads to lycopene, leads to gamma-carotene, and finally leads to β-carotene. This pathway differs from that in higher plants in that the *cis* isomers of phytoene and phytofluene are not on the main pathway of carotene biosynthesis, as they are in higher plants [95]. On the other hand, it has been suggested that bacterioruberin is synthesised by addition of C_5 isoprene units to each end of the lycopene chain, followed by the introduction of four hydroxyl groups. The evidence supporting these suggestions where reported around 40 years ago from experiments where nicotine was used to inhibit the bacterioruberin synthesis [96,97]. The presence of multiple genes for several steps in *Halobacterium* NRC-1 carotenoid production suggests that there may be more than one biosynthetic pathway [98,99].

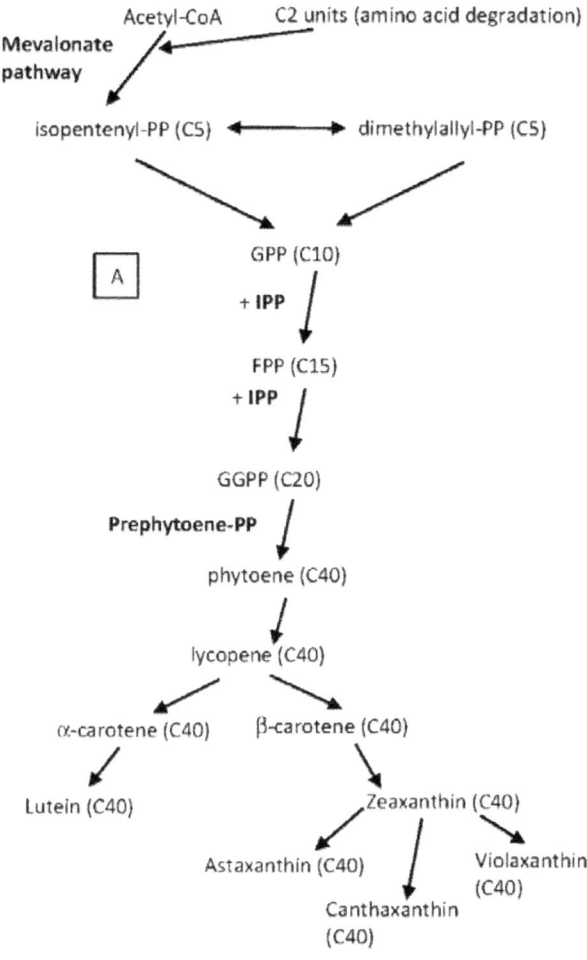

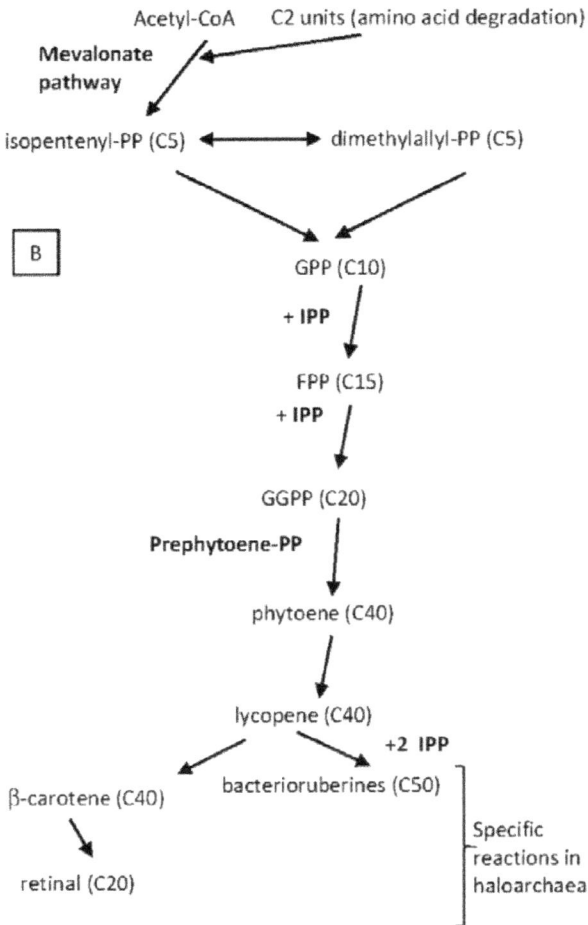

Figure 2. Comparison between the biosynthesis of isoprenoids in photosynthetic organisms (**A**) and the biosynthesis pathway proposed in haloarchaea (**B**). C5 prenyl units are synthesized via the mevalonate pathway starting from two acetyl-CoA molecules and C2 unit arising from amino acid degradation. *Cis*- and *trans*-prenyl chains are derived through head-tail (HT) condensation steps with isopentenyl-diphosphate (IPP). C15 and C20 prenyl chains are modified by head-head condensations and desaturase reactions. Genes coding for the enzymes catalysing those reactions have been identified in the *Halobacterium salinarum* genome. β-Carotene is the precursor for retinal synthesis while lycopene is the precursor for bacterioruberines in the pathway proposed from the *Halobacterium salinarum*'s genomic analysis [99]. Preliminary evidence from other genomic analysis (*Haloferax* sp.) also support this proposal [104]. Details about genes, enzymes and chemical reactions involved in retinal and

bacterioruberines synthesis are far from known. There are not reports about the biosynthesis reactions of other carotenoids such as zeaxanthin, canthaxanthin, astaxanthin, *etc.* in haloarchaea. GPP = geranyl diphosphate; FPP = farnesyl diphosphate; GGPP = geranylgeranyl diphosphate.

Computational genome and pathway analysis of halophilic Archaea done by Falb and co-workers [100] suggested that phytoene is reduced to lycopene by phytoene desaturase. Lycopene is the branching point for the synthesis of bacterioruberins (C_{50}) and β-carotene (C_{40}) [101,102]. Although the reactions leading from lycopene to bacterioruberins have not been elucidated in detail yet, there is some evidence supporting that the lycopene cyclase (OE3983R) converts lycopene to β-carotene in *Halobacterium salinarum* str. NRC-1 [98]. More recently, studies carried out in *Haloarcula japonica* have clearly identified that the genes named *c0507, c0506,* and *c0505* encoded a carotenoid 3,4-desaturase (CrtD), a bifunctional lycopene elongase and 1,2-hydratase (LyeJ), and a C_{50} carotenoid 2",3"-hydratase (CruF), respectively. The above three carotenoid biosynthetic enzymes catalyse the reactions that convert lycopene to bacterioruberin in *Haloarcula japonica* [103]. Figure 2compares the biosynthesis of carotenoids in photosynthetic organisms and in haloarchaea (on the basis of the results reported from *Halobacterium, Haloarcula* and preliminary evidence from *Haloferax* genomic analysis).

Bacterioruberin Is One of the Major Carotenoids Produced by Haloarchaea

As it can be concluded from the previous section, bacterioruberin is the main carotenoid component responsible for the colour of the red archaea of the family *Halobacteriaceae*. This pigment has a rather different molecular structure. It has a primary conjugated isoprenoid chain length of 13 C=C units with no subsidiary conjugation arising from terminal groups, which contain four –OH group functionalities only [37,78]. Table 1 summarises the bacterioruberin chemical structure as well as its derivatives.

This pigment protects the cells against damage produced by high intensities of light in the visible and ultraviolet range of the spectrum and provides aid in photoreactivation [106,107]. It is also involved in the reinforcement of the cell membrane. It was described for the first time from cells of *Halobacterium* species [88,108,109]. The byosinthesis of C_{50} carotenoids in general terms, and the effect of several chemical compounds on this biosynthesis were first described from *Halobacterium cutirubrum* (*Halobacteriaceae* family) [88,97,110].

Table 1. Chemical structures of bacterioruberin and its derivatives [79,105].

Name	Chemical Structure
Bacterioruberin	
Monoanhydrobacterioruberin	
Bisanhydrobacterioruberin	
Trisanhydrobacterioruberin	
2-isopentenyl-3,4-dehydrorhodopin	
5-cis-bacterioruberin	
9-cis-bacterioruberin	
13-cis-bacterioruberin	

A few years later, it was described that bacterioruberin is synthesized from other C_{50} carotenoids, such as isopentenyldehydrorhodopin, bisanhydrobacterioruberin, and monoanhydrobacterioruberin [5] and the synthesis is induced by (i) low oxygen tension and high light intensity [111,112]; (ii) osmotic stress [113]; and (iii) the presence of different compounds such as aniline [114] (Figure 2). However, this general pattern has some exceptions, for *example Haloquadratum walsbyi*: cells grown under osmotic stress did not experience changes in terms of either membrane lipid composition or carotenoids production [94]. Composition of the total carotenoids fraction in haloarchaea can also change on the basis of the nutritive factors within the culture media [105], the light intensity, oxygen tension, NaCl concentration [91,105,111], and other physical-chemical parameters such as pH value of the culture media. Figure 3 shows the effect of pH on carotenoids profile in *Haloferax mediterranei* cells grown in aerobic complex media. It has been reported that pH significantly influences cell growth and total carotenoid production in a lot of microorganism [115]. Hamidi *et al.*reported an analysis of pH and other environmental factors through response surface methodology on the total carotenoid production of extremely halophilic archaeon *Halorubrum* sp. TBZ126 [116]. They have found that optimum conditions for biomass and total carotenoid production occurred between pH 7 and 10 and biomass and total carotenoid production in pH 10 was about 93% and 90% respectively, compared to data reached from optimum conditions.

More recently, bacterioruberin has been used for the detection of extremely halophilic archaea embedded in halite in terrestrial and possibly extra-terrestrial samples [117].

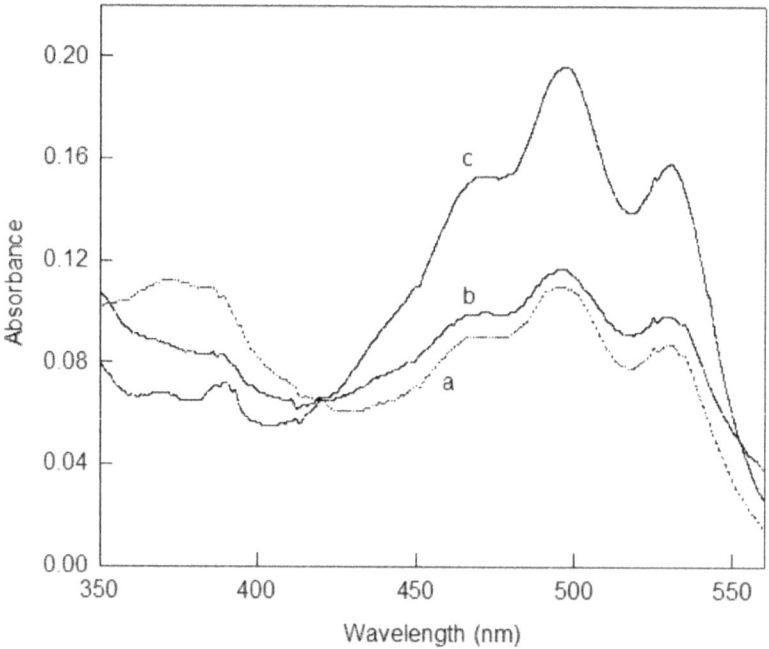

Figure 3. Absorption spectra of acetone extracts of *Haloferax mediterranei* cells grown in complex medium (**a**) pH 5; (**b**) pH 7 and (**c**) pH 9. *Hfx. mediterranei* was grown in complex medium pH 7 (0.5% yeast extract and 25% salted water) until shortly before the culture entered the stationary phase, after which cells were transferred to fresh

complex medium pH 5, 7 or 9. Although bacterioruberin and its derivatives possess extraordinary biological functions, the research regarding the biosynthesis regulation or practical applications of C_{50} carotenoids produced by halophilic archaea is still scarce.

Bacterioruberin Biological Roles

Bacterioruberin as antioxidant compound: The scavenging capacity of the oxygen reactive species (ROS) is dependent on the carotenoid concentration as it has been described so far. On the other hand, the antioxidant capacity of carotenoids in general is related to the length of the carbon chain, the number of pairs of conjugated double bonds and the carotenoids concentration [118,119,120]. As mentioned before, bacterioruberin contains 13 pairs of conjugated double bonds *versus* the nine pairs of conjugated double bonds of the β-carotene. Therefore, bacterioruberin is much better than β-carotene as radical scavenger [5,121]. It has been demonstrated that it protects the cells against oxidative damage. As a consequence of this important biological role, haloarchaea escape from fatal injury under strong light, and resist oxidative DNA damage resulting from radiography, UV irradiation, high doses (5 kGy) of gamma irradiation and H_2O_2 exposure [107,122]. What is clearly stated up to now is that the carotenoids of halophilic microorganisms present higher antioxidant capacity than the carotenoids produced by the other microorganism (extremophilic or not extremophilic).

Bacterioruberin controls membrane rigidity: With 4 hydroxyl substitutes in this dipolar C_{50} carotenoid, bacterioruberin was suggested to act as a "rivet" in the membrane cells. This carotenoid has some effect on fluidity of the membrane, acts as a barrier to water and allows permeability to oxygen and other molecules, so strains can survive in hypersaline or low-temperature conditions [105,123].

Bacterioruberin as part of the rhodopsin complexes: Archaerhodopsin-2 (aR2) is a retinal protein-carotenoid complex found in the claret membrane of *Halorubrum* sp. as well as in other species [124,125,126]. It functions as a light-driven proton pump highly important for haloarchaea cells to obtain energy. Using crystallographic studies it has been demonstrated that bacterioruberin binds to crevices between the subunits of the archaerhodopsin structure, which is a trimer. So, bacterioruberin sustains structural support related to the archaerhodopsin structure [127]. Bacterioruberin is also part of a complex constituted by this carotenoid and halorhodopsin in haloarchaea membranes such as those from *Natronomonas pharaonis*. Halorhodopsin is a retinal protein with a seven-transmembrane helix and acts as an inward light-driven Cl(−) pump [128].

BIOTECHNOLOGICAL USES AND PRODUCTION POTENTIALITY OF CAROTENOIDS FROM HALOARCHAEA

Biotechnological Uses of Carotenoids from Haloarchaea

Carotenoids have numerous applications as colorants (in food products and cosmetics), feed additives for poultry, livestock, fish, and crustaceans (Patent ES2324077 A1. See Table 2), antioxidants, antitumor and heart disease prevention agents, precursors of vitamin A and enhancers of *in vitro* antibody production. Hence, they are widely applied in the food, medical, pharmaceutical, and cosmetic industries as dyes and functional ingredients [3,6,58].

Table 2. Patents (last 20 years) related to carotenoids production by haloarchaea or its biotechnological uses (as pigments or as antioxidants). The key words used to find out the patents were: bacterioruberin, halobacteria, haloarchaea and carotenoids. Data obtained from different websites [129,130,131,132,133,134,135].

Publication Number	Publication Date	Title	International Application Number
WO/2009/042734	02.04.2009	Radiation-resistant mutants of a halophilic archaeon and uses thereof	PCT/US2008/077596
ES2324077 A1	29.07.2009	Compuesto a base de membranas celulares liofilizadas	
US 7939220 B2	10.05.2011	Proton-translocating retinal protein	PCT/EP2001/008715
WO2011133907 A2	27.10.2011	Methods to increase and harvest desired metabolite production in algae	PCT/US2011/033637
WO2012169623	13.12.2012	Method for producing carotenoid each having 50 carbon atoms	PCT/JP2012/064817
WO2014045280 A1	27.03.2014	Topical halobacteria extract composition for treating radiation skin tissue damage	PCT/IL2013/050786
WO/2014/045279	27.03.2014	Halobacteria extracts composition for tumour reduction	PCT/IL2013/050785
US 20140356854 A1	4.10.2014	Methods and compositions relating to mevalonate phosphate decarboxylase	
07-132096	23.05.1995	Production of C_{50} Carotenoid	

In some of the patents, the authors use the term Halobacteria, which was the first name used to identify what it is now call haloarchaea (Families *Halobacteriaceae* and *Haloferacaceae*).).

More than 600 carotenoids are known to occur naturally and many of them are still being identified. Although the carotenoids' market is highly segmented it has grown significantly in the last few years and this growth is projected to continue. β-Carotene and the xanthophylls astaxanthin, cantaxanthin, and lutein are the major carotenoids with commercial interest, and Europe is currently the largest market for this kind of compounds with nearly 45% of worldwide sales [136]. There are several advantages and disadvantages of chemical synthesis for carotenoids production. Chemical synthesis technology has been developed so far for many carotenoids (mainly for all of those most demanded by the market). This synthesis procedures produce carotenoids of exceptional purity and consistency at high concentration, and usually the overall cost of the production is relatively low. However, the chemical synthesis of certain carotenoids is very complex, and as a consequence of that, it is slow and expensive. Besides, the chemical synthesis of a new carotenoid usually requires the development of a new chemical route *in vitro*. Finally, some stereoisomers may not be active as the naturally occurring carotenoids isomers, or may have undesired side effects. As a consequence of all those aspects, the production of carotenoids from biological sources has been an area of intensive investigation. Also the consumer preference for natural products, as well as high costs, presence of by-products and damaging effects on the environment have together intensified efforts to identify alternative sources for chemical method. The production of natural colorants through fermentation has a number of advantages, such as cheaper production, higher yields, possibly easier extraction, less batch-to-batch variations and no seasonal variations. The production is flexible and can easily be controlled. Furthermore, the collection of microbial organisms is sustainable and usually microbial engineering has no negative impact on the environment [58]. Accordingly for all the previous reasons, different species of bacteria, moulds, yeasts and algae have attracted a great interest as alternative biosources for high-scale production of carotenoids [137,138]. Carotenoid-producing microorganisms have biotechnological attributes proper of microbial cells (fast growth in liquid culture and ability to accumulate or secrete some metabolites). Besides, the use of several molecular biology techniques enables the production of mutant strains able to overproduce carotenoids of interest. For these reasons, microorganisms have become excellent tools to look for new applied processes to obtain biomolecules of high interest in biotechnology and biomedicine and represent the basis of biotechnology-based companies [34].

New research results highlight the possibility of using halophilic microorganisms (mainly halophilic archaea) as natural sources for carotenoids production [19] thanks to the simplicity in increasing carotenoid production by culture conditions and genetic manipulation, and the feasibility of downstream processes of the cells to isolate the carotenoids. The extremely halophilic archaea has unique features making them suitable potential sources for carotenoids production, including: (i) the high-salt tolerance of haloarchaea enables their cultivation under non-sterile conditions because high salt concentrations prevent contamination by other organisms. This feature makes cultivation of haloarchaea advantageous if compared to cultivation of other microorganisms; (ii) the process to obtain the carotenoids is simple because in lower NaCl concentrations cell lysis is induced and consequently, carotenoids extraction could be conducted directly from the cells without any mechanical operation which is required in case of plants and (iii) the procedures for pigments extraction and purification seem to be simpler than those from other sources. Therefore, production potentiality of carotenoids from halophilic haloarchaea should be studied in order to assess alternative commercial sources for carotenoids [58].

Production Potentiality of Carotenoids from Haloarchaea

There are few examples of studies about haloarchaea carotenoids accumulation supporting the idea that these microorganisms might be considered good carotenoids producers [113,137], specifically for bacterioruberin and its C_{50}-related pigments. The culture conditions that are required to promote fast growth of halophilic archaea include high salt concentration (from 20% to 25% w/v). However, promoting massive carotenoid accumulation of these halophilic microorganisms generally requires much lower concentrations or NaCl, normally below 16% w/v [113,116]. Such lower salt concentrations address slower growth rates or even cell lysis. Therefore, carotenoid accumulation and growth of halophilic archaea are often opposite events. Moreover, besides the culture medium salinity, other key factors as temperature and pH may affect carotenoids accumulation and also growth rates of halophilic microorganisms tremendously [116]. In some cases, changes in the nutrient composition of the culture medium might result in enhanced C_{50} carotenoid accumulation [105]. Consequently, if cultivated under suitable conditions halophilic microorganisms may accumulate carotenoids.

So far, only few studies have been reported on the accumulation of carotenoids in halophilic archaea, and all of them at laboratory scale systems. The data obtained so far help clarify that the carotenoid accumulation potential of halophilic archaea is worth being studied in terms of bioprocess engineering. The maximal intracellular concentration of carotenoids reported so far for halophilic archaea were obtained in small flasks during laboratory experiments and ranges from 20 to 25 mg·g^{-1} dry weight, with maximal volumetric productions of about 10 mg·L^{-1} [116]. The intracellular concentration of 20–25 mg·g^{-1} dry weight means an accumulation rate per biomass unit of 2.0%–2.5%. These data compare well to those of carotenoid producing microalgae, many of which are below 1% [139] per biomass unit with the exception of *Dunaliella salina*, obviously the most efficient carotenoid producer microorganism [140].

Productivity is the key parameter to understand the potential of a microorganism for production of whatever value compound. For biomass, volumetric productivity is calculated in terms of g·L^{-1}·day^{-1}, and productivity of a desirable compound can also be calculated per reactor volume (g·L^{-1}·day^{-1}) or as mg·g^{-1} dry weight·day^{-1}. At large scale, surface is included as a factor to which productivity is referred. For instance, biomass areal productivity is expressed as g·m^{-2}·day^{-1}. Independently of how productivity is expressed, the rate at which biomass is produced in time obviously determines the potential of the microorganism for accumulation of desirable compounds. To our knowledge, no data have been published on both biomass and carotenoid productivities of halophilic archaea at scales other than just few examples in laboratory flasks. Data such these are needed to approach the potential of halophilic archaea for carotenoids production. An approach to the biomass productivity of halophilic archaea can be done simply with the data above. For instance, the average time taken for a culture of *Halorubrum* sp. and other halophilic archaea species to grow to the end of the exponential phase is about 10 days [116,141], in batch systems. Maximal biomass concentration reported is about 0.8 g·L^{-1}. Simple calculations give an average biomass productivity of 0.08 g (dry biomass)·L^{-1}·day^{-1}. This productivity data should be higher if the halophilic archaea were cultured in continuous production systems at optimal growing conditions. Accordingly, there should still be significant room for improving such laboratory biomass productivity data. Combining the average biomass productivity data (0.08 g (dry biomass)·L^{-1}·day^{-1}) and the maximal intracellular concentration of carotenoids found for *Halorubrum* sp. under specific conditions, 25 mg·g^{-1} dry biomass, a maximal productivity of carotenoids of about 2 mg·L^{-1}·day^{-1} is obtained. Microalgae, natural carotenoid producers, can be produced at biomass volumetric productivities likely between 0.1 and 0.3 g·L^{-1}·day^{-1} [139,142], which means maximal carotenoids productivity of about 1 to 3 mg·L^{-1}·day^{-1} for a microalga that accumulates carotenoids at 1% (w/w), therefore giving haloarchaea a chance to be considered in studies for assessing potential of carotenoid production. Of course, these productivity data are far from those carotenoid productivities obtained with *Dunaliella* species, which are 5 to 10-fold higher. Microalgae have the advantage of using natural light as energy source and carbon dioxide, and haloarchaea, which do not need light to grow, should have the advantage of being produced at a larger volume per surface ratio; this might help to overcome lower volumetric biomass productivities.

These aspects might at least be promising enough to study continuous processes of carotenoids production from halophilic archaea in laboratory and pre-pilot scale, particularly for production of other carotenoids than those typically obtained from microalgae (C_{50}-bacterioruberin and derivate C_{50} pigments). Therefore, it is suggested that haloarchaea might become complementary to those already known to be good carotenoid producers, namely microalgae, in the panel of potential producers. Of course this should just be the starting point of a challenging subject. The use of cheap, raw suitable carbon sources, the economics of cultivation—particularly energy costs for mixing and harvesting—biomass processing and carotenoid purification costs are among those key factors that should be extensively studied.

Although the biological roles of the carotenoids produced by haloarchaea in haloarchaeal cells are known (see Section 5), potential benefits of those carotenoids on animal cells (including human beings) have not been tested yet. However, there is some evidence supporting that carotenoids from haloarchaea are, at least, as efficient as those antioxidant compounds produced by other microorganism [79,143]. For instance, the halophilic bacteria *Halobacterium salinarum* produces various pigments such as phytoene, β-carotene, lycopene and derivatives of bacterioruberin and salinixanthin. These pigments have been tested for their free radical scavenging activity by DPPH (di(phenyl)-(2,4,6-trinitrophenyl)iminoazanium) assay and the results validated the known antioxidant activity of carotenoids. A further analysis of the cytotoxic properties against human liver cancer cell lines showed dose-dependent increase in cytotoxicity of the carotenoids on these cells, suggesting the probable anti-cancer properties [143,144]. This is the reason why several research groups around the world as well as some I&D (innovation and development) companies focused on secondary metabolites production have focused their attention on haloarchaea as carotenoids source. This interest is not only supported by the huge amount of publications on that subject, but also by the patents related to carotenoid production by haloarchaea (wild types as well as genetically modified strains) or methods/technologies to isolate and to purify those carotenoids. Table 2 summarises the patents focused on those subjects during the last 20 years.

CONCLUSIONS

On the basis of carotenoid production by haloarchaea in terms of quantity and variety, these microorganisms are revealed as good candidates to produce carotenoids at high-scale following cheap and quick culture and downstream processes. The most interesting carotenoids from a commercial point of view nowadays are not the major ones produced by haloarchaea as it can be concluded from the previous sections, with the exception of β-carotene, which is produced at significant concentration by several species. Since there are not studies on the potential benefits of the carotenoids produced by haloarchaea on human health reported in the scientific literature, more efforts should be made to properly address this question. Studies about carotenoid metabolism in haloarchaea are also required to provide further insights into the mechanisms controlling localized and context-specific carotenoid synthesis and degradation; such analysis would lead to a better understanding of the spatial distribution and function of different carotenoids and their derivatives in response to environmental and developmental signals. This knowledge may facilitate further progress in the field of carotenoid metabolic engineering in haloarchaea.

ACKNOWLEDGMENTS

This work was funded by research grant from the MINECO Spain (CTM2013-43147-R).

CONFLICTS OF INTEREST

The authors declare no conflict of interest. The founding sponsors had no role in the design of the study; in the collection, analyses, or interpretation of data; in the writing of the manuscript, and in the decision to publish the results.

REFERENCES

1. Zhang, J.; Sun, Z.; Sun, P.; Chen, T.; Chen, F. Microalgal carotenoids: Beneficial effects and potential in human health. *Food Funct.* 2014, *5*, 413–425.

2. Fiedor, J.; Burda, K. Potential role of carotenoids as antioxidants in human health and disease. *Nutrients* 2014, *6*, 466–488.

3. Vílchez, C.; Forján, E.; Cuaresma, M.; Bédmar, F.; Garbayo, I.; Vega, J.M. Marine carotenoids: Biological functions and commercial applications. *Mar. Drugs* 2011, *9*, 319–333.

4. Nisar, N.; Li, L.; Lu, S.; Khin, N.C.; Pogson, B.J. Carotenoid metabolism in plants. *Mol. Plant* 2015, *8*, 68–82.

5. Yatsunami, R.; Ando, A.; Yang, Y.; Takaichi, S.; Kohno, M.; Matsumura, Y.; Ikeda, H.; Fukui, T.; Nakasone, K.; Fujita, N.; *et al.* Identification of carotenoids from the extremely halophilic archaeon *Haloarcula japonica*. *Front. Microbiol.* 2014, *5*, 100–105.

6. Mata-Gómez, L.C.; Montañez, J.C.; Méndez-Zavala, A.; Aguilar, C.N.

Biotechnological production of carotenoids by yeasts: An overview. *Microb. Cell Fact.* 2014, *21*, 12.

7. Goodwin, T.W.; Britton, G. Distribution and analysis of carotenoids. In *Plant Pigments*; Goodwin, T.W., Ed.; Academic Press: London, UK, 1980; pp. 61–132.

8. Cunningham, F.X.; Gantt, E. Genes and enzymes of carotenoid biosynthesis in plants. *Annu. Rev. Plant Physiol. Plant Mol. Biol.* 1998, *49*, 557–583.

9. Blanco, A.M.; Moreno, J.; del Campo, J.A.; Rivas, J.; Guerrero, M.G. Outdoor cultivation of lutein-rich cells of*Muriellopsis* sp. in open ponds. *Appl. Microbiol. Biotechnol.* 2007, *73*, 1259–1266.

10. Nelis, H.J.; de Leenheer, A.P. Microbial sources of carotenoid pigments used in foods and feeds. *J. Appl. Bacteriol.* 1991,*70*, 181–191.

11. Bourgaud, F.; Gravot, A.; Milesi, S.; Gontier, E. Production of plant secondary metabolites: A historical perspective.*Plant Sci.* 2001, *161*, 839–851.

12. Olaizola, M. Commercial development of microalgal biotechnology: From the test tube to the marketplace. *Biomol. Eng.* 2003, *20*, 459–466.

13. Lichtenthaler, H.K.; Buschmann, C. Chlorophylls and Carotenoids: Measurement and Characterization by UV-VIS Spectroscopy. *Curr. Protoc. Food Analyt.Chem.* 2001, *F:F4:F4.3*.

14. Azevedo-Meleiro, C.H.; Rodriguez-Amaya, D.B. Confirmation of the identity of the carotenoids of tropical fruits by HPLC-DAD and HPLC-MS. *J. Food Comp. Anal.* 2004, *17*, 385–396.

15. Jaime, L.; Mendiola, J.; Herrero, M.; Soler-Rivas, C.; Santoyo, S.; Señorans, F.J.; Cifuentes, A.; Ibañez, E. Separation and characterization of antioxidants from *Spirulina platensis* microalga combining pressurized liquid extraction, TLC, and HPLC-DAD. *J. Sep. Sci.* 2005, *28*, 2111–2119.

16. Hengartner, U.; Bernhard, K.; Meyer, K.; Englert, G.; Glinz, E. Synthesis, isolation, and NMR-Spectroscopic characterization of Fourteen (Z)-Isomers of Lycopene and of come AcetylenicDidehydro- and Tetradehydrolycopenes.*Helv. Chim. Acta* 1992, *75*, 1848–1865.

17. Britton, G. Structure and properties of carotenoids in relation to function. *FASEB J.* 1995, *9*, 1551–1558.

18. Meléndez-Martínez, A.J.; Britton, G.; Vicario, I.M.; Heredia, F.J. Relationship between the colour and the chemical structure of carotenoid pigments. *Food Chem.* 2007, *101*, 1145–1150.

19. De Lourdes Moreno, M.; Sánchez-Porro, C.; García, M.T.; Mellado, E. Carotenoids' production from halophilic bacteria. *Methods Mol. Biol.* 2012, *892*, 207–217.
20. Oren, A. A hundred years of *Dunaliella* research: 1905–2005. *Saline Syst.* 2005, *4*, 2.
21. Oren, A. The ecology of *Dunaliella* in high-salt environments. *J. Biol. Res.* 2014, *21*, 23.
22. Hosseini Tafreshi, A.; Shariati, M. *Dunaliella* biotechnology: Methods and applications. *J. Appl. Microbiol.* 2009, *107*, 14–35.
23. Lamers, P.P.; Janssen, M.; de Vos, R.C.H.; Bino, R.J.; Wijffels, R.H. Exploring and exploiting carotenoid accumulation in *Dunaliella salina* for cell-factory applications. *Trends Biotechnol.* 2008, *26*, 631–638.
24. Asker, D.; Ohta, Y. Production of Canthaxanthin by Extremely Halophilic Bacteria. *J. Biosci. Bioeng.* 1999, *88*, 617–621.
25. Asker, D.; Awad, T.; Ohta, T. Lipids of *Haloferax alexandrinus* Strain TMT. An Extremely Halophilic Canthaxanthin-Producing Archaeon. *J. Biosci. Bioeng.* 2002, *93*, 37–43.
26. Ronnekleiv, M.; Liaaen-Jensen, S. Bacterial Carotenoids 53*, C_{50}-Carotenoids 23; Carotenoids of *Haloferax volcanii versus* other Halophilic Bacteria. *Biochem. Syst. Ecol.* 1995, *23*, 627–734.
27. Gupta, R.S.; Naushad, S.; Baker, S. Phylogenomic analyses and molecular signatures for the class *Halobacteria* and its two major clades: A proposal for division of the class *Halobacteria* into an emended order *Halobacteriales* and two new orders, *Haloferacales* ord nov and *Natrialbales* ord. nov., containing the novel families *Haloferacaceae* fam. nov. and*Natrialbaceae* fam. nov. *Int. J. Syst. Evol. Microbiol.* 2015, *65*, 1050–1069.
28. Oren, A. Life at high salt concentrations, intracellular KCl concentrations, and acidic proteomes. *Front. Microbiol.* 2013,*5*, 315.
29. Oren, A. Industrial and environmental applications of halophilic microorganisms. *Environ. Technol.* 2010, *31*, 825–834.
30. Oren, A. Halophilic archaea on Earth and in space: Growth and survival under extreme conditions. *Philos. Trans. A Math. Phys. Eng. Sci.* 2014, *13*, 372.
31. Schwieter, U.; Rüegg, R.; Isler, O. Syntheses in the carotenoid series. 21. Synthesis of 2,2'-diketo-spirilloxanthin (P 518) and 2,2'-diketo-bacterioruberin. *Helv. Chim. Acta* 1966, *49*, 992–996.
32. Kelly, M.; Jensen, S.L. Bacterial carotenoids. XXVI. C_{50}-carotenoids. 2.

Bacterioruberin. *Acta Chem. Scand.* 1967, *21*, 2578–2580.

33. Rao, A.V.; Rao, L.G. Carotenoids and human health. *Pharmacol. Res.* 2007, *55*, 207–216.

34. Del Campo, J.A.; García-González, M.; Guerrero, M.G. Outdoor cultivation of microalgae for carotenoid production: Current state and perspectives. *Appl. Microbiol. Biotechnol.* 2007, *74*, 1163–1174.

35. Rivera, S.M.; Canela-Garayoa, R. Analytical tools for the analysis of carotenoids in diverse materials. *J. Chromatogr. A* 2012, *1224*.

36. Fassett, R.G.; Coombes, J.S. Astaxanthin in cardiovascular health and disease. *Molecules* 2012, *17*, 2030–2048.

37. Jehlička, J.; Oren, A. Raman spectroscopy in halophile research. *Front. Microbiol.* 2013, *10*, 380.

38. Tanaka, T.; Shnimizu, M.; Moriwaki, H. Cancer chemoprevention by carotenoids. *Molecules* 2012, *17*, 3202–3242.

39. Higuera-Ciapara, I.; Félix-Valenzuela, L.; Goycoolea, F.M. Astaxanthin: A review of its chemistry and applications.*Crit. Rev. Food Sci. Nutr.* 2006, *46*, 185–196.

40. Ambati, R.R.; Phang, S.M.; Ravi, S.; Aswathanarayana, R.G. Astaxanthin: Sources, extraction, stability, biological activities and its commercial applications—A review. *Mar. Drugs* 2014, *12*, 128–152.

41. Stutz, H.; Bresgen, N.; Eckl, P.M. Analytical tools for the analysis of β-carotene and its degradation products. *Free Radic. Res.* 2015, *49*, 650–680.

42. Englert, M.; Hammann, S.; Vetter, W. Isolation of β-carotene, α-carotene and lutein from carrots by countercurrent chromatography with the solvent system modifier benzotrifluoride. *J. Chromatogr. A* 2015, *1388*, 119–125.

43. Li, Y.; Liu, S.; Man, Y.; Li, N.; Zhou, Y.U. Effects of vitamins E and C combined with β-carotene on cognitive function in the elderly. *Exp. Ther. Med.* 2015, *9*, 1489–1493.

44. Relevy, N.Z.; Harats, D.; Harari, A.; Ben-Amotz, A.; Bitzur, R.; Rühl, R.; Shaish, A. Vitamin A-Deficient Diet Accelerated Atherogenesis in Apolipoprotein E(−/−) Mice and Dietary β-Carotene Prevents This Consequence. *Biomed. Res. Int.* 2015, *2015*.

45. Tanaka, T.; Makita, H.; Ohnishi, M.; Mori, H.; Satoh, K.; Hara, A. Chemoprevention of rat oral carcinogenesis by naturally occurring xanthophylls, astaxanthin and canthaxanthin. *Cancer Res.* 1995, *55*, 4059–4064.

46. Surai, P.F. The antioxidant properties of canthaxanthin and its potential effects in the poultry eggs and on embryonic development of the chick, Part 1. *World Poult. Sci. J.* 2012, *68*, 465–476.
47. Rostami, F.; Razavi, S.H.; Sepahi, A.A.; Gharibzahedi, S.M. Canthaxanthin biosynthesis by *Dietzia natronolimnaea* HS-1: Effects of inoculation and aeration rate. *Braz. J. Microbiol.* 2014, *45*, 447–456.
48. Hojjati, M.; Razavi, S.H.; Rezaei, K.; Gilani, K. Stabilization of canthaxanthin produced by *Dietzia natronolimnaea* HS-1 with spray drying microencapsulation. *J. Food Sci. Technol.* 2014, *51*, 2134–2140.
49. Heying, E.K.; Tanumihardjo, J.P.; Vasic, V.; Cook, M.; Palacios-Rojas, N.; Tanumihardjo, S.A. Biofortified orange maize enhances β-cryptoxanthin concentrations in egg yolks of laying hens better than tangerine peel fortificant. *J. Agric. Food Chem.* 2014, *62*, 11892–11900.
50. Burri, B.J. β-Cryptoxanthin as a source of vitamin A. *J. Sci. Food Agric.* 2015, *95*, 1786–1794.
51. Granado-Lorencio, F.; de Las Heras, L.; Millán, C.S.; Garcia-López, F.J.; Blanco-Navarro, I.; Pérez-Sacristán, B.; Domínguez, G. β-Cryptoxanthin modulates the response to phytosterols in post-menopausal women carrying NPC1L1 L272L and ABCG8 A632 V polymorphisms: An exploratory study. *Genes Nutr.* 2014, *9*, 428.
52. Chisté, R.C.; Freitas, M.; Mercadante, A.Z.; Fernandes, E. Carotenoids are effective inhibitors of *in vitro* hemolysis of human erythrocytes, as determined by a practical and optimized cellular antioxidant assay. *J. Food Sci.* 2014, *79*, H1841–H1877.
53. Ghodratizadeh, S.; Kanbak, G.; Beyramzadeh, M.; Dikmen, Z.G.; Memarzadeh, S.; Habibian, R. Effect of carotenoid β-cryptoxanthin on cellular and humoral immune response in rabbit. *Vet. Res. Commun.* 2014, *38*, 59–62.
54. Li, D.; Xiao, Y.; Zhang, Z.; Liu, C. Light-induced oxidation and isomerization of all-*trans*-β-cryptoxanthin in a model system. *J. Photochem. Photobiol. B Biol.* 2015, *142*, 51–58.
55. Riccioni, G.; D'Orazio, N.; Franceschelli, S.; Speranza, L. Marine carotenoids and cardiovascular risk markers. *Mar. Drugs.* 2011, *9*, 1166–1175.
56. Igielska-Kalwat, J.; Gościańska, J.; Nowak, I. Carotenoids as natural antioxidants. *Postepy Hig. Med. Dosw.* 2015, *69*, 418–428.
57. Pirayesh Islamian, J.; Mehrali, H. Lycopene as a carotenoid provides radioprotectant and antioxidant effects by quenching radiation-induced

free radical singlet oxygen: An overview. *Cell J.* 2015, *16*, 386–391.

58. Naziri, D.; Hamidi, M.; Hassanzadeh, S.; Tarhriz, V.; Maleki Zanjani, B.; Nazemyieh, H.; Hejazi, M.A.; Hejazi, M.S. Analysis of Carotenoid Production by *Halorubrum.* sp. TBZ126: An Extremely Halophilic Archeon from Urmia Lake.*Adv. Pharm. Bull.* 2014, *4*, 61–67.

59. Flaks, B.; Bresloff, P. Some observations on the fine structure of the lutein cells of X-irradiated rat ovary. *J. Cell Biol.*1966, *30*, 227–236.

60. Altemimi, A.; Lightfoot, D.A.; Kinsel, M.; Watson, D.G. Employing Response Surface Methodology for the Optimization of Ultrasound Assisted Extraction of Lutein and β-Carotene from Spinach. *Molecules* 2015, *20*, 6611–6625.

61. Huang, Y.M.; Dou, H.L.; Huang, F.F.; Xu, X.R.; Zou, Z.Y.; Lin, X.M. Effect of supplemental lutein and zeaxanthin on serum, macular pigmentation, and visual performance in patients with early age-related macular degeneration.*Biomed. Res. Int.* 2015.

62. Costa, S.; Giannantonio, C.; Romagnoli, C.; Barone, G.; Gervasoni, J.; Perri, A.; Zecca, E. Lutein and zeaxanthin concentrations in formula and human milk samples from Italian mothers. *Eur. J. Clin. Nutr.* 2015, *69*, 531–532.

63. Li, X.R.; Tian, G.Q.; Shen, H.J.; Liu, J.Z. Metabolic engineering of *Escherichia coli* to produce zeaxanthin. *J. Ind. Microbiol. Biotechnol.* 2015, *42*, 627–636.

64. Yamamoto, H.Y.; Chang, J.L.; Aihara, M.S. Light-induced interconversion of violaxanthin and zeaxanthin in New Zealand spinach-leaf segments. *Biochim. Biophys. Acta* 1967, *141*, 342–347.

65. Yamamoto, H.Y.; Kamite, L.; Wang, Y.Y. An Ascorbate-induced Absorbance Change in Chloroplasts from Violaxanthin De-epoxidation. *Plant Physiol.* 1972, *49*, 224–228.

66. Sapozhnikov, D.I. Investigation on the violaxanthin cycle. *Pure Appl. Chem.* 1973, *35*, 47–61.

67. Soontornchaiboon, W.; Joo, S.S.; Kim, S.M. Anti-inflammatory effects of violaxanthin isolated from microalga *Chlorella ellipsoidea* in RAW 264.7 macrophages. *Biol. Pharm Bull.* 2012, *35*, 1137–1144.

68. Hallin, E.I.; Guo, K.; Åkerlund, H.E. Violaxanthin de-epoxidase disulphides and their role in activity and thermal stability. *Photosynth. Res.* 2015, *124*, 191–198.

69. Burton, G.W.; Foster, D.O.; Perly, B.; Slater, T.F.; Smith, I.C.; Ingold, K.U. Biological antioxidants. *Philos. Trans. R. Soc. Lond. B Biol. Sci.*

1985, *311*, 565–578.
70. Gammone, M.A.; Riccioni, G.; D'Orazio, N. Carotenoids: Potential allies of cardiovascular health? *Food Nutr. Res.* 2015,*59*.
71. LaFountain, A.M.; Prum, R.O.; Frank, H.A. Diversity, physiology, and evolution of avian plumage carotenoids and the role of carotenoid-protein interactions in plumage colour appearance. *Arch. Biochem. Biophys.* 2015, *572*, 201–212.
72. Namitha, K.K.; Negi, P.S. Chemistry and biotechnology of carotenoids. *Crit. Rev. Food Sci. Nutr.* 2010, *50*, 728–760.
73. Zile, M.H. Vitamin A and embryonic development: An overview. *J. Nutr.* 1998, *128*, 455S–458S.
74. Kaulmann, A.; Bohn, T. Carotenoids, inflammation, and oxidative stress—Implications of cellular signaling pathways and relation to chronic disease prevention. *Nutr. Res.* 2014, *34*, 907–929.
75. Gupta, C.; Prakash, D. Phytonutrients as therapeutic agents. *J. Complement. Integr. Med.* 2014, *11*, 151–169.
76. Ascenso, A.; Ribeiro, H.; Marques, H.C.; Oliveira, H.; Santos, C.; Simões, S. Chemoprevention of photocarcinogenesis by lycopene. *Exp. Dermatol.* 2014, *23*, 874–878.
77. Amundsen, C.R.; Nordeide, J.T.; Gjøen, H.M.; Larsen, B.; Egeland, E.S. Conspicuous carotenoid-based pelvic spine ornament in three-spined stickleback populations—Occurrence and inheritance. *Peer J.* 2015, *3*.
78. Jehlicka, J.; Edwards, H.G.; Oren, A. Bacterioruberin and salinixanthin carotenoids of extremely halophilic Archaea and Bacteria: A Raman spectroscopic study. *Spectrochim. Acta A Mol. Biomol. Spectrosc.* 2013, *106*, 99–103.
79. Mandelli, F.; Miranda, V.S.; Rodrigues, E.; Mercadante, A.Z. Identification of carotenoids with high antioxidant capacity produced by extremophile microorganisms. *World J. Microbiol. Biotechnol.* 2012, *28*, 1781–1790.
80. Othman, R.; Mohd Zaifuddin, F.A.; Hassan, N.M. Carotenoid biosynthesis regulatory mechanisms in plants. *J. Oleo Sci.* 2014, *63*, 753–760.
81. Palczewski, G.; Amengual, J.; Hoppel, C.L.; von Lintig, J. Evidence for compartmentalization of mammalian carotenoid metabolism. *FASEB J.* 2014, *28*, 4457–4469.
82. Giuliano, G. Plant carotenoids: Genomics meets multi-gene engineering. *Curr. Opin. Plant Biol.* 2014, *19*, 111–117.
83. Parker, R.S. Absorption, metabolism, and transport of carotenoids. *FASEB J.* 1996, *10*, 542–551.

84. Reboul, E.; Borel, P. Proteins involved in uptake, intracellular transport and basolateral secretion of fat-soluble vitamins and carotenoids by mammalian enterocytes. *Prog. Lipid Res.* 2011, *50*, 388–402.
85. Desmarais, D.; Jablonski, P.E.; Fedarko, N.S.; Roberts, M.F. 2-Sulfotrehalose, a novel osmolyte in haloalkaliphilic archaea. *J. Bacteriol.* 1997, *179*, 3146–3153.
86. Madern, D.; Camacho, M.; Rodríguez-Arnedo, A.; Bonete, M.J.; Zaccai, G. Salt-dependent studies of NADP-dependent isocitrate dehydrogenase from the halophilic archaeon *Haloferax volcanii*. *Extremophiles* 2004, *8*, 377–384.
87. Bonete, M.J.; Martínez-Espinosa, R.M. Enzymes from Halophilic Archaea: Open Questions. In *Halophiles and Hypersaline Environments: Current Research and Future Trends*; Ventosa, A., Oren, A., Eds.; Springer-Verlag GmbH: Berlin, Germany, 2011; pp. 358–370.
88. Kushwaha, S.C.; Kramer, J.K.; Kates, M. Isolation and characterization of C_{50}-carotenoid pigments and other polar isoprenoids from *Halobacterium cutirubrum*. *Biochim. Biophys. Acta* 1975, *398*, 303–314.
89. Bidle, K.A.; Hanson, T.E.; Howell, K.; Nannen, J. HMG-CoA reductase is regulated by salinity at the level of transcription in *Haloferax volcanii*. *Extremophiles* 2007, *11*, 49–55.
90. Oren, A.; Gurevich, P. Dynamics of a bloom of halophilic archaea in the Dead Sea. *Hydrobiologia* 1995, *315*, 149–158.
91. Asker, D.; Awad, T.; Ohta, Y. Lipids of *Haloferax. alexandrinus* strain TM^T: An extremely halophilic canthaxanthin-producing archaeon. *J. Biosci. Bioeng.* 2002, *93*, 37–43.
92. Marshall, C.P.; Leuko, S.; Coyle, C.M.; Walter, M.R.; Burns, B.P.; Neilan, B.A. Carotenoid analysis of halophilic archaea by resonance Raman spectroscopy. *Astrobiology* 2007, *7*, 631–643.
93. Jehlička, J.; Edwards, H.G.; Oren, A. Raman spectroscopy of microbial pigments. *Appl. Environ. Microbiol.* 2014, *80*, 3286–3295.
94. Lobasso, S.; Lopalco, P.; Mascolo, G.; Corcelli, A. Lipids of the ultra-thin square halophilic archaeon *Haloquadratum walsbyi*. *Archaea* 2008, *2*, 177–183.
95. Kushwaha, S.C.; Kates, M.; Porter, J.W. Enzymatic synthesis of C_{40} carotenes by cell-free preparation from *Halobacterium cutirubrum*. *Can. J. Biochem.* 1976, *54*, 816–823.
96. Kushwaha, S.C.; Kates, M. Effect of nicotine on biosynthesis of C_{50} carotenoids in *Halobacterium cutirubrum*. *Can. J. Biochem.* 1976, *54*,

824–829.

97. Kushwaha, S.C.; Kates, M. Effect of glycerol on carotenogenesis in the extreme halophile, *Halobacterium cutirubrum.Can. J. Microbiol.* 1979, *25*, 1288–1291.

98. Peck, R.F.; Echavarri-Erasun, C.; Johnson, E.A.; Ng, W.V.; Kennedy, S.P.; Hood, L.; DasSarma, S.; Krebs, M.P. brp and*blh* are required for synthesis of the retinal cofactor of bacteriorhodopsin in *Halobacterium salinarum. J. Biol. Chem.*2001, *276*, 5739–5744.

99. Dassarma, S.; Kennedy, S.P.; Berquist, B.; Victor, N.W.; Baliga, N.S.; Spudich, J.L.; Krebs, M.P.; Eisen, J.A.; Johnson, C.H.; Hood, L. Genomic perspective on the photobiology of *Halobacterium* species NRC-1, a phototrophic, phototactic, and UV-tolerant haloarchaeon. *Photosynth. Res.* 2001, *70*, 3–17.

100. Falb, M.; Müller, K.; Königsmaier, L.; Oberwinkler, T.; Horn, P.; von Gronau, S.; Gonzalez, O.; Pfeiffer, F.; Bornberg-Bauer, E.; Oesterhelt, D. Metabolism of halophilic archaea. *Extremophiles* 2008, *12*, 177–196.

101. Oesterhelt, D. Bacteriorhodopsin as an example of a light-driven proton pump. *Angew. Chem. Int. Ed. Engl.* 1976, *15*, 17–24.

102. Sumper, M.; Reitmeier, H.; Oesterhelt, D. Biosynthesis of the purple membrane of halobacteria. *Angew. Chem. Int. Ed. Engl.* 1976, *15*, 187–194.

103. Yang, Y.; Yatsunami, R.; Ando, A.; Miyoko, N.; Fukui, T.; Takaichi, S.; Nakamura, S. Complete Biosynthetic Pathway of the C_{50} Carotenoid Bacterioruberin from Lycopene in the extremely halophilic archaeon *Haloarcula japonica. J. Bacteriol.* 2015, *197*, 1614–1623.

104. Rodrigo-Baños, M.; Garbayo, I.; Vilchez, C.; Bonete, M.J.; Martínez. Espinosa, R.M. Genomic analysis of the biosynthesis of isoprenoids in *Haloferax* genus, to be submitted for publication.

105. Fang, C.J.; Ku, K.L.; Lee, M.H.; Su, N.W. Influence of nutritive factors on C_{50} carotenoids production by *Haloferax mediterranei* ATCC 33500 with two-stage cultivation. *Bioresour. Technol.* 2010, *101*, 6487–6493.

106. Dundas, I.D.; Larsen, H. A study on the killing by light of photosensitized cells of *Halobacterium salinarium. Arch. Mikrobiol.* 1963, *46*, 19–28.

107. Shahmohammadi, H.R.; Asgarani, E.; Terato, H.; Saito, T.; Ohyama, Y.; Gekko, K.; Yamamoto, O.; Ide, H. Protective roles of bacterioruberin and intracellular KCl in the resistance of *Halobacterium salinarium* against DNA-damaging agents. *J. Radiat. Res.* 1998, *39*, 251–262.

108. Kelly, M.; Norgard, S.; Liaaen-Jensen, S. Bacterial carotenoids. 31. C_{50}-

carotenoids 5. Carotenoids of *Halobacterium. salinarium*, especially bacterioruberin. *Acta Chem. Scand.* 1970, *24*, 2169–2182.

109. Becher, B.M.; Cassim, J.Y. Improved isolation procedures for the purple membrane of *Halobacterium halobium*. *Prep. Biochem.* 1975, *5*, 161–178.

110. Kushwaha, S.C.; Kates, M. Studies of the biosynthesis of C_{50} carotenoids in *Halobacterium cutirubrum*. *Can. J. Microbiol.* 1979, *25*, 1292–1297.

111. Shand, R.F.; Betlach, M.C. Expression of the bop gene cluster of *Halobacterium halobium* is induced by low oxygen tension and by light. *J. Bacteriol.* 1991, *173*, 4692–4699.

112. El-Sayed, W.S.; Takaichi, S.; Saida, H.; Kamekura, M.; Abu-Shady, M.; Seki, H.; Kuwabara, T. Effects of light and low oxygen tension on pigment biosynthesis in *Halobacterium salinarum*, revealed by a novel method to quantify both retinal and carotenoids. *Plant Cell Physiol.* 2002, *43*, 379–383.

113. D'Souza, S.E.; Altekar, W.; D'Souza, S.F. Adaptive response of *Haloferax mediterranei* to low concentrations of NaCl (<20%) in the growth medium. *Arch. Microbiol.* 1997, *168*, 68–71.

114. Raghavan, T.M.; Furtado, I. Expression of carotenoid pigments of haloarchaeal cultures exposed to aniline. *Environ. Toxicol.* 2005, *20*, 165–169.

115. Raghavan, T.; Furtado, I. Occurrence of extremely halophilic Archaea in sediments from the continental shelf of west coast of India. *Curr. Sci.* 2004, *86*, 1065–1067.

116. Hamidi, M.; Abdin, M.Z.; Nazemyieh, H.; Hejazi, M.A.; Hejazi, M.S. Optimization of Total Carotenoid Production by*Halorubrum* sp. TBZ126 using response surface methodology. *J. Microb. Biochem. Technol.* 2014, *6*, 286–294.

117. Fendrihan, S.; Musso, M.; Stan-Lotter, H. Raman spectroscopy as a potential method for the detection of extremely halophilic archaea embedded in halite in terrestrial and possibly extraterrestrial samples. *J. Raman Spectrosc.* 2009, *40*, 1996–2003.

118. Miller, N.J.; Sampson, J.; Candeias, L.P.; Bramley, P.M.; Rice-Evans, C.A. Antioxidant activities of carotenes and xanthophylls. *FEBS Lett.* 1996, *384*, 240–242.

119. Albrecht, M.; Takaichi, S.; Steiger, S.; Wang, Z.Y.; Sandmann, G. Novel hydroxycarotenoids with improved antioxidative properties produced by gene combination in *Escherichia coli*. *Nat. Biotechnol.* 2000, *18*, 843–

846.

120. Tian, B.; Xu, Z.; Sun, Z.; Lin, J.; Hua, Y. Evaluation of the antioxidant effects of carotenoids from *Deinococcus radiodurans* through targeted mutagenesis, chemiluminescence, and DNA damage analyses. *Biochim. Biophys. Acta* 2007, *1770*, 902–911.

121. Saito, T.; Miyabe, Y.; Ide, H.; Yamamoto, O. Hydroxyl radical scavenging ability of bacterioruberin. *Radiat. Phys. Chem.* 1997, *50*, 267–269.

122. Kottemann, M.; Kish, A.; Iloanusi, C.; Bjork, S.; DiRuggiero, J. Physiological responses of the halophilic archaeon*Halobacterium* sp. strain NRC1 to desiccation and gamma irradiation. *Extremophiles* 2005, *9*, 219–227.

123. Lazrk, T.; Wolff, G.; Albrecht, A.M.; Nakatani, Y.; Ourisson, G.; Kates, M. Bacterioruberins reinforce reconstituted halobacterium lipid-membranes. *Biochim. Biophys. Acta* 1988, *939*, 160–162.

124. Cao, Z.; Ding, X.; Peng, B.; Zhao, Y.; Ding, J.; Watts, A.; Zhao, X. Novel expression and characterization of a light driven proton pump archaerhodopsin-4 in a *Halobacterium salinarum* strain. *Biochim. Biophys. Acta* 2015, *1847*, 390–398.

125. Feng, J.; Liu, H.C.; Chu, J.F.; Zhou, P.J.; Tang, J.A.; Liu, S.J. Genetic cloning and functional expression in *Escherichia coli*of an archaerhodopsin gene from *Halorubrum xinjiangense*. *Extremophiles* 2006, *10*, 29–33.

126. Li, Q.; Sun, Q.; Zhao, W.; Wang, H.; Xu, D. Newly isolated archaerhodopsin from a strain of Chinese halobacteria and its proton pumping behavior. *Biochim. Biophys. Acta* 2000, *1466*, 260–266.

127. Yoshimura, K.; Kouyama, T. Structural role of bacterioruberin in the trimeric structure of archaerhodopsin-2. *J. Mol. Biol.* 2008, *375*, 1267–1281.

128. Sasaki, T.; Razak, N.W.; Kato, N.; Mukai, Y. Characteristics of halorhodopsin-bacterioruberin complex from*Natronomonas pharaonis* membrane in the solubilized system. *Biochemistry* 2012, *51*, 2785–2794.

129. Google (Key words: caroten and haloarchaea). Available online: https://www.google.es/?tbm=pts&gws_rd=cr,ssl&ei=md8wVcesMYyCPeKdgdgC#tbm=pts&q=caroten+%26+haloarchaea+patents (accessed on 13 April 2015).

130. Pantentscope (Key words: halobacteria, carotenoids and haloarchaea). Available online: https://patentscope.wipo.int/search/en/result.jsf (accessed on 14 April 2015).

131. Oficina Española de patentes y marcas - Invenciones. Available online: http://www.oepm.es/es/invenciones/resultados.html?field=TITU_RESU&bases=0&keyword=carotenoid (accessed on 15 April 2015).

132. World Intellectual Property Organization Global Brand Database. Available online: http://www.wipo.int/branddb/en (accessed on 21 April 2015).

133. Japan Platform for Patent information website. Available online: https://www.j-platpat.inpit.go.jp/web/all/top/BTmTopEnglishPage (accessed on 22 April 2015).

134. Espacenet Patent search. Available online: http://worldwide.espacenet.com/?locale=en_EP (accessed on 23 April 2015).

135. European patent register. Available online: https://register.epo.org/regviewer (accessed on 23 April 2015).

136. Markets and markets website—New market reports. Available online: http://www.marketsandmarkets.com/search.asp?Search=carotenoid&x=0&y=0 (accessed on 24 April 2015).

137. Yachai, M. Carotenoid Production by Halophilic Archaea and Its Applications. Ph.D. Thesis, Prince of Songkla University, Songkhla, Thailand, 2009.

138. Varela, J.C.; Pereira, H.; Vila, M.; León, R. Production of carotenoids by microalgae: Achievements and challenges. *Photosynth. Res.* 2015, *125*.

139. Norsker, N.; Barbosa, M.; Vermue, M.; Wijffels, R. Microalgal production: A close look at the economics. *Biotechnol. Adv.* 2001, *29*, 24–27.

140. Wichuk, K.; Brynjolfsson, S.; Fu, W. Biotechnological production of value-added carotenoids from microalgae: Emerging technology and prospects. *Bioengineered* 2014, *5*, 204–208.

141. Zeng, C.; Zhu, J.C.; Liu, Y.; Yang, Y.; Zhu, J.Y.; Huang, Y.P.; Shen, P. Investigation of the influence of NaCl concentration on *Halobacterium salinarum* growth. *J. Therm. Anal. Calorim.* 2006, *84*, 625–630.

142. Mata, T.; Martins, A.; Caetano, N. Microalgae for biodiesel production and other applications: A review. *Renew. Sustain. Energy Rev.* 2010, *14*, 217–232.

143. Abbes, M.; Baati, H.; Guermazi, S.; Messina, C.; Santulli, A.; Gharsallah, N.; Ammar, E. Biological properties of carotenoids extracted from *Halobacterium halobium* isolated from a Tunisian solar saltern. *BMC Complement. Altern. Med.* 2013, *13*, 255.

144. Sikkandar, S.; Murugan, K.; Al-Sohaibani, S.; Rayappan, F.; Nair, A.; Tilton, F. Halophilic bacteria-A potent source of carotenoids with

antioxidant and anticancer potentials. *J. Pure Appl. Microbiol.* 2013, 7, 2825–2830.

Chapter 7

THE BIFUNCTIONAL DIHYDROFOLATE REDUCTASE THYMIDYLATE SYNTHASE OF TETRAHYMENA THERMOPHILA PROVIDES A TOOL FOR MOLECULAR AND BIOTECHNOLOGY APPLICATIONS

Lutz Herrmann[1], Ulrike Bockau[1,2], Arno Tiedtke[2], Marcus WW Hartmann[1] and Thomas Weide[1]

[1]Cilian AG, Johann-Krane Weg 42, D-48149 Münster, Germany
[2]Institut für allgemeine Zoologie und Genetik, Universität Münster, Schloßplatz 5, D-48149 Münster, Germany

ABSTRACT

Background

Dihydrofolate reductase (DHFR) and thymidylate synthase (TS) are crucial enzymes in DNA synthesis. In alveolata both enzymes are expressed as one bifunctional enzyme.

Results

Loss of this essential enzyme activities after successful allelic assortment of knock out alleles yields an auxotrophic marker in ciliates. Here the cloning, characterisation and functional analysis of *Tetrahymena thermophila's* DHFR-TS is presented. A first aspect of the presented work relates to destruction of DHFR-TS enzyme function in an alveolate thereby causing an auxotrophy for thymidine. A second aspect is to knock in an expression cassette encoding for a foreign gene with subsequent expression of the target protein.

BACKGROUND

Tetrahymena thermophila is a ciliated eukaryotic unicellular organism belonging to the regnum of protozoa and bearing two nuclei, a transcriptionally silent, diploid germline micronucleus (MIC) and a transcriptionally active, polyploid somatic macronucleus (MAC)[1]. In 1923, when Nobel Laureate Andre Lwoff succeeded in growing *Tetrahymena* in pure culture, the basis for exploiting this alveolate as a model organism was laid. Milestone discoveries made in *T. thermophila* are the discovery of dynein motors[2], telomeres[3], RNA-mediated catalysis[4], telomerase[5] and the function of histone acetyltransferases in transcription regulation[6]. Within the last decades molecular biological techniques have been developed to alter *T. thermophila*'s genome and proteome: DNA transfection methods range from microinjection[7] and electroporation[8] into the MAC to biolistic bombardment of MIC and MAC[9]. Episomal plasmids based on an rDNA-replicon are available[10], as well as knock out/-in techniques based on homologous recombination[11, 12].

On protein level heterologous expression of related species has been performed[13, 14] and also endogenous proteins were silenced by a novel antisense-ribosome-technique[15]. The biotechnological potential of *T. thermophila* has been proven in numerous publications, demonstrating fast growth, high biomass, fermentation in ordinary bacterial/yeast equipment, up-scalability, existence of cheap and chemical defined media [16–18].

Although known and analysed for decades, only a few markers have been described for *T. thermophila*. So far there are ribosomal point mutation mediated resistances, a plasmid based neomycin resistance[19] and a beta-tubulin selection marker making use of an inducible promoter in combination with mutated tubulins being resistant or sensitive to the mitotic drug taxol[20]. Yet no true auxotrophic marker is available that permits selection without the use of antibiotics or drugs.

Critical enzymes in pyrimidine biosynthesis are dihydrofolate reductase (DHFR) and thymidylate synthase (TS). DHFR catalyses the production of tetrahydrofolate from dihydrofolate; TS is in charge of transferring a methyl-group from N5, N10-methylene-tetrahydrofolate to dUMP thereby generating dTMP and tetrahydrofolate. These enzymes being crucial for pyrimidine synthesis have been used as auxotrophic markers in various systems by targeted gene disruption but also a number of inhibitors (antifolates) have been developed as anti-cancer drugs[21]. In animals, fungi and eubacteria the DHFR and TS gene are separately translated, whereas plants, alveolata and euglenozoa have a bifunctional fusion gene with both enzyme activities combined in one protein ("DHFR-TS").

The occurrence of the bifunctional enzyme in *T. pyriformis* has been postulated in 1984 and 1985 but no functional or even molecular biological analysis had been performed. A partial amino acid sequence of DHFR-TS of a non determined «*T. pyriformis*-like strain" has been published in

2002[22], but this work is lacking any proof of linkage to the described partial cDNA to enzyme function. Here we present a first characterization of the *T. thermophila* DHFR-TS gene including gene structure and functional data on the enzyme and data on *in vivo* function. For the first time we show that the *T. thermophila* DHFR-TS locus provides an auxotrophic marker system that enables monitoring of allelic assortment processes. As PCR based approaches always have to cope with wildtype alleles present in the MIC, the DHFR-TS auxotrophy system is able to deliver direct proof of the allelic assortment to be completed: Only in the case of all wildtype alleles in the MAC having been substituted by the knock out construct the auxotrophy will occur. Combining this methodology with a knock in, a new, stable and useful strain to express recombinant enzymes or proteins has been generated.

RESULTS

Tetrahymenathermophila is a free-living ciliated protozoan that has become increasingly interesting as an excellent expression system. It is one of the best characterised unicellular eukaryotes and its genome has been sequenced in its entirety (The Institute for Genomic Research[23]). These features form the basis for using *T. thermophila* in future biotechnology applications. However, appropriate marker systems that are essential tools for genetic manipulations are limited. The enzymes dihydrofolate reductase (DHFR) and thymidylate reductase (TS) play a crucial for pyrimidine synthesis, therefore both enzymes have been used as auxotrophic marker-systems in various species by targeted gene disruption. The aim was to establish such a DHFR marker system for *T. thermophila*. Interestingly, the bifunctional DHFR-TS has been as taxonomic tool to set up phylogenetic trees for years. These approaches led to a very rudimental DHFR-TS amino acid fragment sequence of a non determined "*T. pyriformis*-like strain". By using this incomplete information and data from the entire *T. thermophila* genome it was possible to determine the whole sequence of the bifunctional DHFR-TS MAC gene in *T. thermophila*. The *T. thermophila* DHFR-TS gene and its flanking regions were amplified by using the primer pairs DHFR 5'1F NotI, DHFR 5›2R BamHI for the 5›region non coding region, the primer pair DHFR 3›1F XhoI: 5, DHFR 3›1R Acc65I for the 3›region and the primers DHFR CDS-F, DHFR CDS-F for amplification of the coding region.

The *T. thermophila*DHFR-TS gene structure

The comparison of the amino acid sequences of related species to the translated coding region of the DHFR-TS structure gene revealed the corresponding cDNA (see figure 1). Alignment with the cDNA with the gene revealed the intron exon boundaries. Figure 1 summarises the results and shows the MAC gene structure with two introns, the corresponding DNA sequence and the amino acid sequence of the *T. thermophila* DHFR-TS.

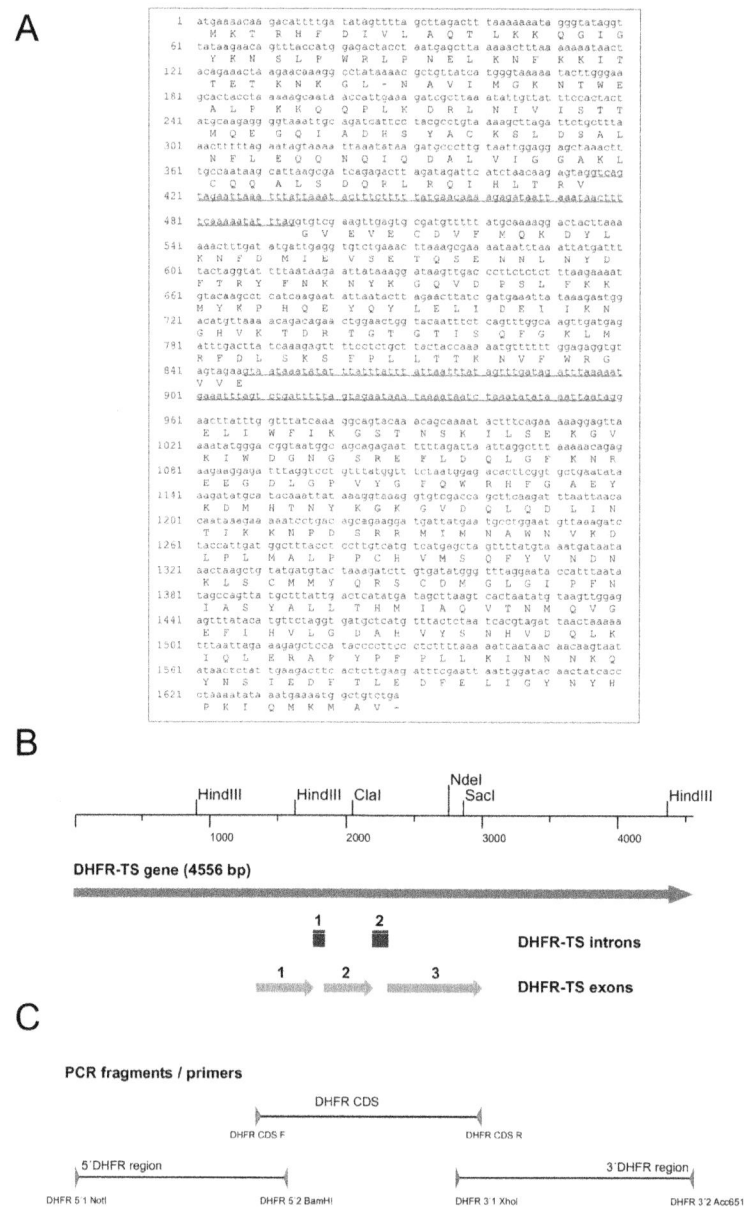

Figure 1: Genomic structure of *T. thermophila* bifunctional DHFR-TS enzyme. **A:** The DNA sequences of the gene and the cDNA as well as the deduced

amino acid sequence are shown on the left (introns underlined, blue). **B:** The DHFR-TS gene structure of *T. thermophila* consists of three exons (green) and two introns (blue). **C:** Overview of primer pairs to amplify DNA for homologous integration and to amplify the CDS.

Sequence alignment of the *T. thermophila*DHFR-TS enzyme to other alveolates

The alignment of the DHFR-TS between the alveolates *Tetrahymena thermophila*, *Plasmodium falciparum*, *Plasmodium vivax*, *Cryptosporidium hominis* and *Toxoplasma gondii* (Tt, Pf, Pv, Ch, Tg in figure 2) illustrates a 50–60% degree of identity between the proteins. This is also the case if the ciliate DHFR-TS is compared to the analogous enzyme of *Leishmania major* and *Trypanosoma cruzi* (Lm, Tc in figure 2).

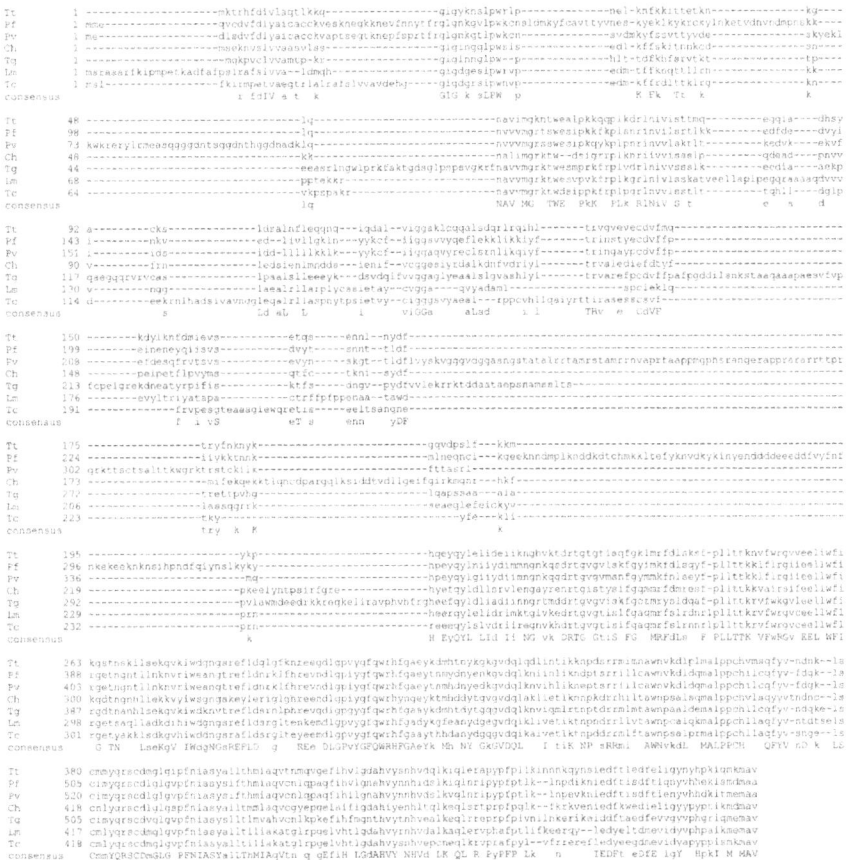

Figure 2: Comparison of the *T. thermophila* DHFR-TS bifunctional enzyme

to other protozoa. Amino acid sequence comparison of *Tetrahymena thermophila* (Tt), *Plasmodium falciparum* (Pf), *Plasmodium vivax* (Pv), *Cryptosporidium hominis* (Ch), *Toxoplasma gondii* (Tg), *Leishmania major* (Lm) and *Trypanosoma cruzi* (Tc) DHFR-TS bifunctional enzyme.

This is almost due to the highly conserved thymidylate synthase part of the bifunctional enzymes. A highly conserved histidine is found at position 198, suggesting that the amino acids 198–485 represent the thymidylate synthase part of the bifunctional enzyme. Much more heterogeneity is found in the N-terminal part of the bifunctional enzyme that consists of the DHFR part and a linker site. The alignment reveals that almost all enzymes bear one or more individual, non-homologue amino acid stretches (figure 2).

Proper integration of the knock out construct into the DHFR-TS locus and auxotrophy for thymidine

In order to verify that only one DHFR-TS activity is present in *T. thermophila* and that the sequence given in figure 1 corresponds to this activity a construct for knocking out the DHFR-TS gene was made. By inserting the flanking regions of the DHFR-TS gene into a pBS II SK backbone we created a plasmid for stable integration into the ciliate genome. The *neo2* cassette of pH4T2[19] was used to monitor the successful uptake of this plasmid by selection against paromomycin. Figure 3 shows that the *neo2* cassette is flanked by the ~1.5 kb fragments of the 5' and 3' parts of the non-coding regions of the DHFR-TS gene, respectively. Because the pBS backbone lacks an appropriate origin of replication, paromomycin resistant *T. thermophila* clones argue for proper homologous recombination in the DHFR-TS gene locus.

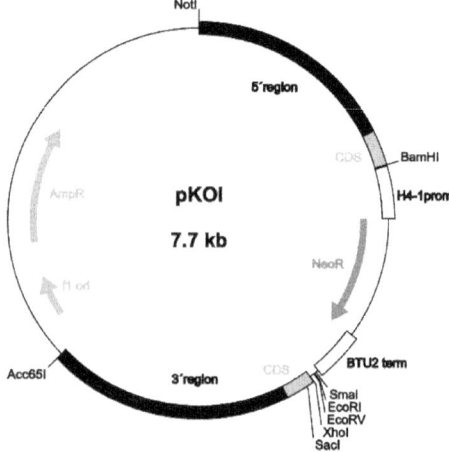

Figure 3: pKOI: The basis for a DHFR-TS knock out construct. Figure 3 shows

the knock out construct used for a targeted knock out of the DHFR-TS gene in *T. thermophila*. It consists of 3' and 5' flanking regions of the *T. thermophila* DHFR-TS gene and parts of its coding sequence (CDS), disrupted by a functional neomycin cassette conferring resistance to paromomycin. For more details refer to material and methods section.

The most convincing evidence for the correct integration into the DHFR-TS locus is a loss of the DHFR-TS activity in the transformed strain. However, in the case of *T. thermophila* this requires the complete replacement of all chromosomal DHFR-TS wildtype alleles (~45 ARPs) by the ARPs that includes the knock out cassette. We achieved this by allelic assortment, a phenomenon summarised in figure 4. This allelic or phenotypic assortment is based on randomised distribution of the MAC chromosomes units (ARPs) during mitosis. In order to force the assortment process into the desired direction – namely into the recombinant resistance gene – the transformed cells were cultivated for at least 3–4 weeks using increasing concentrations of the drug paromomycin. Single clones were isolated and tested for DHFR-TS deficiency by using a minimal chemical defined medium (CDM) with (+T) and without thymidine (-T). It could be shown that DHFR-TS knock out clones are real auxotrophic strains. The mutants are able to grow in CDM with thymidine like wildtype strains or strains with an incomplete allelic assortment. In CDM lacking thymidine they are unable to grow (figure 5).

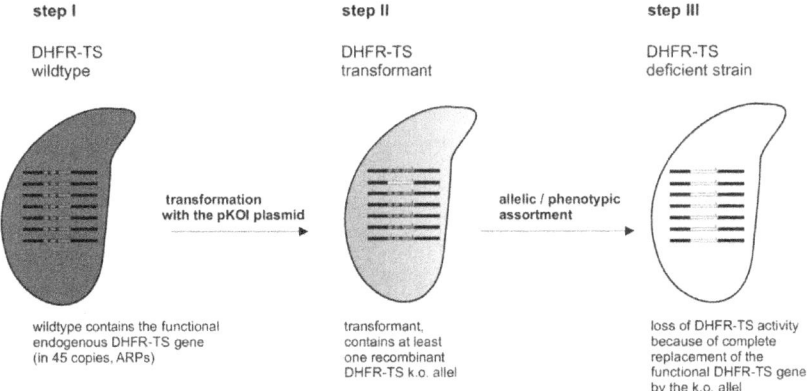

Figure 4: Generation of DHFR-TS deficient strains by allelic assortment. Wildtype strains are transfected with pKOI. In one copy of the 45 ARPs the endogenous DHFR-TS gene is substituted by the knock out construct (step 2). By amitotic division of the MAC and increasing selection pressure clones will arise, that have sorted out all endogenous DHFR-TS genes and retain only recombinant and defect DHFR-TS genes (step 3).

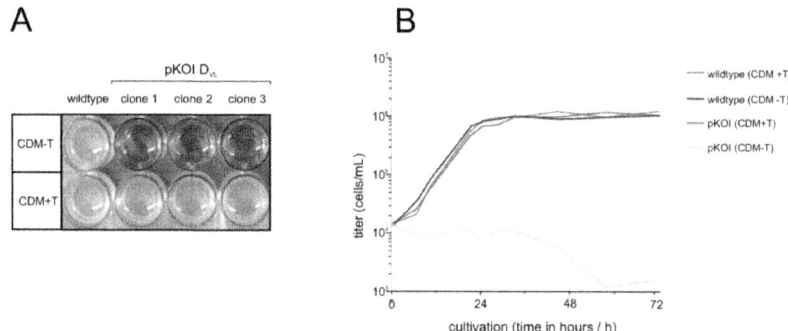

Figure 5: Growth of DHFR-TS deficient *T. thermophila* compared to wildtype. A: Selection of DHFR-TS knock out cells by growth on thymidine. *T. thermophila* cells with disrupted DHFR-TS gene (clone 1–3) do not grow without the presence of thymidine (CDM-T), whereas wildtype cells do. Addition of thymidine to the medium (CDM+T) recovers growth. **B:** Growth kinetics of DHFR-TS knock out cells compared to wildtype cells in media with or without thymidine show, that the knock out strain (pKOI) is growing as fast as wildtype cells on thymidine supplemented media (CDM+T). Knock out cells die without thymidine present (CDM-T). The curves are calculated by mean values of at least three independent experiments.

Functional addition of an expression cassette to the knock out construct

The *knockout* of the endogenous DHFR-TS gene of *T. thermophila* also provides the possibility to *knockin* a further foreign gene that can be expressed heterologously in the DHFR-TS knock out strains. Therefore we named our constructs pKOI (pKOI: knock out and knock in). To illustrate this knock out/-in concept we constructed the pKOI D_{VL} plasmid. It consists of a pKOI backbone with an additional expression cassette that encodes the first 115 amino acids (aa) of the precursor sequence of the PLA_1 gene and the mature human DNase I (aa 23 to 281). The PLA_1 pre/pro-peptide (aa 1 to 110) has significant similarity to members of the cathepsin L family and mediates secretion into the medium[24]. The five additional amino acids (aa 111 to 115) should ensure an optimal cleavage of the pro PLA_1-DNaseI fusion protein by endogenous pro-peptidases. In contrast to the *neo2* cassette the expression of the ppPLA_{115}-DNase I fusion protein is regulated by the inducible *MTT1* promoter[25]. The inducible system was selected because it allows a clear discrimination between the DNase activity of heterologously expressed recombinant human DNase I and the basal activity due to at least two endogenous DNases. The transformation, selection of positive clones and the directed allelic assortment were done as described in the material and methods section. Furthermore, we tested the correct and complete integration of both expression cassettes (*neo2* and the DNase I) in the DHFR-TS locus by a PCR approach (figure 6). In order to demonstrate the pKOI concept, cells of these DHFR-TS knock out strains carrying the ppPLA$_{115}$-DNase expression cassette were treated with and without Cadmium. Only induced strains showed an elevated DNase activity in the supernatant (Figure 7). To confirm this enzymatic data a specific antiserum against human DNase I was used to analyse the cell extracts and the supernatant of these human DNase expressing DHFR-TS knock out strains by Western blot. The results illustrate that the DHFR-TS knock out strains are capable of expressing and secreting the functional recombinant human protein (Figure 8).

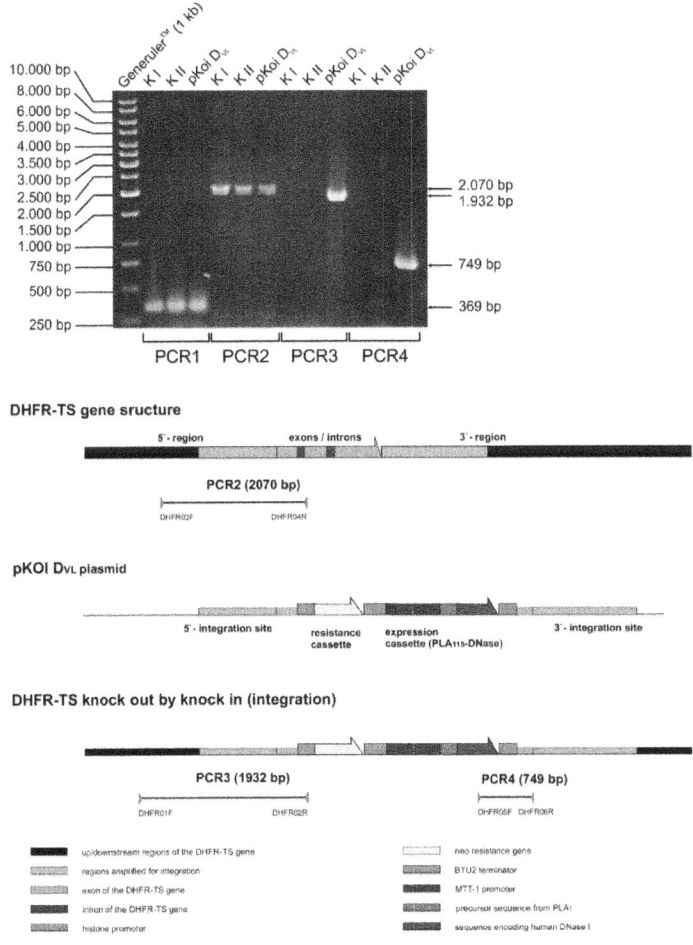

Figure 6: Proper integration of DHFR-TS knock out construct. This figure illus-

trates the PCR approach to determine that the knock out/-in construct has integrated into the DHFR-TS gene locus. Three different cells were tested: KI is a wildtype control, KII are cells transfected with a plasmid carrying only the disrupting cassettes but no DHFR-TS gene sequences and pKOI D_{VL} are cells transformed with the pKOI D_{VL} plasmid. PCR1 is a control reaction amplyfing 369 bp of the beta-hexosaminidase gene. PCR2 is to detect endogenous DHFR-TS (note that in the pKOID$_{VL}$ cells there still is a wildtype gene in the MIC). PCR3 only yields PCR-product for correctly integrated pKOI D_{VL} DNA. PCR4 shows that the full-length expression cassette has integrated. The first scheme illustrate the overall structure of the DHFR-TS gene, exons are coloured in green and introns in blue. The parts of the non-coding 5› and 3› regions of the DHFR-TS structure gene that we used as integration sites in the pKOI/ pKOI D_{VL} plasmids are coloured in grey. The next scheme gives on overall structure of the pKOID$_{VL}$ expression construct that was used for transformation and integration into the MAC. The third scheme illustrates the disruption of the endogenous DHFR-TS structure gene by proper integration of the resistance and expression cassettes into the MAC.

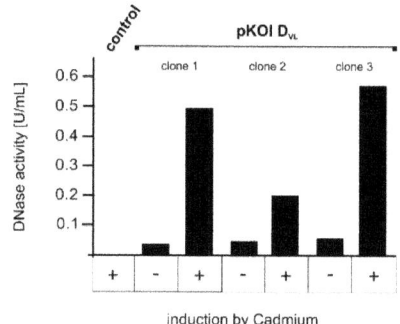

Figure 7: Enzyme function of knock in target. Supernatants of three clones transformed with pKOI D_{VL} were assayed for DNase activity. Only heavy metal induced cells (+) show high levels of DNase activity. Transformed, but non-induced cells show a slightly elevated enzyme activity compared to wildtype cells due to low basal promoter activity.

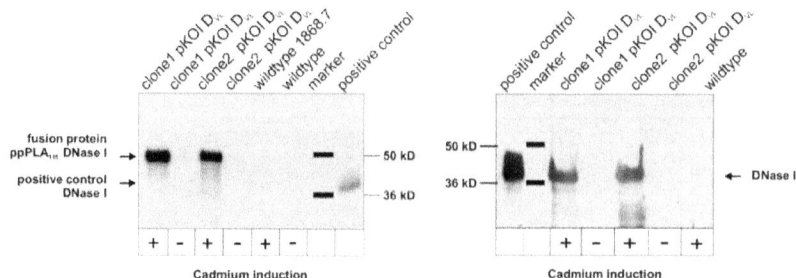

Figure 8: Expression of knock in target. Western blots show expression of recom-

binant human DNase I: Only transformed and heavy metal induced cells (+) show strong signals due to anti-DNase I antibodies. The intracellular PLA_1-DNase-fusion protein is visualized on the left blot by samples of cell lysates. Bands are running at higher molecular weight than the mature and processed positive control due to the pre/pro-peptide. On the right, supernatants were subjected to Western blot; the size of the secreted protein of *T. thermophila* argues for a correct processing when compared to the positive control.

DISCUSSION

The ciliated protozoan *T. thermophila* features two nuclei, a somatic macronucleus (MAC) and a genetic micronucleus (MIC) offering different possibilities of manipulating the organism›s properties (for a short and comprehensive review see[26]).

It is well known that altering the cell›s phenotype ultimately needs direct or indirect (by MIC transformation and conjugation) genetic engineering of the vegetative MAC. The first approaches for heterologous expression were done by using plasmids that use the vast amplification of the rDNA gene during *anlagen*/MAC development. However, the episomal presence of these plasmids depends on the drug concentration and the plasmid may recombine homologously and non-directionally into endogenous rDNA units.

Stable integration of expression cassettes into diploid MIC can be achieved by biolistic bombardment. After conjugation of two different mating types the old MACs disappear and new ones form that carry the new information derived from the recombinant MIC. The whole process follows the statistics of the Mendelian genetics. The advantage of this approach is that one obtains stable clones that maintain the genetic properties and that can be crossed via classical genetics to combine various properties of different *T. thermophila* strains. This approach is elaborative and time consuming. Furthermore, it has been shown recently that scan RNAs (snRNA) derived from the old MAC play an important role in DNA elimination during the development of the somatic MAC from the germline MIC (see [27] for review). The primary sequence of these small RNAs explains how the parental MAC epigenetically controls the genome rearrangement in the new MAC. In the case of stable MIC transformants this RNAi-like mechanism may cause partial deletion of foreign expression cassettes in the developing new MAC [28].

Therefore instead of episomal transformation by rDNA based plasmids or the stable transformation of the MIC we used a shortcut by combining MAC transformation with allelic assortment. This combination has several advantages. Firstly, the MAC transformation is much more efficient because there are at least about 45 potential integration sites per gene locus. This increases the probability of integration to at least one order of magnitude. Secondly, not only conjugating but also non-conjugating and therefore defined strains can be transformed. This is very important, because it allows the stepwise improvement of strains by maintaining defined genetic properties. Thirdly after completed allelic assortment the use of any antibiotic is obsolete as wildtype alleles have vanished in a one-way-manner. This allows cheap cultivation at large scales for e.g. biotechnical production processes.

CONCLUSION

In summary use of the DHFR-TS gene locus as a target for homologous genomic integration does not only provide a robust and simple marker system to assure complete allelic assortment of altered ARPs in the MAC but also the pKOI concept combining the DHFR-TS knock out with an additional knock in and subsequent expression of foreign genes.

METHODS

Cells and cell culture

Tetrahymena thermophila strains B 1868/4, B 1868/7 and B 2068/1 were cultivated in skimmed milk medium[29] in SPP (0.5% proteose peptone, 0.5% yeast extract, 0.1% ferrous sulphate chelate solution and 1% glucose) or in CDM medium[18].

Amplification of the DHFR-TS gene of *T. thermophila*

The DHFR-TS cDNA, gene including 5' and 3' flanking sites can be amplified using the following primer pairs. Nucleotides in small letters encode sites for restriction endonucleases. Amplification of the DHFR-TS 5' flanking region:

DHFR 5'1 F NotI: 5'-cccgcggccgcACAGAGTTAATGGAAATGGAGC-3'

DHFR 5'2R BamHI: 5'-gggggatccATATTTAAGCGATCTTTCAATGG-3'

Amplification of the DHFR-TS cDNA and gene with introns:

DHFR CDS-F:

5'-cgcGAATTCATGAAAACAAGACATTTTGATATAGTTTTAGC-3'

DHFR CDS-R:

5'-gcgCTCGAGTCAGACAGCCATTTTCATTTATATTTTAGGG-3'

Amplification of the DHFR-TS 3' flanking region:

DHFR 3'1F XhoI: 5'-gggctcgagATGCTCATGTTTACTCTAATCACG-3'

DHFR 3'1R Acc65I: 5'-gggggtaccAGTAAAAATAGAGTAGAAGGAG-3'.

Construction of plasmids

Construction of pKOI

The pKOI (knock out/-in) plasmid was constructed as follows: As backbone for selection and propagation in *E. coli.* the pBlueScript II SK plasmid was used. The 1.5 kb 5'-DHFR-TS integration site was amplified using the primer pair DHFR 5' 1 F NotI and DHFR 5' 2 R BamHI cloned into pBS II SK by using NotI and BamHI sites. Next the 1.4 kb paromomycin selection cassette from the pH4T2 *(neo2)* was cloned into the intermediate pBS IISK by BamHI and SmaI sites. Finally, the 3›-DHFR-TS integration site was amplified by primers DHFR 3› 1 F XhoI and DHFR 3› 1 R Acc65I and cloned by using the XhoI and Acc65I sites to finish the DHFR-TS knock out cassette. The SacI site of the pBS II SK backbone had been destroyed by site directed mutagenesis to facilitate the use of the endogenous SacI site in the 3›-DHFR-TS integrating sequence and the XhoI site as unique cloning site in pKOI. The whole pKOI basis vector is ~7.7 kb in size and contains a multiple cloning site (figure 3).

Construction of pKOI D_{VL}

The unique XhoI and SacI sites were used to insert the ppPLA$_{115}$-DNase knock in expression cassette (~2.5 kb). This cassette encodes a fusion protein of the first 115 aa of the endogenous PLA$_1$ precursor and the aa 23–281 of the mature human DNaseI, flanked by a ~1 kb *MTT1* promotor active sequence and the ~0.4 kb *BTU2* terminator, leading to a ~10.2 kb vector. To ensure proper translation of this fusion protein a codon optimised synthetic human DNase I gene was used (submitted at BMC Biotechnol.)

The junction sequence between the *MTT1* promoter and the precursor sequence of PLA$_1$ is given with the sequence ATGgatatcAAC, using a EcoRV site (gatatc) between the initial ATG of the *MTT1* gene and the second codon (AAC) of the PLA$_1$ precursor sequence. The cDNA that encodes the ppPLA$_{115}$-DNase I fusion protein ends with the TGA stop codon followed by a BglII site (agatct). Therefore the "knock in" cassette (generated in an intermediate vector) offers a modular structure that allows the simple replacement of the promoter (by XhoI/EcoRV), the coding sequence (by EcoRV/BglII) and the terminator DNA sequences (by BglII/SacI). Further details of the primers and constructs are available from the authors.

Transformation of pKOI plasmids (biolistic bombardment)

We used conjugating cells, as well as vegetative, growing or stationary *T. thermophila* strains. The transformation of the *T. thermophila* cells was performed as previously described in[9].

Selection, allelic assortment and DHFR-TS knock out assay

T. thermophila cell proliferation assay: For the first ca 16 h after biolistic bombardment transformants were grown in skimmed milk medium. After that transformed cells were grown on SPP medium with in increasing concentrations of paromomycin (from 100 µg/mL to 1000 µg/mL) to support the allelic assortment process. After 3–4 weeks each clone was cultivated on CDM replica plates with or without thymidine (10 mg/mL). Functional DHFR-TS knock out clones are only able to grow in CDM medium supplemented with thymidine. The viability of the DHFR-TS knock out strains was monitored by determining the growth kinetic. The complete integration of the DHFR-TS knock out and DNase I knock in cassette was confirmed by PCR using the following primers:

DHFR01F: 5›-CTTTTTAACAGCCTGCTGCTCG-3›

DHFR02R: 5›-GATTTTGATGCTTCAATAAGGTTG-3›

DHFR03F: 5›-TTATTTGTTTTATCATAGTGGAAAAGG-3›

DHFR04R: 5›-CAGACACCTCAATCATATCAAAG-3›

DHFR05F: 5›-GGTCCTCCATCAGATTGTGG-3›

DHFR06R: 5›-CGCGTCGAGTCAGACAGCCATTTTCATTTA-3›

Hex01F: 5›-ATGCAAAAGATACTTTTAATTACTTTC-3›

Hex02R: 5›-TATATTTTAGGAATGTTGTAATC-3›

A pH4T2 plasmid carrying the same *neo2* and DNase I expression/secretion cassettes was used as PCR control.

SDS-PAGE and Western blot

Aliquots of transformed cells and of SPP supernatants were resuspended in sample buffer and separated on 15 % SDS-PAGE. The gels were blotted onto nitrocellulose membranes and blocked in PBS containing 0.05 % Tween 20 and 5 % skim milk (PBS-TM). The expression of recombinant human DNase I in transformed Ciliates was detected by two specific anti sera from rabbit against human DNase I (antigen: recombinant human DNase I, Pulmozyme,

Roche). Both sera detected the recombinant DNase I antigen. The serum was used in a 1:500 dilution in PBS-TM. After washing with PBS/T and applying an HRP-conjugated anti rabbit serum. The blots were developed by using chemiluminescence.

DNase I activity assay

The methyl green based DNase activity assay was performed as already published [30]. Samples were incubated at 37°C for 24 h on a microtiter plate. Absorbance was measured at 620 nm. Calibration of the assay was achieved by different amounts of defined DNase I Units of Pulmozyme from Roche (CHO derived) in each experiment and linear regression. These results combined with semi-quantitative Western blotting were used to calculate the specific activity of expressed DNase I.

Sequence alignments

DHFR-TS sequences were aligned with Sci Ed Central's Clonemanger Professional Suite v6.0 using the BLOSUM 62 score matrix.

Declarations

Acknowledgements

We would like to thank Angelika Kronenfeld for excellent technical assistance and commitment. This work represents main parts of the PhD thesis of UB.

Authors' contributions

LH evaluated the DHFR-TS targeting sequences, setup and cloned the pKOI vector system and participated in manuscript drafting. UB participated in performing the experiments (cloning of pKOI D_{VL} expression constructs, transformation and screening of the ciliates, including monitoring of foreign gene expression) and in figure preparation. MWWH conceived of the study and participated in its design and coordination. TW participated in project conception, preparation of figures, did sequence alignments and was involved in manuscript drafting. All authors read and approved the final manuscript.

REFERENCES

1. Karrer KM: Tetrahymena genetics: two nuclei are better than one. Methods Cell Biol. 2000, 62: 127-186.
2. Gibbons IR, A.J R: Dynein: a protein with adenosine triphosphatase activity from cilia. Science. 1965, 149: 424-426.

3. Blackburn EH, Gall JG: A tandemly repeated sequence at the termini of the extrachromosomal ribosomal RNA genes in Tetrahymena. J Mol Biol. 1978, 120: 33-53. 10.1016/0022-2836(78)90294-2.

4. Cech TR, Zaug AJ, Grabowski PJ: In vitro splicing of the ribosomal RNA precursor of Tetrahymena: involvement of a guanosine nucleotide in the excision of the intervening sequence. Cell. 1981, 27: 487-496. 10.1016/0092-8674(81)90390-1.

5. Greider CW, Blackburn EH: Identification of a specific telomere terminal transferase activity in Tetrahymena extracts. Cell. 1985, 43: 405-413. 10.1016/0092-8674(85)90170-9.

6. Brownell JE, Zhou J, Ranalli T, Kobayashi R, Edmondson DG, Roth SY, Allis CD: Tetrahymena histone acetyltransferase A: a homolog to yeast Gcn5p linking histone acetylation to gene activation. Cell. 1996, 84: 843-851. 10.1016/S0092-8674(00)81063-6.

7. Pan WC, Blackburn EH: Single extrachromosomal ribosomal RNA gene copies are synthesized during amplification of the rDNA in Tetrahymena. Cell. 1981, 23: 459-466. 10.1016/0092-8674(81)90141-0.

8. Gaertig J, Gorovsky MA: Efficient mass transformation of Tetrahymena thermophila by electroporation of conjugants. Proc Natl Acad Sci U S A. 1992, 89: 9196-9200.

9. Cassidy-Hanley D, Bowen J, Lee JH, Cole E, VerPlank LA, Gaertig J, Gorovsky MA, Bruns PJ: Germline and somatic transformation of mating Tetrahymena thermophila by particle bombardment. Genetics. 1997, 146: 135-147.

10. Larson DD, Blackburn EH, Yaeger PC, Orias E: Control of rDNA replication in Tetrahymena involves a cis-acting upstream repeat of a promoter element. Cell. 1986, 47: 229-240. 10.1016/0092-8674(86)90445-9.

11. Hai B, Gaertig J, Gorovsky MA: Knockout heterokaryons enable facile mutagenic analysis of essential genes in Tetrahymena. Methods Cell Biol. 2000, 62: 513-531.

12. Gaertig J, Kapler G: Transient and stable DNA transformation of Tetrahymena thermophila by electroporation. Methods Cell Biol. 2000, 62: 485-500.

13. Clark TG, Gao Y, Gaertig J, Wang X, Cheng G: The I-antigens of Ichthyophthirius multifiliis are GPI-anchored proteins. J Eukaryot Microbiol. 2001, 48: 332-337. 10.1111/j.1550-7408.2001.tb00322.x.

14. Peterson DS, Gao Y, Asokan K, Gaertig J: The circumsporozoite protein

of Plasmodium falciparum is expressed and localized to the cell surface in the free-living ciliate Tetrahymena thermophila. Mol Biochem Parasitol. 2002, 122: 119-126. 10.1016/S0166-6851(02)00079-8.

15. Sweeney R, Fan Q, Yao MC: Antisense ribosomes: rRNA as a vehicle for antisense RNAs. Proc Natl Acad Sci U S A. 1996, 93: 8518-8523. 10.1073/pnas.93.16.8518.

16. Guberman A, Hartmann M, Tiedtke A, Florin-Christensen J, Florin-Christensen M: A method for the preparation of Tetrahymena thermophila phospholipase A1 suitable for large-scale production. J Appl Microbiol. 1999, 86: 226-230. 10.1046/j.1365-2672.1999.00651.x.

17. Wheatley DN, Rasmussen L, Tiedtke A: Tetrahymena: a model for growth, cell cycle and nutritional studies, with biotechnological potential. Bioessays. 1994, 16: 367-372. 10.1002/bies.950160512.

18. Hellenbroich D, Valley U, Ryll T, Wagner R, Tekkanat N, Kessler W, Ross A, Deckwer WD: Cultivation of Tetrahymena thermophila in a 1.5-m3 airlift bioreactor. Appl Microbiol Biotechnol. 1999, 51: 447-455. 10.1007/s002530051415.

19. Gaertig J, Gu L, Hai B, Gorovsky MA: High frequency vector-mediated transformation and gene replacement in Tetrahymena. Nucleic Acids Res. 1994, 22: 5391-5398.

20. Gaertig J, Thatcher TH, Gu L, Gorovsky MA: Electroporation-mediated replacement of a positively and negatively selectable beta-tubulin gene in Tetrahymena thermophila. Proc Natl Acad Sci U S A. 1994, 91: 4549-4553.

21. Huennekens FM: The methotrexate story: a paradigm for development of cancer chemotherapeutic agents. Adv Enzyme Regul. 1994, 34:397-419.: 397-419. 10.1016/0065-2571(94)90025-6.

22. Stechmann A, Cavalier-Smith T: Rooting the eukaryote tree by using a derived gene fusion. Science. 2002, 297: 89-91. 10.1126/science.1071196.

23. The Institute for Genomic Research. [http://www.tigr.org]

24. Weide T, Herrmann L, Bockau U, Niebur N, Aldag I, Laroy W, Contreras A, Tiedtke A, Hartmann MWW: Secretion of functional human enzymes by Tetrahymena thermophila. BMC Biotechnol. 2006, 6: 19-

25. Shang Y, Song X, Bowen J, Corstanje R, Gao Y, Gaertig J, Gorovsky MA: A robust inducible-repressible promoter greatly facilitates gene knockouts, conditional expression, and overexpression of homologous and heterologous genes in Tetrahymena thermophila. Proc Natl Acad Sci U S A. 2002, 99: 3734-3739. 10.1073/pnas.052016199.

26. Collins K, Gorovsky MA: Tetrahymena thermophila. Curr Biol. 2005, 15: R317-R318. 10.1016/j.cub.2005.04.039.
27. Yao MC, Chao JL: RNA-Guided DNA Deletion in Tetrahymena: An RNAi-Based Mechanism for Programmed Genome Rearrangements. Annu Rev Genet. 2005, .:
28. Yao MC, Fuller P, Xi X: Programmed DNA deletion as an RNA-guided system of genome defense. Science. 2003, 300: 1581-1584. 10.1126/science.1084737.
29. Kiy T, Tiedtke A: Continuous high-cell-density fermentation of the ciliated protozoon Tetrahymena in a perfused bioreactor. Appl Microbiol Biotechnol. 1992, 38: 141-146. 10.1007/BF00174458.
30. Sinicropi D, Baker DL, Prince WS, Shiffer K, Shak S: Colorimetric determination of DNase I activity with a DNA-methyl green substrate. Anal Biochem. 1994, 222: 351-358. 10.1006/abio.1994.1502.

Chapter 8

FUNCTION AND BIOTECHNOLOGY OF EXTREMOPHILIC ENZYMES IN LOW WATER ACTIVITY

Ram Karan[1,2], Melinda D Capes[1,2] and Shiladitya DasSarma[1,2]

[1]Department of Microbiology and Immunology, University of Maryland School of Medicine
[2]Institute of Marine and Environmental Technology, University System of Maryland

ABSTRACT

Enzymes from extremophilic microorganisms usually catalyze chemical reactions in non-standard conditions. Such conditions promote aggregation, precipitation, and denaturation, reducing the activity of most non-extremophilic enzymes, frequently due to the absence of sufficient hydration. Some extremophilic enzymes maintain a tight hydration shell and remain active in solution even when liquid water is limiting, e.g. in the presence of high ionic concentrations, or at cold temperature when water is close to the freezing point. Extremophilic enzymes are able to compete for hydration via alterations especially to their surface through greater surface charges and increased molecular motion. These properties have enabled some extremophilic enzymes to function in the presence of non-aqueous organic solvents, with potential for design of useful catalysts. In this review, we summarize the current state of knowledge of extremophilic enzymes functioning in high salinity and cold temperatures, focusing on their strategy for function at low water activity. We discuss how the understanding of extremophilic enzyme function is leading to the design of a new generation of enzyme catalysts and their applications to biotechnology.

INTRODUCTION

Enzymes are nature's biocatalysts endowed with high catalytic power, remarkable substrate specificity, and ability to work under mild reaction conditions. These unique features led to enzyme applications in competitive bioprocesses as one of the foremost areas of biotechnology research. Most enzymes are active within a defined set of standard conditions close to what is considered normal for mesophilic terrestrial organisms. However, much of the biosphere is extreme by comparison (e.g. cold oceans and dry, salty deserts). Not surprisingly, the biosphere contains a very large number of extremophilic microorganisms with enzymes capable of functioning in unusual conditions [1, 2].

The discovery of thermostable DNA polymerases and their impact on research, medicine, and industry has underscored the potential benefits of enzymes from extreme environments [3]. Since that time, the biotechnological and industrial demand for stable enzymes functioning in harsh operational conditions has surged. A great deal of current effort is aimed at screening for new sources of novel enzymes capable of functioning in extreme conditions. The parallel development of sophisticated molecular biology tools has also enabled engineering of enzymes with novel properties using techniques such as site-directed mutagenesis, gene shuffling, directed evolution, chemical modifications and immobilization [4–6].

Microorganisms which grow in extreme conditions have been an important source of stable and valuable enzymes [1, 7, 8]. Their enzymes, sometimes called "extremozymes", perform the same enzymatic functions as their non-extreme counterparts, but they can catalyze such reactions in conditions which inhibit or denature the less extreme forms. Interestingly, some of the enzymes derived from extremophiles display polyextremophilicity, i.e. stability and activity in more than one extreme condition, including high salt, alkaline pH, low temperature, and non-aqueous medium [2, 9–11]. A basic understanding of the stability and function of extremozymes under extreme conditions is important for innovations in biotechnology.

One of the underlying reasons for limited enzyme activity in extreme conditions is their effects on water structure and dynamics. When water activity is perturbed by extreme temperatures, high salinity, or other extreme conditions, normally structured liquid water may become limiting to enzymes, with deleterious consequences to enzyme structure and/or function. For example, at high salinity, water is sequestered in hydrated ionic structures, limiting the availability of free water molecules for protein hydration [12, 13]. An analogous effect is felt by enzymes in cold temperatures due to the freezing of water molecules, forming structured ice-like lattices that are less available to interact with proteins [14]. Therefore, improved hydration characteristics in some extremozymes are critical for their function in their natural conditions.

An interesting and potentially useful consequence of the hydration properties of such enzymes may be in extending their range of function to non-aqueous environments [5]. Enzymes capable of functioning in the presence of organic solvents may permit their use in some specialized applications, such as for catalysis of reactions using novel substrates. As a result, a better understanding of molecular mechanisms used by such extremozymes for improved solubility and hydration is of substantial biotechnological interest.

Salt adapted enzymes

Water molecules are known to play a critical role in biological functions of proteins by binding to the surface and incorporating into the interior of protein molecules [15–18]. Water has a tendency to form ordered cages around hydrophobic groups on the protein surface [19]. Salt ions are known to disrupt the local water structure, diminishing the number of intermolecular hydrogen bonds [20–22]. High salt concentrations critically affect the solubility, binding, stability, and crystallization of proteins [23]. The interactions between proteins and protein subunits in solution are also altered by salts. The electrostatic interactions between charged amino acids are also perturbed with significant consequences for protein structure and function [24]. The effects depend on the chemical nature of the salts, generally following the position of ions in the Hofmeister series [25, 26].

Water is necessary for native structure, proper function, and to prevent aggregation of proteins. As salt ions inside the cell increases, water is removed from hydrophobic regions of protein surfaces, until proteins are no longer sufficiently hydrated [17, 18, 27] (Figure 1). Non-halophilic proteins are generally less able to compete with salts and lose their structure and activity at relatively lower ionic concentration. However, halophilic proteins are able to successfully compete with salt ions for hydration and maintain their functional conformation in the presence of high ionic concentration [28–30] (Table 1). This is especially true for proteins from halophilic microorganisms which use the salt-in mechanism for osmotic stabilization (e.g. halophilic archaea and some halophilic bacteria) [8]. Such halophilic proteins have evolved a specific set of molecular features that help them to compete with ions for water and maintain a stable hydration shell. In fact, in comparison to non-halophilic enzymes, halophilic enzymes are found to have multilayered hydration shells that are of considerably greater size and order (Figure 1).

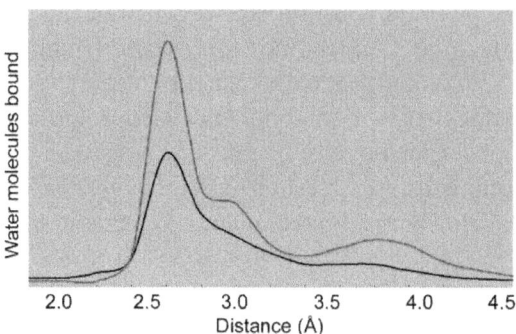

Figure 1: Distribution of the water molecules near the protein surface predicted from high resolution structures (adapted from ref. 30). The relative number of water molecules versus distance is plotted for a halophilic glucose dehydrogenase enzyme active at low water activity (red) [30] and for non-halophilic enzymes (black). Multiple hydration layers may surround extremophilic proteins as a result of their ability to bind more tightly to water than non-extremophilic proteins.

Table 1: Extremophilic enzymes studied for function in low water activity

Name	Organism(s)	Method(s)	Reference(s)
Salt adapted			
α-amylase	*Halothermothrix* sp.	CD spectroscopy, sedimentation velocity, crystal structure	[59]
α-amylase	*Pseudoalteromonas* sp.	CD and fluorescence spectroscopy	[53]
Carbonic anhydrase	*Dunaliella* sp.	Crystal structure	[125]
Cysteinyl tRNA synthetase	*Halobacterium* sp.	Mutagenesis	[126]
Dihydrofolate reductase	*Halobacterium* sp.	Homology modeling	[127]
Dihydrofolate reductase	*Haloferax* sp.	Crystal structure	[41]
Dihydrofolate reductase	*Haloarcula* sp., *Halobacterium* sp., *Haloquadratum* sp., *Natrosomonas* sp.	Homology modeling	[38]
Dihydrolipoamide dehydrogenase	*Haloferax* sp.	Homology modeling, site-directed mutagenesis	[50]
DNA ligase	*Haloferax* sp.	Mutagenesis	[128]

DNA ligase	*Haloferax* sp.	Site-directed mutagenesis, CD, fluorescence and NMR spectroscopy	[35]
DNA ligase	*Haloferax* sp.	Homology modeling, CD spectroscopy	[26]
Esterase	*Haloarcula* sp.	Homology modeling, CD spectroscopy	[43]
Ferredoxin [2Fe-2S]	*Haloarcula* sp.	Crystal structure	[29]
Ferredoxin [2Fe-2S]	*Halobacterium* sp.	Fluorescence and CD spectroscopy	[48, 129]
Glutamate dehydrogenase	*Halobacterium* sp.	Homology modeling	[130]
Glucose dehydrogenase	*Haloferax* sp.	Crystal structure	[30]
Glutaminase	*Micrococcus* sp.	Crystallization and X-ray crystallography	[131]
Malate dehydrogenase	*Halobacterium* sp.	Fluorescence spectroscopy	[132]
Malate dehydrogenase	*Halobacterium* sp.	Neutron scattering, ultracentrifugation and quasi-elastic light-scattering	[46]
Malate dehydrogenase	*Haloarcula* sp.	Densitometry and neutron scattering	[133]
Malate dehydrogenase	*Haloarcula* sp.	X-ray crystallography	[28]
Malate dehydrogenase	*Haloarcula* sp.	Site-directed mutagenesis	[134]
Malate dehydrogenase	*Haloarcula* sp.	Mutagenesis, crystal structure	[135]
Malate dehydrogenase	*Haloarcula* sp.	Crystal structure, neutron scattering	[23]
Malate dehydrogenase	*Salinibacter* sp.	Analytical centrifugation, CD spectroscopy	[136]
Malate dehydrogenase	*Haloarcula* sp.	Neutron diffraction, CD and neutron spectroscopy	[19]
Malate dehydrogenase	*Haloarcula* sp.	Neutron spectroscopy	[137]
Nucleoside diphosphate kinase	*Haloarcula* sp.	CD spectroscopy, crystal structure	[52]
Proliferating cell nuclear antigen	*Haloferax* sp.	Crystal structure	[42]

Protease	*Halobacillus sp.*	Fluorescence resonance energy transfer	[51]
TATA-box binding protein	*Pyrococcus sp.*	Analytical ultracentrifugation, isothermal titration calorimetry	[54]
TATA-box binding protein	*Pyrococcus sp.*	Site-directed mutagenesis, isothermal titration calorimetry	[55–57]
Xylanase	*Bacillus sp.*	Crystal structure	[138]
Cold active			
Adenylate kinase	*Bacillus sp.*	Crystal structure	[85]
Adenylate kinase	*Marinibacillus sp.*	Crystal structure, CD spectroscopy	[139]
Alkaline phosphatase	*Gadus sp.*	Fluorescence spectroscopy	[140]
Alkaline phosphatase	Antarctic strain TAB5	Site-directed mutagenesis	[84, 141]
Alkaline phosphatase	*Vibrio sp.*	Mutagenesis, CD spectroscopy	[142]
Aminopeptidase	*Colwellia sp.*	Crystal structure	[143]
Aminopeptidase	*Colwellia sp.*	Differential scanning calorimetry, fluorescence spectroscopy	[144]
α-amylase	*Alteromonas sp.*	Crystal structure	[86]
α-amylase	*Pseudoalteromonas sp.*	Mutagenesis, differential scanning calorimetry, fluorescence spectroscopy	[145]
α-amylase	*Pseudoalteromonas sp.*	Differential scanning calorimetry, fluorescence spectroscopy	[79]
α-amylase	*Pseudoalteromonas sp.*	Matrix assisted laser desorption ionization time-of-flight mass spectrometry	[146]
α-amylase	*Alteromonas sp.*	Mutagenesis, crystal structure, molecular dynamics simulations	[147]
Aspartate aminotransferase	*Pseudoalteromonas sp.*	Homology modeling, CD and fluorescence spectroscopy	[148]
β-galactosidase	*Arthrobacter sp.*	Crystal structure	[149]
β-lactamase	*Pseudomonas sp.*	Crystal structure	[87]

Catalase	*Vibrio* sp.	Differential scanning calorimetry, fluorescence spectroscopy	[150]
Catalase	*Vibrio* sp.	Crystal structure	[151]
Chitinase	*Arthrobacter* sp.	Homology-modeling, mutagenesis, fluorescence spectroscopy	[77]
Chitobiase	*Arthrobacter* sp.	Differential scanning calorimetry	[152]
Citrate synthase	Antarctic bacterium DS2-3R	Crystal structure	[153]
Citrate synthase	*Arthrobacter* sp.	Site-directed mutagenesis	[154]
Citrate synthase	*Sulfolobus* sp.	Crystal structure	[155]
Citrate synthase	*Arthobacter* sp., *Pyrococcus* sp.	Homology modeling	[156]
Endonuclease I	*Vibrio* sp.	Crystal structure	[59]
Esterase	*Pseudoalteromonas* sp.	Fourier transform infrared spectroscopy, molecular dynamics simulation	[78]
Iron superoxide	*Pseudoalteromonas* sp.	Crystal structure, CD and fluorescence spectroscopy	[75]
Lipase	*Photobacterium* sp.	Crystal structure	[157]
Malate dehydrogenase	*Aquaspirillium* sp.	Crystal structure	[88]
Nitrate reductase	*Shewanella* sp.	Homology modeling	[158]
Pepsin	*Trematomus* sp.	Homology modeling	[159]
Protease	*Bacillus* sp.	Homology modeling, mutagenesis, CD spectroscopy	[160]
Protease	*Pseudomonas* sp.	Crystal structure	[161]
Protease	*Pseudoalteromonas* sp.	Homology modeling, CD, fluorescence spectroscopy	[162]
Protease	*Bacillus* sp.	Homology modeling, mutagenesis	[163]
Protease	*Vibrio* sp.	Site-directed mutagenesis	[164]
Protease	*Bacillus* sp.	Crystal structure	[165]
Protease	*Geomicrobium* sp.	Homology modeling, CD and fluorescence spectroscopy	[166, 167]

Ribonuclease	*Shewanella* sp.	Site-directed mutagenesis, CD spectroscopy	[168]
Superoxide dismutase	*Aliivibrio* sp.	Crystal structure, differential scanning calorimetry	[93]
Subtilisin	*Bacillus* sp.	Site-directed mutagenesis	[169]
Triose phosphate isomerase	*Vibrio* sp.	Crystal structures, calorimetry	[98]
Organic solvent active			
Alcohol dehydrogenase	*Rhodococcus* sp.	Crystal structure	[119]
Protease	*Pseudomonas* sp.	Site-directed and random mutagenesis	[116, 117]
Protease	*Pseudomonas* sp.	Homology modeling	[118]

In order to enhance activity in high salt concentrations, an increase in the number of charged amino acids, especially acidic residues at the protein surface, is observed in halophilic proteins [31–35] (Figure 2). Bioinformatic studies of the extreme halophile *Halobacterium* sp. NRC-1 and other species have shown that an increase in the number of acidic (glutamic acid, and to a lesser extent, aspartic acid) over basic residues is a general property of proteins predicted from the genomes of halophilic microorganisms [13, 27] (Figure 2). Glutamate residues have superior water binding capacity over all other amino acids and are generally found in excess on the surface of halophilic proteins [15, 16, 28, 30]. Acidic amino acids can constitute a high fraction of an individual protein, with up to 20-23% having been reported [36, 37]. The negatively charged amino acids in halophilic proteins bind hydrated cations and help maintain a surface hydration layer, reducing their surface hydrophobicity, and contributing to mutual electrostatic repulsion [29, 35, 38]. These properties prevent aggregation at high salt concentrations [39].

X-ray and neutron diffraction structures have confirmed that the high content of acidic residues play significant roles in binding of essential water molecules and salt ions, preventing protein aggregation and providing flexibility to protein structure through electrostatic repulsion (Figure 3). For example, the structure of malate dehydrogenase from the extremely halophilic archaeaon *Haloarcula marismortui* received considerable attention from Mevarech and co-workers [40] and the group of Zaccai [19].

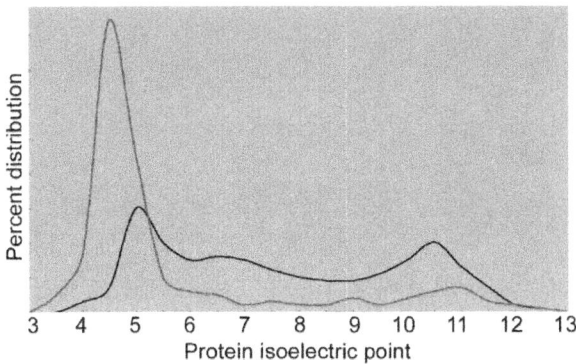

Figure 2: Distribution of protein isoelectric point in halophilic and non-halophilic organisms predicted from genome sequences (adapted from ref. 13). The percent of all predicted proteins is plotted versus their calculated isoelectric points. The distribution of protein isoelectric points for the halophile *Halobacterium* sp. NRC-1 (red) is skewed towards acidic range while those of non-halophiles (black) have a broader distribution of isoelectric points with an average of neutrality in most cases.

The presence of clusters of acidic residues has been observed in the crystal structure of dihydrofolate reductase (DHFR) and proliferating cell nuclear antigen (PCNA) from the extremely halophilic archaeaon*Haloferax volcanii* [41, 42]. Crystal structure of the glucose dehydrogenase of the extremely halophilic archaeaon *H. mediterranei* has also contributed much information about halophilic adaptation and concluded that the surface of enzyme was predominantly acidic in nature and contributed to the halophilic characters of the enzyme [30]. In another study, Tadeo *et al.* [35] reported that by altering the amino acid composition at the protein surface, it is possible to modify the salt dependence of proteins and interconvert salt tolerant and non-tolerant proteins. Through the analysis of a large number of mutants, they concluded that the effect of salt on protein stability is largely independent of the total protein charge. In a recent study, a model of the recombinant esterase from *H. marismortui*, cloned and overexpressed in *Escherichia coli*, confirmed the enrichment of acidic residues and showed a high negative potential from clusters of aspartate and glutamate residues, with most acidic residues confined on the surface [43].

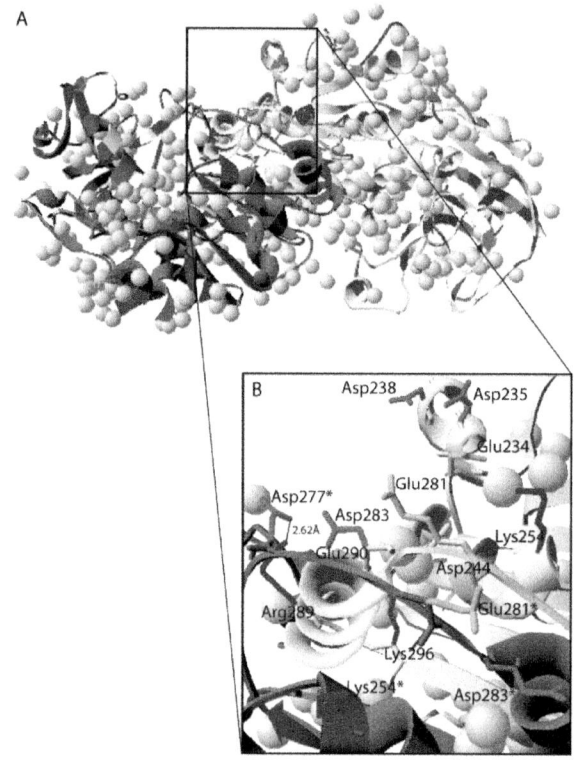

Figure 3: Structural features of an extremophilic glucose dehydrogenase. The protein structure (PDB ID:2B5V) [30] was downloaded from RCSB Protein Data Bank [123] and illustrated using DeepView Swiss-PdbViewer [124]. (A) Ribbon structure is shown with one subunit colored light gray and one subunit colored dark gray. Boxed region encompassing three α-helices of one subunit and two partial α-helices of the other subunit are shown in detail in part B. Acidic residues (aspartic acid and glutamic acid) are colored red and pink respectively, and basic residues (arginine and lysine) are colored dark blue and medium blue, respectively. Water molecules are colored light purple. (B) Expanded region showing a portion with side chains of exposed acidic residues and buried basic residues. Asterisk indicates residues of the dark gray subunit. An inter-subunit ion pair between Arg289 of one subunit and Asp277 of the other subunit is shown by a line labeled 2.62 Å, the distance between interacting atoms of the two residues.

An interesting finding also apparent from the genome-wide bioinformatic analysis of predicted proteins was the deficit of protein surface lysine residues in halophilic proteins [13, 27, 33]. This observation is consistent with the reduction in hydrophobic surfaces in halophilic proteins, resulting in increased hydration at the protein surface. This prediction was confirmed directly via the structure of glucose dehydrogenase from the extreme halophile *H. mediterranei* solved by Britton *et al.* [30], where lysine residues on the enzyme surface have their side chains buried and better ordered than those from non-halophiles (Figure 3). In halophilic proteins, lysine residues are often replaced by arginines, likely due to the greater hydrophilicity of the guanidinyl side chain, with a substantial role in maintaining the active protein structure [11, 30,35, 38, 44].

Halophilic proteins have also been found to contain a low content of bulky hydrophobic side chains on their surface, compared to non-halophilic proteins. The number of larger hydrophobic amino acid residues (phenylalanine, isoleucine, and leucine) is reduced compared to small (glycine and alanine) and borderline hydrophobic (serine and threonine) amino acid residues [27, 31–34, 38, 45]. These findings are also consistent with increased flexibility, increased surface hydration, and reduced surface hydrophobicity of halophilic proteins.

In halophilic proteins, oppositely charged neighboring residues often interact to form salt bridges and these salt bridges play significant roles in protein folding, structure, and oligomerization (Figure 3). For example, the crystal structure of malate dehydrogenase from *H. marismortui* showed an increase in the number of salt bridges compared with the non-halophilic homologs, which enhanced enzyme stability at high salt concentrations [28]. This enzyme exists as a tetramer at high salt concentrations and dissociates into monomers as the salinity is reduced [46]. Similarly Madern *et al.* [47] have shown that isocitrate dehydrogenase from the halophilic archaeon *H. volcanii* exists as a dimer at high salt concentration but at low salt concentration it is irreversibly deactivated, due to dissociation of the dimer towards an inactive, partially folded monomeric species. *Halobacterium* sp. ferredoxin studied using CD and fluorescence techniques showed that the increase in salt concentration decreased electrostatic repulsion by ion binding, likely stabilizing oligomerization necessary for catalytic activity [48].

High salt concentrations generally enhance native conformation and functionality in halophilic proteins [31, 49, 50]. Salt concentrations may significantly affect the folding, conformation, subunit structure, and kinetics of halophilic proteins. Withdrawal of salt generally results in the gradual loss of protein structure and unfolding of halophilic proteins. Okamoto*et al.* [51] studied the role of salt on the kinetics of a high salt-active extracellular protease from the moderately halophilic bacterium *Halobacillus* sp. with fluorescence resonance energy transfer (FRET) and found that NaCl and other kosmotropic salts have a positive effect on catalytic activity of the enzyme. Their results suggested that these salts are excluded from the solvation shell of proteins because they have higher affinity for water than for the protein surface. Using far-ultraviolet circular dichroism (farUV-CD), Müller-Santos *et al.* [43] showed that in salt-free medium, an esterase from *H. marismortui* was completely unfolded and secondary structures appeared only in the presence of high concentrations of salt. The salt-dependent activity profiles of nucleoside diphosphate kinases from extremely halophilic archaea *Haloarcula quadrata* and *H. sinaiiensis*, have also been studied by farUV-CD spectroscopy, with salt-dependent oligomerization observed only for the latter [52]. Srimathi *et al.* [53] investigated a cold adapted amylase from the psychrophile *Pseudoalteromonas haloplanktis* by CD and fluorescence techniques. This cold-active amylase showed increased activity and improved folding at higher concentrations of salt similar to halophilic enzymes, indicating similar mechanisms of enhanced activity in both high salt and low temperature conditions.

Salt is also known to play a critical role in protein-DNA interactions. O'Brien *et al.* [54] studied the effect of salt on the thermodynamic-structural relationship of the binding of TATA box-binding protein (TBP) from *Pyrococcus woesei*, a moderately halophilic and hyperthermophilic organism, to its DNA binding site. This group hypothesized that uptake of cations and discharge of water accompanies protein-DNA complex formation. Subsequently Bergqvist *et al.* [55] used site-directed mutagenesis to change cation binding sites, i.e. negatively charged, acidic glutamate residues on the protein surface. Consistent with the hypothesis, they found that some of the mutants were able to convert the halophilic, relatively salt insensitive TBP into non-halophilic, salt sensitive variants [56, 57].

While the underlying molecular mechanisms of halophilic protein function are still not fully understood, available studies have begun to shed considerable light on their strategies for adaptation to high salinity and relatively low water activity. Based on the many available studies, clustered surface negative charges, decreased hydrophobicity at the surface of the protein, and enrichment of salt bridges appear to be general strategies for improving the function of halophilic proteins in high salt, low water conditions. However, these mechanisms may not be universal [58, 59], and additional research will continue to provide further insights into their function and activity. It is clear, however, that enzymes isolated from halophiles possess extraordinary structural and catalytic properties that allow function at low water activity. These exceptional biomolecules have great potential for applications to many biotechnological and industrial processes (Table 2).

Table 2: Extremophilic enzymes in biotechnology

Name	Organism	Activity			Application(s)	Reference(s)
		Salt	Cold	Organic solvent		
Amylase	*Pseudoalterimonas* sp.	+			*Saccharification of marine microalgae, saccharification of marine microalgae producing ethanol	[170]
Amylase	*Halococcus* sp.	+		+	Starch hydrolysis in industrial processes in saline and organic solvent medium	[171]
Amylase	*Streptomyces* sp.	+			Detergent formulations	[172]
Alcohol dehydrogenase	*Rhodococcus* sp.			+	*Enantioselective oxidation of sec-alcohol and the asymmetric reduction of ketones	[173]
Alkaline phosphatase	Antarctic bacteria strain HK47		+		*Radioactive end-labeling of nucleic acids	[174]
Alkaline phosphatase	Antarctic strain TAB5		+		*Dephosphorylation of DNA vectors	[84]
β-galactosidase	*Pseudoalteromonas* sp.		+		*Lactose hydrolysis	[175]
β-galactosidase	*Kluyveromyces* sp.			+	*Synthesis of galacto-oligosaccharides from lactose	[176]
β-galactosidase	*Arthrobacter* sp.		+		Lactose hydrolysis at low temperature, production of ethanol from lactose-based feedstock	[177]
β-galactosidase	*Bacillus* sp.			+	*Synthesis of N-acetyl-lactosamine	[178]
Chitinase	*Halobacterium* sp.	+		+	Oligosaccharide synthesis	[179]

Enzyme	Organism			Application	Reference
Chitinase	*Virgibacillus* sp.	+		Bioconversion of chitin from fish, crab or shrimp; treatment of chitinous waste	[180]
Cholesterol oxidase	*Pseudomonas* sp.		+	Organic synthesis	[181]
Glutaminase	*Micrococcus* sp.	+		Flavor-enhancing in food industries, antileukaemic agent	[182]
Esterase	*Pyrobaculum* sp.		+	Organic synthesis	[183]
Esterase	*Pseudoalteromonas* sp.		+	Hydrolyzing esters of medical relevance	[184]
Lipase	*Candida* sp.		+	*Organic synthesis related to food/feed processing, pharmaceuticals or cosmetics	[185]
Lipase	*Rhizopus* sp., *Candida* sp.		+	Biodiesel production	[101–103]
Lipase	*Candida* sp.		+	Biodiesel production	[104, 105]
Lipase	*Pseudoalteromonas* sp., *Psychrobacter* sp., *Vibrio* sp.		+	Detergent formulations and bioremediation of fat-contaminated aqueous systems	[186]
Lipase	*Staphylococcus* sp.		+	Detergent formulations	[187]
Lipase	*Psuedomonas* sp.		+	Biodiesel production	[106, 107]
Lipase	*Salinivibrio* sp.	+		Detergent formulations and fatty acid degradations	[188]
Lipase	*Psuedomonas* sp.		+	Solvent bioremediation, biotransformations and detergent formulations	[189]
Lipase	*Marinobacter* sp.	+		*Hydrolysis of fish oil into free eicosapentaenoic acid	[190]

Enzyme	Organism			Application	Ref.
Nuclease	*Micrococcus* sp.	+		*Production of the flavoring agent 5'-guanylic acid	[191]
Pectinase	*Pseudoalteromonas* sp.		+	Enhancing extraction yield, clarification, and taste of fruit juices	[192]
Protease	*Halobacterium* sp.	+		*Peptide synthesis	[193, 194]
Protease	*Pseudomonas* sp.		+	*Synthesis of N-carbobenzoxy-L-arginine-L-leucine amide, N-carbobenzoxy-L-alanine-L-leucine amide and aspartame precursor	[195, 196]
Protease	*Pseudomonas* sp.		+	Peptide synthesis	[197]
Protease	*Bacillus* sp.		+	*Cleansing of contact lenses	[198]
Protease	*Natrialba* sp.	+		*Synthesis of tripeptide Ac-Phe-Gly-Phe-NH$_2$	[199]
Protease	*Halobacterium* sp.	+		*Fish sauce preparation	[200]
Protease	*Geomicrobium* sp.	+	+	Peptide synthesis, detergent formulations	[201, 202]
Xylanase	*Pseudoalteromonas* sp.		+	*Baking industry for increasing loaf volume	[203, 204]
Xylanase	*Bacillus* sp.	+		Xylan biodegradation in pulp and paper industry	[205, 206]

* applications established in laboratory and/or industry

Cold Active Enzymes

Like high salinity, cold temperatures also critically affect the properties and structures of enzymes as well as the surrounding water. Cold temperatures affect the dynamic activity of bulk water as well as the spheres of hydration surrounding the protein surface [60]. Since water acts as a lubricant, easing the essential peptide amide-carbonyl hydrogen bonding dynamic, the effects of water are highly dependent on the temperature [61–63]. Dependence of the strength of aqueous hydrogen bonds on temperature is neither linear nor monotonic but unimodal, with the maximum density for pure water at approximately 4°C. Broken hydrogen bonds are found in high-density water while strong networks of hydrogen bonds are found in low-density water [63]. Deficiency or disturbance of hydrogen bonds and water networks around the protein may be linked to the loss of biological activity and protein denaturation at low temperatures [64].

As temperature decreases, the water molecules surrounding the protein surface become more ordered and thereby less associated with the protein, eventually pushing the system equilibrium toward the unfolded or denatured state. This change in protein structure is driven by an increase in hydration energies of non-polar groups at lower temperature [65,66]. The hydration energies of cold-active enzymes are generally less affected by lower temperature, and their lower inherent surface hydrophobicity is less sensitive than mesophilic proteins, keeping their structures more intact [67]. In effect, cold active proteins are able to hold on more tightly to the available water, similar to salt adapted proteins.

Temperature and viscosity of the medium are inversely related, with viscosity halved when the temperature is reduced from 37°C to 0°C [68]. Low temperatures therefore reduce the speed of reactions at least in part due to the strong effect of temperature on viscosity of the medium [68–71]. Based on biophysical considerations, reaction rates are predicted to decrease 2-3 fold for every 10°C decrease in temperature, according to the Arrhenius equation [72]. As a result, the effect of reduced temperature on enzyme activity is very significant, and the design of cold-active enzymes must have some structural adaptations to maintain the level of 'breathing' necessary for catalysis [73, 74].

Studies of cold active enzymes have suggested that both an increase in interactions with the solvent and an increase in structural flexibility contribute to maintaining catalytic activity at lower temperatures [14, 73, 75–83] (Table 1). Aurilia *et al.* [78] used Fourier transform infrared spectroscopy (FTIR) and molecular dynamics simulations to investigate a cold active esterase from *P. haloplanktis* and found that the flexibility of the protein loop near the active site plays a crucial role in its function. An α-amylase from the same organism was investigated by D›Amico *et al.* [79] and shown using differential scanning calorimetry and fluorescence spectroscopy to have high conformational flexibility and electrostatic potential at the protein surface, which lowers activation energy for hydrolysis. The temperature dependence of alkaline phosphatase from the Antarctic strain TAB5 was studied by selection of thermostable and thermolabile variants. The study showed that cold activity is sensitive to slight changes in the protein sequence, particularly in residues located within or close to the active site of the enzyme [84].

There are relatively few solved three-dimensional structures of cold-active proteins compared to mesophilic or thermophilic proteins. While overall structures of cold adapted proteins are generally almost identical to mesophilic and thermophilic homologs [76], variations that are observed confirm an overall increase in protein flexibility and solvent interactions. For example, the crystal structures of cold active superoxide dismutase from *P. haloplanktis* and *Aliivibrio salmonicida* were compared with the mesophilic homolog from *E. coli*. Both cold-active superoxide dismutases showed an increased flexibility of the active site residues with respect to their mesophilic homologue [75]. Bae and Phillips [85] compared the crystal structures of adenylate kinases from the psychrophile *B. globisporus* and the mesophile *B. subtilis* with the thermophilic *B. stearothermophilus* enzyme. They concluded that the maintenance of proper flexibility is critical for the cold active proteins to function at their environmental temperatures. Similarly the crystal structure of *Alteromonas haloplanctis* α-amylase and *Pseudomonas fluorescens* β-lactamase showed a decrease in the number of hydrogen-bonds favoring more flexibility in these cold-active enzymes [86, 87]. The crystal structure of a cold active malate dehydrogenase from *Aquaspirillium arcticum* showed similar features to be responsible for cold activity, including an increased flexibility around the active site region, more favorable surface charge distribution for substrate and cofactor interactions, and reduced intersubunit ion pair interactions [88].

A bioinformatic study of amino acid contacts that differ between proteins adapted to different temperatures, which included nearly 400 psychrophilic proteins, found that interactions with the solvent at the protein surface play an important role in temperature adaptation [89]. Additional bioinformatic and experimental studies have also suggested that the temperature-dependent activity of cold active enzymes may be altered by changing the amino acid composition, especially the overall protein charge, decreasing the hydrophobicity in the core of the enzyme, or decreasing the number of hydrogen bonds, salt bridges, or bound ions at the surface [72, 90–92].

Amino acids present on the protein surface of cold active enzymes have been shown to play critical roles in both activity in cold and in high salinity, with increased activity and improved folding at higher concentrations of salt [53, 59]. Moreover, the crystal structure of a cold-active iron superoxide dismutase from the *A. salmonicida* also showed an increase in the net negative charge on the surface of cold-active iron superoxide dismutase [93]. These findings and others [94–97] suggest that solvent interactions of cold active enzymes display remarkable similarity to salt adapted enzymes.

While the adaptive mechanisms of cold active proteins are still under investigation, the best understood mechanisms include increased conformational flexibility at the expense of stability and enhanced interactions with the solvent. To increase flexibility, many structural modifications and the disappearance of discrete stabilizing interactions (electrostatic interactions and an improved interaction of surface side chains with the solvent) have been observed. While nearly all cold active enzymes studied to date have shown these tendencies, there are examples of unusual enzymes, *e.g.* enzymes displaying rapid kinetics, which display temperature independent characteristics. Triosephosphate isomerase from *Vibrio marinus* is an example of a temperature independent enzyme [98]. The number of cold active proteins studied is expanding and additional future research including comparisons with closely related mesophilic homologs will provide further insights into enzymology at low temperature. The novel properties of cold active enzymes are likely to be valuable for a variety of applications in biotechnology and industry [99].

Enzyme Function in Organic Solvents

One of the most useful outcomes of a better understanding of enzyme-solvent interactions is the potential engineering of new and more effective catalysts functioning in non-aqueous environments. Such enzymes may be useful for both biofuel and bioenergy applications, where large quantity of ethanol or other organic solvents are produced [100–107], and for synthetic chemistry, especially when catalysis of desired chemical reactions requires the presence of organic solvents [108,109]. Organic solvents in the reaction mixture increase the solubility of hydrophobic substrates, and have the potential to improve the kinetic equilibrium and increase the yield and specificity of the product. However, limiting the disruption of molecular interactions in enzymes by organic solvents and the concomitant loss of activity remains a significant challenge.

A main factor responsible for loss of enzyme activity in organic solvents is the loss of crucial water molecules [109, 110]. The low water content restrains protein conformation mobility and affects K_m and V_{max} values [111]. This rigidity increases resistance to thermal vibrations and reduces the enzyme-substrate interactions, leading to a reduced catalytic rate [112,113]. Loss of water can also disrupt hydrogen bond formation between protein subunits on the exterior surface and active site interactions in the interior of proteins may be weakened. The presence of organic solvents can cause disruption of the forces important to the hydrophobic core due to the increased hydrophobicity of the medium. Low water activity may limit diffusion of substrates and stabilize the ground state of the enzyme or change the enzyme conformation altogether. Enzymes in non-aqueous systems can be active provided that the enzyme surface and the active site region are well hydrated [114].

The polarity of organic solvents is the most important factor in the balance between stabilization and inactivation of enzymes. Co-solvents systems (water plus water-miscible organic solvents), organic aqueous biphasic systems (water plus water-immiscible organic solvents), nearly anhydrous systems, and reverse micelles may be the result of addition of organic solvents with water. The relative proportion of organic solvent and water depends on the miscibility of the components [109–115]. Highly polar, miscible organic solvents may strip the hydration layer from the enzyme surface, affecting enzyme flexibility and catalytic activity. Hydrophobic solvents, in contrast, may form a two-phase non-homogeneous system, leaving the hydration shell of the protein intact. However, they may sequester substrate away from the enzyme, depending on solubility and partitioning between phases. For improved activity in two-phase systems, a microenvironment or surface where favorable conditions (high enzyme activity, high substrate concentration, and low product solubility) driving high reaction rates may be desirable.

Published studies of the mechanism of adaptation of enzymes to function in organic solvent are relatively few. Ogino *et al.* [116, 117] investigated the mechanism of organic solvent tolerance in a *Pseudomonas aeruginosa* PST-01 protease by site-directed and random mutagenesis. They reported that the disulfide bonds and amino acid residues located on the surface of the molecule play important roles in organic solvent stability of the enzymes. The structural analysis of organic solvent effects on a protease from a similar *P. aeruginosa* strain also identified two disulfide bonds as well as a number of hydrophobic clusters at the protein surface. These hydrophobic patches and disulfide bonds were proposed to be responsible for the solvent-stable nature of the enzyme [118]. Recently, Karabec *et al.* [119] solved the crystal structure of alcohol dehydrogenase from *Rhodococcus ruber* DSM 44541 and suggested that salt-bridges play a significance role in the stability of this enzyme in non-aqueous media.

Organic solvent mediated enzymatic reactions have many advantages over aqueous enzymatic reactions: (i) increased solubility of apolar substrate and alteration in substrate specificity, (ii) enhanced regio-and stereo-selectivity, (iii) absence of racemization, (iv) lack of requirement of side chain protection, (v) reduced water activity altering the hydrolytic equilibrium, (vi) elimination of microbial contamination, and (vii) suppression of unwanted water-dependent side reactions [108, 109,114, 115]. Additionally, enzymes in organic solvents tend to be more rigid than in water (due to increased electrostatic and hydrogen bonding interactions among the surface residues of enzyme) and provide the possibility of techniques such as molecular imprinting [109]. In molecular imprinting, the enzyme solution is freeze-dried with a ligand («imprinter») that locks the enzymes into a catalysis favorable condition during lyophilization and enhances the enzyme activity in organic solvents [120]. In some cases, it has been found that molecular imprinting increases enzyme activity in organic solvents when co-lyophilized with an inorganic salt such as KCl. KCl prevents the reversible denaturation of proteins and produces a strong additive activation effect during the drying process [121]. Among the disadvantages of non-aqueous organic enzyme catalysis, the loss of catalytic activity is the most common. Many enzymatic reactions are also susceptible to substrate and/or product inhibition, and expensive natural cofactors for enzyme activity may be required for full activity. Finally, labor and cost-intensive preparation of biocatalysts and enzymes may require narrow operation parameters, limiting the value of some enzymes that may be active in the presence of organic solvents [109, 122]. However, the potential for organic solvent mediated enzyme catalysis to enable desirable biotechnological aims remains a major motivating factor for additional research.

CONCLUSION

Extremophilic salt adapted and cold active enzymes have expanded our understanding of enzyme stability and activity mechanisms, protein structure-function relationships, and enzyme engineering and evolution. The still emerging understanding of protein-solvent interactions are likely to aid in development of new catalysts for use in novel synthetic applications, including enzymes operating in low water activity and organic solvents, and in the development of efficient catalytic systems active in organic solvents for applications in bioenergy and biotechnology.

Acknowledgements

This work was funded by NSF grant MCB-0450695, Henry M Jackson Foundation grant HU0001-09-1-0002-660883, and the National Aeronautics and Space Administration grant NNX10AP47G to SD. RK was partially supported by an ASM International Fellowship for Asia.

Competing interests

The authors declare that they have no competing interests.

Authors' contributions

The manuscript was drafted by RK and revised and finalized together with MDC and SD. All authors have read and approved the final manuscript.

REFERENCES

1. Hough DW, Danson MJ: Extremozymes. Curr Opin Chem Biol. 1999, 1: 39-46.
2. Gomes J, Steiner W: The biocatalytic potential of extremophiles and extremozymes. Food Technol Biotechnol. 2004, 42: 223-235.
3. Vieille C, Zeikus GJ: Hyperthermophilic enzymes: sources, uses, and molecular mechanisms for thermostability. Microbiol Mol Biol Rev. 2001, 1: 1-43.
4. Bull AT, Ward AC, Goodfellow M: Search and discovery strategies for biotechnology: the paradigm shift. Microbiol Mol Biol Rev. 2000, 3: 573-606.
5. Iyer PV, Ananthanarasyan L: Enzyme stability and stabilization-aqueous and non-aqueous environment. Process Biochem. 2008, 43: 1019-1032.
6. Kaul P, Asano Y: Strategies for discovery and improvement of enzyme function: state of the art and opportunities. Microb Biotechnol. 2011, doi: 10.1111/j.1751-7915.2011.00280.x
7. Adams MWW, Perler FB, Kelly RM: Extremozymes: Expanding the limits of biocatalysis. Nat Biotechnol. 1995, 13: 662-668.
8. DasSarma P, Coker JA, Huse V, DasSarma S: Halophiles, Biotechnology. Encyclopedia of Industrial Biotechnology, Bioprocess, Bioseparation, and Cell Technology. Edited by: Flickinger MC. 2010, Hoboken, NJ: John Wiley and Sons, 2769-2777.
9. Bowers KJ, Mesbah NM, Wiegel J: Biodiversity of poly-extremophilic bacteria: Does combining the extremes of high salt, alkaline pH and elevated temperature approach a physico-chemical boundary for life?. Saline Syst. 2009, 5: 9-17.
10. Marhuenda-Egea FC, Bonete MJ: Extreme halophilic enzymes in organic solvents. Curr Opin Biotechnol. 2002, 13: 385-389.
11. Pire C, Marhuenda-egea FC, Esclapez J, Alcaraz L, Ferrer J, Bonete MJ: Stability and enzymatic studies of glucose dehydrogenase from

the Archaeon *Haloferax mediterranei* in reverse micelles. Biocatal Biotransform. 2004, 22: 17-23.

12. Danson MJ, Hough DW: The structural basis of protein halophilicity. Comp Biochem Physiol A Physiol. 1997, 117: 307-312.

13. Kennedy SP, Ng WV, Salzberg SL, Hood L, DasSarma S: Understanding the adaptation of *Halobacterium* species NRC-1 to its extreme environment through computational analysis of its genome sequence. Genome Res. 2001, 11: 1641-1650.

14. Siddiqui KS, Cavicchioli R: Cold-adapted enzymes. Annu Rev Biochem. 2006, 75: 403-433.

15. Kuntz ID: Hydration of macromolecules IV Polypeptide conformation in frozen solutions. J Am Chem Soc. 1971, 93: 516-518.

16. Saenger W: Structure and dynamics of water surrounding biomolecules. Annu Rev Biophys Biophys Chem. 1987, 16: 93-114.

17. Persson E, Halle B: Cell water dynamics on multiple time scales. Proc Natl Acad Sci USA. 2008, 17: 6266-6271.

18. Spitzer J: From water and ions to crowded biomacromolecules: In vivo structuring of a prokaryotic cell. Microbiol Mol Biol R. 2011, 3: 491-506.

19. Zaccai G: The effect of water on protein dynamics. Philos Trans R Soc Lond B Biol Sci. 2004, 359: 1269-1275.

20. Mountain RD, Thirumalai D: Alterations in water structure induced by guanidinium and sodium ions. J Phys Chem. 2004, 108: 19711-19716.

21. Mancinelli R, Botti A, Bruni F, Ricci MA, Soper AK: Hydration of sodium, potassium, and chloride ions in solution and the concept of structure maker/breaker. Phys Chem. 2007, 111: 13570-13577.

22. Bakker HJ: Water dynamics: Ion-ing out the details. Nature Chemistry. 2009, 1: 24-25.

23. Irimia A, Ebel C, Madern D, Richard SB, Cosenza LW, Zaccaï G, Vellieux FM: The oligomeric states of *Haloarcula marismortui* malate dehydrogenase are modulated by solvent components as shown by crystallographic and biochemical studies. J Mol Biol. 2003, 3: 859-873.

24. Dennis PP, Shimmin LC: Evolutionary divergence and salinity-mediated selection in halophilic archaea. Microbiol Mol Biol Rev. 1997, 61: 90-104.

25. Sedlák E, Stagg L, Wittung-Stafshede P: Role of cations in stability of acidic protein *Desulfovibrio desulfuricans* apoflavodoxin. Arch Biochem Biophys. 2008, 1: 128-135.

26. Ortega G, Laín A, Tadeo X, López-Méndez B, Castaño D, Millet O: Halophilic enzyme activation induced by salts. Scientific Reports. 2011, 1: 6-
27. Paul S, Bag SK, Das S, Harvill ET, Dutta C: Molecular signature of hypersaline adaptation: Insights from genome and proteome composition of halophilic prokaryotes. Genome Biol. 2008, 9: R70-
28. Dym O, Mevarech M, Sussman JL: Structural features that stabilize halophilic malate dehydrogenase from an archaebacterium. Science. 1995, 267: 1344-1346.
29. Frolow F, Harel M, Sussman JL, Mevarech M, Shoham M: Insights into protein adaptation to a saturated salt environment from the crystal structure of a halophilic 2Fe-2S ferredoxin. Nature Struct Biol. 1996, 3: 452-458.
30. Britton KL, Baker PJ, Fisher M, Ruzheinikov S, Gilmour DJ, Bonete MJ, Ferrer J, Pire C, Esclapez J, Rice DW: Analysis of protein solvent interactions in glucose dehydrogenase from the extreme halophile *Haloferax mediterranei*. Proc Natl Acad Sci USA. 2006, 103: 4846-4851.
31. Lanyi JK: Salt dependent properties of proteins from extremely halophilic bacteria. Bacteriol Rev. 1974, 38: 272-290.
32. Madern D, Ebel C, Zaccai G: Halophilic adaptation of enzymes. Extremophiles. 2000, 4: 91-98.
33. Fukuchi S, Yoshimune K, Wakayama M, Moriguchi M, Nishikawa K: Unique amino acid composition of proteins in halophilic bacteria. J Mol Biol. 2003, 327: 347-357.
34. Bolhuis A, Kwan D, Thomas JR: Halophilic adaptations of proteins. Protein adaptation in extremophiles. Edited by: Siddiqui KS, Thomas T. 2008, New York: Nova Science Publishers Inc USA, 71-104.
35. Tadeo X, López-Méndez B, Trigueros T, Laín A, Castaño D, Millet O: Structural basis for the amino acid composition of proteins from halophilic archaea. PLoS Biol. 2009, 7: e1000257-
36. Ishibashi M, Tokunaga H, Hiratsuka K, Yonezawa Y, Tsurumaru H, Arakawa T, Tokunaga M: NaCl-activated nucleoside diphosphate kinase from extremely halophilic archaeon, *Halobacterium salinarum*, maintains native conformation without salt. FEBS Lett. 2001, 493: 134-138.
37. De Castro RE, Ruiz DM, Giménez MI, Silveyra MX, Paggi RA, Maupin-Furlow JA: Gene cloning and heterologous synthesis of a haloalkaliphilic extracellular protease of *Natrialba magadii* (Nep). Extremophiles. 2008, 5: 677-687.

38. Kastritis PL, Papandreou NC, Hamodrakas SJ: Haloadaptation: insights from comparative modeling studies of halophilic archaeal DHFRs. Int J Biol Macromol. 2007, 41: 447-453.

39. Elcock AH, McCammon JA: Electrostatic contributions to the stability of halophilic proteins. J Mol Biol. 1998, 4: 731-748.

40. Mevarech M, Frolow F, Gloss LM: Halophilic enzymes: Proteins with a grain of salt. Biophys Chem. 2000, 86: 155-164.

41. Pieper U, Kapadia G, Mevarech M, Herzberg O: Structural features of halophilicity derived from the crystal structure of dihydrofolate reductase from the Dead Sea halophilic archaeon, *Haloferax volcanii*. Structure. 1998, 6: 75-88.

42. Winter JA, Christofi P, Morroll S, Bunting KA: The crystal structure of *Haloferax volcanii* proliferating cell nuclear antigen reveals unique surface charge characteristics due to halophilic adaptation. BMC Struct Biol. 2009, 9: 55-

43. Müller-Santos M, de Souza EM, Pedrosa Fde O, Mitchell DA, Longhi S, Carrière F, Canaan S, Krieger N: First evidence for the salt-dependent folding and activity of an esterase from the halophilic archaea *Haloarcula marismortui*. Biochim Biophys Acta. 2009, 1791: 719-729.

44. Esclapez J, Pire C, Bautista V, Martínez-Espinosa RM, Ferrer J, Bonete MJ: Analysis of acidic surface of *Haloferax mediterranei* glucose dehydrogenase by site-directed mutagenesis. FEBS Lett. 2007, 581: 837-842.

45. Rao JKM, Argos P: Structural stability of halophilic proteins. Biochemistry. 1981, 20: 6536-6543.

46. Zaccai G, Cendrin F, Haik Y, Borochov N, Eisenberg H: Stabilization of halophilic malate dehydrogenase. J Mol Biol. 1989, 208: 491-500.

47. Madern D, Camacho M, Rodríguez-Arnedo A, Bonete MJ, Zaccai G: Salt-dependent studies of NADP-dependent isocitrate dehydrogenase from the halophilic archaeon *Haloferax volcanii*. Extremophiles. 2004, 5: 377-384.

48. Bandyopadhyay AK, Sonawat HM: Salt dependent stability and unfolding of [Fe2-S2] ferredoxin of *Halobacterium salinarum*: Spectroscopic investigations. Biophys J. 2000, 79: 501-510.

49. Rao L, Zhao X, Pan F, Li Y, Xue Y, Ma Y, Lu JR: Solution behavior and activity of a halophilic esterase under high salt concentration. PLoS One. 2009, 9: e6980-

50. Jolley KA, Russell RJ, Hough DW, Danson MJ: Site-directed mutagenesis

and halophilicity of dihydrolipoamide dehydrogenase from the halophilic archaeon, *Haloferax volcanii*. Eur J Biochem. 1997, 2: 362-368.

51. Okamoto DN, Kondo MY, Santos JA, Nakajima S, Hiraga K, Oda K, Juliano MA, Juliano L, Gouvea IE: Kinetic analysis of salting activation of a subtilisin-like halophilic protease. Biochim Biophys Acta. 2009, 1794: 367-373.

52. Yamamura A, Ichimura T, Kamekura M, Mizuki T, Usami R, Makino T, Ohtsuka J, Miyazono K, Okai M, Nagata K, Tanokura M: Molecular mechanism of distinct salt-dependent enzyme activity of two halophilic nucleoside diphosphate kinases. Biophys J. 2009, 96: 4692-4700.

53. Srimathi S, Jayaraman G, Feller G, Danielsson B, Narayanan PR: Intrinsic halotolerance of the psychrophilic alpha-amylase from *Pseudoalteromonas haloplanktis*. Extremophiles. 2007, 11: 505-515.

54. O'Brien R, DeDecker B, Fleming K, Sigler PB, Ladbury JE: The effects of salt on the TATA binding protein-DNA interaction from a hyperthermophilic archaeon. J Mol Biol. 1998, 279: 117-125.

55. Bergqvist S, O'Brien R, Ladbury JE: Site-specific cation binding mediates TATA binding protein-DNA interaction from a hyperthermophilic archaeon. Biochemistry. 2001, 40: 2419-2425.

56. Bergqvist S, Williams MA, O'Brien R, Ladbury JE: Reversal of halophilicity in a protein-DNA interaction by limited mutation strategy. Structure. 2002, 10: 629-637.

57. Bergqvist S, Williams MA, O'Brien R, Ladbury JE: Halophilic adaptation of protein-DNA interactions. Biochem Soc Trans. 2003, 31: 677-680.

58. Sivakumar N, Li N, Tang JW, Patel BK, Swaminathan K: Crystal structure of AmyA lacks acidic surface and provide insights into protein stability at poly-extreme condition. FEBS Lett. 2006, 11: 2646-2652.

59. Altermark B, Helland R, Moe E, Willassen NP, Smalås AO: Structural adaptation of endonuclease I from the cold-adapted and halophilic bacterium *Vibrio salmonicida*. Acta Crystallogr D Biol Crystallogr. 2008, 64: 368-376.

60. Zhong D, Pal SK, Zewail AH: Biological water: A critique. Chem Phys Lett. 2010, 503: 1-11.

61. Fernández A, Scheraga HA: Insufficiently dehydrated hydrogen bonds as determinants of protein interactions. Proc Natl Acad Sci USA. 2003, 100: 113-118.

62. Kurkal-Siebert V, Daniel RM, L Finney J, Tehei M, Dunn RV, Smith JC: Enzyme hydration, activity and flexibility: A neutron scattering approach.

J Non-Cryst Solids. 2006, 352: 4387-4393.

63. Chaplin M: Do we underestimate the importance of water in cell biology?. Nat Rev Mol Cell Biol. 2006, 11: 861-866.

64. Koizumi M, Hirai H, Onai T, Inoue K, Hirai M: Collapse of the hydration shell of a protein prior to thermal unfolding. J Appl Cryst. 2007, 40: s175-s178.

65. Lopez CF, Darst RK, Rossky PJ: Mechanistic elements of protein cold denaturation. J Phys Chem B. 2008, 112: 5961-5967.

66. Dias CL, Ala-Nissila T, Wong-ekkabut J, Vattulainen I, Grant M, Karttunen M: The hydrophobic effect and its role in cold denaturation. Cryobiol. 2010, 60: 91-99.

67. Fields PA: Protein function at thermal extremes: balancing stability and flexibility. Comp Biochem Physiol Pt A. 2001, 129: 417-431.

68. D'Amico S, Collins T, Marx JC, Feller G, Gerday C: Psychrophilic microorganisms: challenges for life. EMBO Rep. 2006, 7: 385-389.

69. Demchenko AP, Ruskyn OI, Saburova EA: Kinetics of the lactate dehydrogenase reaction in high-viscosity media. Biochim Biophys Acta. 1989, 998: 196-203.

70. Siddiqui KS, Bokhari SA, Afzal AJ, Singh S: A novel thermodynamic relationship based on Kramers theory for studying enzyme kinetics under high viscosity. IUBMB Life. 2004, 56: 403-407.

71. Marx JC, Collins T, D'Amico S, Feller G, Gerday C: Cold-adapted enzymes from marine Antarctic microorganisms. Mar Biotechnol (NY). 2007, 9: 293-304.

72. Georlette D, Blaise V, Collins T, D'Amico S, Gratia E, Hoyoux A, Marx JC, Sonan G, Feller G, Gerday C: Some like it cold: biocatalysis at low temperatures. FEMS Microbiol Rev. 2004, 28: 25-42.

73. Rasmussen BF, Stock AM, Ringe D, Petsko GA: Crystalline ribonuclease: A loses function below the dynamical transition at 220 K. Nature. 1992, 357: 423-424.

74. Siglioccolo A, Gerace R, Pascarella S: "Cold spots" in protein cold adaptation: Insights from normalized atomic displacement parameters (B'-factors). Biophys Chem. 2010, 153: 104-114.

75. Merlino A, Russo Krauss I, Castellano I, De Vendittis E, Rossi B, Conte M, Vergara A, Sica F: Structure and flexibility in cold-adapted iron superoxide dismutases: the case of the enzyme isolated from *Pseudoalteromonas haloplanktis*. J Struct Biol. 2010, 172: 343-352.

76. Margesin R, Feller G: Biotechnological applications of psychrophiles.

Environ Technol. 2010, 31: 835-844.
77. Mavromatis K, Feller G, Kokkinidis M, Bouriotis V: Cold adaptation of a psychrophilic chitinase: a mutagenesis study. Protein Eng. 2003, 16: 497-503.
78. Aurilia V, Rioux-Dubé JF, Marabotti A, Pézolet M, D'Auria S: Structure and dynamics of cold-adapted enzymes as investigated by FT-IR spectroscopy and MD. The case of an esterase from *Pseudoalteromonas haloplanktis*. J Phys Chem B. 2009, 113: 7753-7761.
79. D'Amico S, Gerday C, Feller G: Activity stability relationships in extremophilic enzymes. J Biol Chem. 2003, 278: 7891-7896.
80. Somero GN: Temperature as a selective factor in protein evolution: The adaptational strategy of "compro-mise". J exp Zool. 1975, 194: 175-188.
81. Feller G, Narinx E, Arpigigny JL, Aittaleb M, Baise E, Genicot S, Gerday C: Enzymes from psychrophilic organisms. FEMS Microbiol Rev. 1996, 18: 189-202.
82. Feller G, Gerday C: Psychrophilic enzymes: molecular basis of cold adaptation. Cell Mol Life Sci. 1997, 53: 830-841.
83. Fields PA, Somero GN: Hot spots in cold adaptation: localized increases in conformational flexibility in lactate dehydrogenase A4 orthologs of Antarctic notothenioid fishes. Proc Natl Acad Sci USA. 1998, 95: 11476-11481.
84. Koutsioulis D, Wang E, Tzanodaskalaki M, Nikiforaki D, Deli A, Feller G, Heikinheimo P, Bouriotis V: Directed evolution on the cold adapted properties of TAB5 alkaline phosphatase. Protein Eng Des Sel. 2008, 21: 319-327.
85. Bae E, Phillips GN: Structures and analysis of highly homologous psychrophilic, mesophilic, and thermophilic adenylate kinases. J Biol Chem. 2004, 27: 28202-28208.
86. Aghajari N, Feller G, Gerday C, Haser R: Structures of the psychrophilic *Alteromonas haloplanctis* α-amylase give insights into cold adaptation at a molecular level. Structure. 1998, 6: 1503-1516.
87. Michaux C, Massant J, Kerff F, Frère JM, Docquier JD, Vandenberghe I, Samyn B, Pierrard A, Feller G, Charlier P, Van Beeumen J, Wouters J: Crystal structure of a cold-adapted class C beta-lactamase. FEBS J. 2008, 8: 1687-1697.
88. Kim SY, Hwang KY, Kim SH, Sung HC, Han YS, Cho Y: Structural basis for cold adaptation sequence, biochemical properties, and crystal structure of malate dehydrogenase from a psychrophile *Aquaspirillium*

arcticum. J Biol Chem. 1999, 274: 11761-11767.

89. Sælensminde G, Halskau Ø, Jonassen I: Amino acid contacts in proteins adapted to different temperatures: hydrophobic interactions and surface charges play a key role. Extremophiles. 2009, 1: 11-20.

90. Russell NJ: Toward a molecular understanding of cold activity of enzymes from psychrophiles. Extremophiles. 2000, 4: 83-90.

91. Zartler ER, Jenney FE, Terrell M, Eidsness MK, Adams MW, Prestegard JH: Structural basis for thermostability in aporubredoxins from *Pyrococcus furiosus* and *Clostridium pasteurianum.* Biochemistry. 2001, 40: 7279-7290.

92. Feller G: Molecular adaptations to cold in psychrophilic enzymes. Cell Mol Life Sci. 2003, 60: 648-662.

93. Pedersen HL, Willassen NP, Leiros I: The first structure of a cold-adapted superoxide dismutase (SOD): biochemical and structural characterization of iron SOD from *Aliivibrio salmonicida.* Acta Crystallogr Sect F Struct Biol Cryst Commun. 2009, 2: 84-92.

94. Davail S, Feller G, Narinx E, Gerday C: Cold adaptation of proteins. Purification, characterization, and sequence of the heat-labile subtilisin from the antarctic psychrophile *Bacillus* TA41. J Biol Chem. 1994, 26: 17448-17453.

95. Feller G, Zekhnini Z, Lamotte-Brasseur J, Gerday C: Enzymes from cold-adapted microorganisms. The class C beta-lactamase from the antarctic psychrophile *Psychrobacter immobilis* A5. Eur J Biochem. 1997, 1: 186-191.

96. Russell RJ, Gerike U, Danson MJ, Hough DW, Taylor GL: Structural adaptations of the cold-active citrate synthase from an Antarctic bacterium. Structure. 1998, 3: 351-361.

97. Feller G, D›Amico D, Gerday C: Thermodynamic stability of a cold-active alpha-amylase from the Antarctic bacterium *Alteromonas haloplanctis.* Biochemistry. 1999, 14: 4613-4619.

98. Alvarez M, Zeelen JP, Mainfroid V, Rentier-Delrue F, Martial JA, Wyns L, Wierenga RK, Maes D: Triose-phosphate isomerase (TIM) of the psychrophilic bacterium *Vibrio marinus* kinetic and structural properties. J Biol Chem. 1998, 273: 2199-206.

99. Cavicchioli R, Charlton T, Ertan H, Omar SM, Siddiqui KS, Williams TJ: Biotechnological uses of enzymes from psychrophiles. Microb Biotechnol. 2011, 4: 449-460.

100. Fjerbaek L, Christensen KV, Norddahl B: A review of the current state

of biodiesel production using enzymatic transesterification. Biotechnol Bioeng. 2009, 5: 1298-1315.

101. Lee K-T, Foglia TA, Chang KS: Production of alkyl ester as biodiesel from fractionated lard and restaurant grease. J AmOil Chem Soc. 2002, 2: 191-195.

102. Lee DH, Kim JM, Shin HY, Kang SW, Kim SW: Biodiesel production using a mixture of immobilized *Rhizopus oryzae* and *Candida rugosa*lipases. Biotechnol Bioprocess Eng. 2006, 6: 522-525.

103. Lee JH, Kim SB, Kang SW, Song YS, Park C, Han SO, Kim SW: Biodiesel production by a mixture of *Candida rugosa* and *Rhizopus oryzae*lipases using a supercritical carbon dioxide process. Bioresour Technol. 2011, 2: 2105-2108.

104. Deng L, Xu XB, Haraldsson GG, Tan TW, Wang F: Enzymatic production of alkyl esters through alcoholysis: A critical evaluation of lipases and alcohols. J Am Oil Chem Soc. 2005, 5: 341-347.

105. Tan TW, Nie KL, Wang F: Production of biodiesel by immobilized *Candida* sp. lipase at high water content. Appl Biochem Biotechnol. 2006, 2: 109-116.

106. Kumari V, Shah S, Gupta MN: Preparation of biodiesel by lipasecatalyzed transesterification of high free fatty acid containing oil from*Madhuca indica*. Energy Fuels. 2007, 1: 368-372.

107. Shah S, Gupta MN: Lipase catalyzed preparation of biodiesel from *Jatropha* oil in a solvent-free system. Process Biochem. 2007, 3: 409-414.

108. Gupta A, Khare SK: Enzymes from solvent-tolerant microbes: Useful biocatalysts for non-aqueous enzymology. Crit Rev Biotechnol. 2009, 29: 44-54.

109. Doukyua N, Ogino H: Organic solvent-tolerant enzymes. Biochem Eng J. 2010, 48: 270-282.

110. Klibanov AM: Why are enzymes less active in organic solvents than in water?. Trends Biotechnol. 1997, 15: 97-101.

111. Zhu X, Zhou T, Wu X, Cai Y, Yao D, Xie C, Liu D: Covalent immobilization of enzymes within micro-aqueous organic media. J Mol Catal B: Enzym. 2011, 72: 145-149.

112. Ru MT, Dordick JS, Reimer JA, Clark DS: Optimizing the salt-induced activation of enzymes in organic solvents: Effects of lyophilization time and water content. Biotechnol Bioeng. 1999, 63: 233-241.

113. Torres S, Castro GR: Non-aqueous biocatalysis in homogeneous solvent

systems. Food Technol Biotechnol. 2004, 42: 271-277.
114. Gupta MN, Roy I: Enzymes in organic media. Forms, functions and applications. Eur J Biochem. 2004, 13: 2575-2583.
115. Gupta MN: Enzyme function in organic solvents. Eur J Biochem. 1992, 203: 25-32.
116. Ogino H, Uchiho T, Yokoo J, Kobayashi R, Ichise R, Ishikawa H: Role of intermolecular disulfide bonds of the organic solvent-stable PST-01 protease in its organic solvent stability. Appl Environ Microbiol. 2001, 67: 942-947.
117. Ogino H, Uchiho T, Doukyu N, Yasuda M, Ishimi K, Ishikawa H: Effect of exchange amino acid residues of the surface region of the PST-01 protease on its organic solvent-stability. Biochem Biophys Res Commun. 2007, 358: 1028-1033.
118. Gupta A, Ray S, Kapoor S, Khare SK: Solvent-stable *Pseudomonas aeruginosa* PseA protease gene: identification, molecular characterization, phylogenetic and bioinformatic analysis to study reasons for solvent stability. J Mol Microbiol Biotechnol. 2008, 15: 234-243.
119. Karabec M, Łyskowski A, Tauber KC, Steinkellner G, Kroutil W, Grogan G, Gruber K: Structural insights into substrate specificity and solvent tolerance in alcohol dehydrogenase ADH-›A› from *Rhodococcus ruber* DSM 44541. Chem Commun (Camb). 2010, 34: 6314-6316.
120. Rich JO, Dordick JS: Imprinting enzymes for use in organic media. Enzymes in Nonaqueous Media. Edited by: Vulfson J, Halling PJ, Holland HL. 2000, Humana Press: Totowa, NJ
121. Rich JO, Mozhaev VV, Dordick JS, Clark DS, Khmelnitsky YL: Molecular imprinting of enzymes with water-insoluble ligands for nonaqueous biocatalysis. J Am Chem Soc. 2002, 124: 5254-5255.
122. Khmelnitsky Y: Biotransformations in organic chemistry. 1997, A textbook, third edition, by Faber K, Springer-Verlag, Berlin
123. Berman HM, Westbrook J, Feng Z, Gilliland G, Bhat TN, Weissig H, Shindyalov IN, Bourne PE: The Protein Data Bank. Nucleic Acids Res. 2000, 28: 235-242.
124. Guex N, Peitsch MC: SWISS-MODEL and the Swiss-PdbViewer: An environment for comparative protein modeling. Electrophoresis. 1997, 18: 2714-2723.
125. Premkumar L, Greenblatt HM, Bageshwar UK, Savchenko T, Gokhman I, Sussman JL, Zamir A: Three-dimensional structure of a halotolerant algal carbonic anhydrase predicts halotolerance of a mammalian

homolog. Proc Natl Acad Sci USA. 2005, 21: 7493-7498.
126. Evilia C, Ming X, DasSarma S, Hou YM: Aminoacylation of an unusual tRNA(Cys) from an extreme halophile. RNA. 2003, 7: 794-801.
127. Bohm G, Jaenicke R: A structure-based model for the halophilic adaptation of dihydrofolate reductase from *Halobacterium volcanii*. Protein Eng. 1994, 7: 213-220.
128. Poidevin L, MacNeill SA: Biochemical characterisation of LigN, an NAD+-dependent DNA ligase from the halophilic euryarchaeon*Haloferax volcanii* that displays maximal *in vitro* activity at high salt concentrations. BMC Mol Biol. 2006, 7: 44-
129. Bandyopadhyay AK, Krishnamoorthy G, Padhy LC, Sonawat HM: Kinetics of salt-dependent unfolding of [2Fe-2S] ferredoxin of*Halobacterium salinarum*. Extremophiles. 2007, 4: 615-625.
130. Britton KL, Stillman TJ, Yip KSP, Forterre P, Engel PC, Rice DW: Insights into the molecular basis of salt tolerance from the study of glutamate dehydrogenase from *Halobacterium salinarum*. J Biol Chem. 1998, 273: 9023-9030.
131. Chantawannakul P, Yoshimune K, Shirakihara Y, Shiratori A, Wakayama M, Moriguchi M: Crystallization and preliminary X-ray crystallographic studies of salt-tolerant glutaminase from *Micrococcus luteus* K-3. Acta Crystallogr D Biol Crystallogr. 2003, 3: 566-568.
132. Mevarech M, Eisenberg H, Neumann E: Malate dehydrogenase isolated from extremely halophilic bacteria of the Dead Sea. 1. Purification and molecular characterization. Biochemistry. 1977, 17: 3781-3785.
133. Bonnete ÂF, Madern D, Zaccaõ ÈG: Stability against denaturation mechanisms in halophilic malate dehydrogenase ‹›adapt›› to solvent conditions. J Mol Biol. 1994, 244: 436-447.
134. Madern D, Pfister C, Zaccai G: Mutation at a single acidic amino acid enhances the halophilic behaviour of malate dehydrogenase from*Haloarcula marismortui* in physiological salts. Eur J Biochem. 1995, 3: 1088-1095.
135. Richard SB, Madern D, Garcin E, Zaccai G: Halophilic adaptation: novel solvent protein interactions observed in the 2.9 and 2.6 Å resolution structures of the wild type and a mutant of malate dehydrogenase from *Haloarcula marismortui*. Biochemistry. 2000, 5: 992-1000.
136. Madern D, Zaccai G: Molecular adaptation: the malate dehydrogenase from the extreme halophilic bacterium *Salinibacter ruber*behaves like a non-halophilic protein. Biochimie. 2004, 86: 295-303.

137. Tehei M, Zaccai G: Adaptation to extreme environments: macromolecular dynamics in complex systems. Biochim Biophys Acta. 2005, 3: 404-410.
138. Manikandan K, Bhardwaj A, Gupta N, Lokanath NK, Ghosh A, Reddy VS, Ramakumar S: Crystal structures of native and xylosaccharide-bound alkali thermostable xylanase from an alkalophilic *Bacillus* sp. NG-27: structural insights into alkalophilicity and implications for adaptation to polyextreme conditions. Protein Sci. 2006, 8: 1951-1960.
139. Davlieva M, Shamoo Y: Structure and biochemical characterization of an adenylate kinase originating from the psychrophilic organism*Marinibacillus marinus*. Acta Crystallogr Sect F Struct Biol Cryst Commun. 2009, 8: 751-756.
140. Asgeirsson B, Hauksson JB, Gunnarsson GH: Dissociation and unfolding of cold-active alkaline phosphatase from atlantic cod in the presence of guanidinium chloride. Eur J Biochem. 2000, 21: 6403-6412.
141. Tsigos I, Mavromatis K, Tzanodaskalaki M, Pozidis C, Kokkinidis M, Bouriotis V: Engineering the properties of a cold active enzyme through rational redesign of the active site. Eur J Biochem. 2001, 268: 5074-5080.
142. Gudjónsdóttir K, Asgeirsson B: Effects of replacing active site residues in a cold-active alkaline phosphatase with those found in its mesophilic counterpart from *Escherichia coli*. FEBS J. 2008, 1: 117-127.
143. Bauvois C, Jacquamet L, Huston AL, Borel F, Feller G, Ferrer JL: Crystal structure of the cold-active aminopeptidase from *Colwellia psychrerythraea*, a close structural homologue of the human bifunctional leukotriene A_4 hydrolase. J Biol Chem. 2008, 34: 23315-23325.
144. Huston AL, Haeggström JZ, Feller G: Cold adaptation of enzymes: structural, kinetic and microcalorimetric characterizations of an aminopeptidase from the Arctic psychrophile *Colwellia psychrerythraea* and of human leukotriene A(4) hydrolase. Biochim Biophys Acta. 2008, 11: 1865-1872.
145. D›Amico S, Gerday C, Feller G: Structural determinants of cold adaptation and stability in a large protein. J Biol Chem. 2001, 276: 25791-25796.
146. Siddiqui KS, Poljak A, Guilhaus M, Feller G, D›Amico S, Gerday C, Cavicchioli R: Role of disulfide bridges in the activity and stability of a cold-active alpha-amylase. J Bacteriol. 2005, 17: 6206-6212.
147. Papaleo E, Pasi M, Tiberti M, De Gioia L: Molecular dynamics of mesophilic-like mutants of a cold-adapted enzyme: insights into distal effects induced by the mutations. PLoS One. 2011, 9: e24214-
148. Birolo L, Tutino ML, Fontanella B, Gerday C, Mainolfi K, Pascarella

S, Sannia G, Vinci F, Marino G: Aspartate aminotransferase from the Antarctic bacterium *Pseudoalteromonas haloplanktis* TAC 125. Cloning, expression, properties, and molecular modelling. Eur J Biochem. 2000, 9: 2790-2802.

149. Skálová T, Dohnálek J, Spiwok V, Lipovová P, Vondráčková E, Petroková H, Dusková J, Strnad H, Králová B, Hasek J: Cold-active beta-galactosidase from *Arthrobacter* sp. C2-2 forms compact 660 kDa hexamers: crystal structure at 1.9Å resolution. J Mol Biol. 2005, 2: 282-294.

150. Lorentzen MS, Moe E, Jouve HM, Willassen NP: Cold adapted features of *Vibrio salmonicida* catalase: characterisation and comparison to the mesophilic counterpart from *Proteus mirabilis*. Extremophiles. 2006, 5: 427-440.

151. Riise EK, Lorentzen MS, Helland R, Smalås AO, Leiros HK, Willassen NP: The first structure of a cold-active catalase from *Vibrio salmonicida* at 1.96 Å reveals structural aspects of cold adaptation. Acta Crystallogr D Biol Crystallogr. 2007, 2: 135-148.

152. Lonhienne T, Zoidakis J, Vorgias CE, Feller G, Gerday C, Bouriotis V: Modular structure, local flexibility and cold-activity of a novel chitobiase from a psychrophilic Antarctic bacterium. J Mol Biol. 2001, 310: 291-297.

153. Russell RJ, Gerike U, Danson MJ, Hough DW, Taylor GL: Structural adaptations of the cold-active citrate synthase from an Antarctic bacterium. Structure. 1998, 6: 351-361.

154. Gerike U, Danson MJ, Hough DW: Cold-active citrate synthase: mutagenesis of active-site residues. Protein Eng. 2001, 14: 655-661.

155. Bell GS, Russell RJM, Connaris H, Hough DW, Danson MJ, Taylor GL: Stepwise adaptations of citrate synthase to survival at life›s extremes from psychrophile to hyperthermophile. Eur J Biochem. 2002, 269: 6250-6260.

156. Kumar S, Nussinov R: Different roles of electrostatics in heat and in cold: Adaptation by citrate synthase. Chem Bio Chem. 2004, 5: 280-290.

157. Jung SK, Jeong DG, Lee MS, Lee JK, Kim HK, Ryu SE, Park BC, Kim JH, Kim SJ: Structural basis for the cold adaptation of psychrophilic M37 lipase from *Photobacterium lipolyticum*. Proteins. 2008, 1: 476-484.

158. Simpson PJ, Codd R: Cold adaptation of the mononuclear molybdoenzyme periplasmic nitrate reductase from the Antarctic bacterium*Shewanella gelidimarina*. Biochem Biophys Res Commun. 2011, 4: 783-788.

159. Carginale V, Trinchella F, Capasso C, Scudiero R, Parisi E: Gene amplification and cold adaptation of pepsin in Antarctic fish. A possible strategy for food digestion at low temperature. Gene. 2004, 2: 195-205.
160. Miyazaki K, Wintrode PL, Grayling RA, Rubingh DN, Arnold FH: Directed evolution study of temperature adaptation in a psychrophilic enzyme. J Mol Biol. 2000, 297: 1015-1026.
161. Aghajari N, Van Petegem F, Villeret V, Chessa JP, Gerday C, Haser R, Van Beeumen J: Crystal structures of a psychrophilic metalloprotease reveal new insights into catalysis by cold adapted proteases. Proteins. 2003, 50: 636-647.
162. Xie BB, Bian F, Chen XL, He HL, Guo J, Gao X, Zeng YX, Chen B, Zhou BC, Zhang YZ: Cold adaptation of zinc metalloproteases in the thermolysin family from deep sea and arctic sea ice bacteria revealed by catalytic and structural properties and molecular dynamics: new insights into relationship between conformational flexibility and hydrogen bonding. J Biol Chem. 2009, 14: 9257-9269.
163. Zhong CQ, Song S, Fang N, Liang X, Zhu H, Tang XF, Tang B: Improvement of low-temperature caseinolytic activity of a thermophilic subtilase by directed evolution and site-directed mutagenesis. Biotechnol Bioeng. 2009, 5: 862-870.
164. Sigurdardóttir AG, Arnórsdóttir J, Thorbjarnardóttir SH, Eggertsson G, Suhre K, Kristjánsson MM: Characteristics of mutants designed to incorporate a new ion pair into the structure of a cold adapted subtilisin-like serine proteinase. Biochim Biophys Acta. 2009, 3: 512-518.
165. Almog O, González A, Godin N, de Leeuw M, Mekel MJ, Klein D, Braun S, Shoham G, Walter RL: The crystal structures of the psychrophilic subtilisin S41 and the mesophilic subtilisin Sph reveal the same calcium-loaded state. Proteins. 2009, 2: 489-496.
166. Karan R, Khare SK: Stability of haloalkaliphilic *Geomicrobium* sp. protease modulated by salt. Biochemistry (Mosc). 2011, 76: 686-693.
167. Karan R, Singh RK, Kapoor S, Khare SK: Gene identification and molecular characterization of solvent stable protease from a moderately haloalkaliphilic bacterium, *Geomicrobium* sp. EMB2. J Microbiol Biotechnol. 2011, 2: 129-135.
168. Ohtani N, Haruki M, Morikawa M, Kanaya S: Heat labile ribonuclease HI from a psychrotrophic bacterium: gene cloning, characterization and site-directed mutagenesis. Protein Eng. 2001, 12: 975-982.
169. Narinx E, Baise E, Gerday C: Subtilisin from psychrophilic Antarctic bacteria: characterization and site-directed mutagenesis of residues

possibly involved in the adaptation to cold. Protein Eng. 1997, 10: 1271-1279.
170. Matsumoto M, Yokouchi H, Suzuki N, Ohata H, Matsunaga T: Saccharification of marine microalgae using marine bacteria for ethanol production. Appl Biochem Biotechnol. 2003, 108: 247-254.
171. Fukushima T, Mizuki T, Echigo A, Inoue A, Usami R: Organic solvent tolerance of halophilic -amylase from a Haloarchaeon, *Haloarcula*sp. strain S-1. Extremophiles. 2005, 9: 85-89.
172. Chakraborty S, Khopade A, Kokare C, Mahadik K, Chopade B: Isolation and characterization of novel a-amylase from marine*Streptomyces* sp. D1. J Mol Catalysis B Enzymatic. 2009, 58: 17-23.
173. Kosjek B, Stampfer W, Pogorevc M, Goessler W, Faber K, Kroutil W: Purification and characterization of a chemotolerant alcohol dehydrogenase applicable to coupled redox reactions. Biotechnol Bioeng. 2004, 86: 55-62.
174. Kobori H, Sullivan CW, Shizuya H: Heat-labile alkaline phosphatase from antarctic bacteria: rapid 5› end-labeling of nucleic acids. Proc Natl Acad Sci USA. 1984, 81: 6691-6695.
175. Hoyoux A, Jennes I, Dubois P, Genicot S, Dubail F, François JM, Baise E, Feller G, Gerday C: Cold-adapted beta-galactosidase from the Antarctic psychrophile *Pseudoalteromonas haloplanktis*. Appl Environ Microbiol. 2001, 67: 1529-1535.
176. Maugard T, Gaunt D, Legoy MD, Besson T: Microwave-assisted synthesis of galacto-oligosaccharides from lactose with immobilized beta-galactosidase from *Kluyveromyces lactis*. Biotechnol Lett. 2003, 25: 623-629.
177. Hildebrandt P, Wanarska M, Kur J: A new cold-adapted β-D-galactosidase from the Antarctic *Arthrobacter* sp. 32c- gene cloning, overexpression, purification and properties. BMC Microbiology. 2009, 9: 151-
178. Bridiau N, Issaoui N, Maugard T: The effects of organic solvents on the efficiency and regioselectivity of N-acetyl-lactosamine synthesis, using the β-galactosidase from *Bacillus circulans* in hydro-organic media. Biotechnol Prog. 2010, 26: 1278-1289.
179. Hatori Y, Sato M, Orishimo K, Yatsunami R, Endo K, Fukui T, Nakamura S: Characterization of recombinant family 18 chitinase from extremely halophilic archaeon *Halobacterium salinarum* strain NRC-1. Chitin Chitosan Res. 2006, 12: 201-
180. Essghaier B, Hedi A, Bejji M, Jijakli H, Boudabous A, Sadfi-Zouaoui

N: Characterization of a novel chitinase from a moderately halophilic bacterium, *Virgibacillus marismortui* strain M3-23. Annal Microbiol. 2011, DOI: 10.1007/s13213-011-0324-4

181. Doukyu N, Aono R: Purification of extracellular cholesterol oxidase with high activity in the presence of organic solvents from*Pseudomonas* sp ST-200. Appl Environ Microbiol. 1998, 64: 1929-1932.

182. Yoshimune K, Shirakihara Y, Wakayama M, Yumoto I: Crystal structure of salt-tolerant glutaminase from *Micrococcus luteus* K-3 in the presence and absence of its product L-glutamate and its activator Tris. FEBS J. 2010, 3: 738-748.

183. Hotta Y, Ezaki S, Atomi H, Imanaka T: Extremely stable and versatile carboxyl esterase from a hyperthermophilic archaeon. Appl Environ Microbiol. 2002, 68: 3925-3931.

184. Khudary RA, Venkatachalam R, Katzer M, Elleuche S, Antranikian G: A cold-adapted esterase of a novel marine isolate,*Pseudoalteromonas arctica*: gene cloning, enzyme purification and characterization. Extremophiles. 2010, 14: 273-285.

185. Kirk O, Christensen MW: Lipases from *Candida antarctica*: Unique biocatalysts from a unique origin. Org Process Res Dev. 2002, 6: 446-451.

186. Giudice AL, Michaud L, de Pascale D, Domenico MD, di Prisco G, Fani R, Bruni V: Lipolytic activity of Antarctic cold adapted marine bacteria. J Appl Microbiol. 2006, 101: 1039-1048.

187. Joseph B, Ramteke PW, Kumar PA: Studies on the enhanced production of extracellular lipase by *Staphylococcus epidermidis*. J Gen Appl Microbiol. 2006, 52: 315-320.

188. Amoozegar MA, Salehghamari E, Khajeh K, Kabiri M, Naddaf S: Production of an extracellular thermohalophilic lipase from a moderately halophilic bacterium, *Salinivibrio* sp. strain SA-2. J Basic Microbiol. 2008, 48: 160-167.

189. Gaur R, Gupta A, Khare SK: Purification and characterization of lipase from solvent tolerant *Pseudomonas aeruginosa* PseA. Process Biochem. 2008, 43: 1040-1046.

190. Pérez D, Martín S, Fernández-Lorente G, Filice M, Guisán JM, Ventosa A, García MT, Mellado E: A novel halophilic lipase, LipBL, showing high efficiency in the production of eicosapentaenoic acid (EPA). PLoS ONE. 2011, 6: e23325-

191. Kamekura M, Hamakawa T, Onishi H: Application of halophilic nuclease

H of *Micrococcus varians* subsp. *halophilus* to commercial production of flavoring agent 5'-GMP. Appl Environ Microbiol. 1982, 44: 994-995.

192. Truong LV, Tuyen H, Helmke E, Binh LT, Schweder T: Cloning of two pectate lyase genes from the marine Antarctic bacterium *Pseudoalteromonas haloplanktis* strain ANT/505 and characterization of the enzymes. Extremophiles. 2001, 5: 35-44.

193. Ryu K, Kim J, Dordick JS: Catalytic properties and potential of an extracellular protease from an extreme halophile. Enzym Microb Technol. 1994, 16: 266-275.

194. Kim J, Dordick JS: Unusual salt and solvent dependence of a protease from an extreme halophile. Biotechnol Bioeng. 1997, 55: 471-479.

195. Bobe IM, Abdelmoez W, Ogino H, Yasuda M, Ishimi K, Ishikawa H: Kinetics and mechanism of a reaction catalyzed by PST-01 protease from *Pseudomonas aeruginosa* PST-01. Biotechnol Bioeng. 2004, 86: 365-373.

196. Tsuchiyama S, Doukyu N, Yasuda M, Ishimi K, Ogino H: Peptide synthesis of aspartame precursor using organic solvent-stable PST-01 protease in monophasic aqueous organic solvent systems. Biotechnol Prog. 2007, 23: 820-823.

197. Gupta A, Roy I, Khare SK, Gupta MN: Purification and characterization of a solvent stable protease from *Pseudomonas aeruginosa* PseA. J Chrom A. 2005, 1069: 155-161.

198. Pawar R, Zambare V, Barve S, Paratkar G: Application of protease isolated from *Bacillus* sp. in enzymatic cleansing of contact lenses. Biotechnol. 2009, 8: 276-280.

199. Ruiz DM, Iannuci NB, Cascone O, De Castro RE: Peptide synthesis catalysed by a haloalkaliphilic serine protease from the archaeon *Natrialba magadii* (Nep). Lett Appl Microbiol. 2010, 51: 691-696.

200. Akolkar AV, Durai D, Desai AJ: *Halobacterium* sp. SP1 (1) as a starter culture for accelerating fish sauce fermentation. J Appl Microbiol. 2010, 109: 44-53.

201. Karan R, Khare SK: Purification and characterization of a solvent stable protease from *Geomicrobium* sp EMB2. Environ Technol. 2010, 10: 1061-1072.

202. Karan R, Singh SP, Kapoor S, Khare SK: A novel organic solvent tolerant protease from a newly isolated *Geomicrobium* sp. EMB2 (MTCC 10310): production optimization by response surface methodology. N Biotechnol. 2011, 2: 136-145.

203. Collins T, Hoyoux A, Dutron A, Georis J, Genot B, Dauvrin T, Arnaut F, Gerday C, Feller G: Use of glycoside hydrolase family 8 xylanases in baking. J Cereal Sci. 2006, 43: 79-84.
204. Dornez E, Verjans P, Arnaut F, Delcour JA, Courtin CM: Use of psychrophilic xylanases provides insight into the xylanase functionality in bread making. J Agric Food Chem. 2011, 17: 9553-9562.
205. Wang K, Li G, Yu SQ, Zhang CT, Liu YH: A novel metagenome-derived beta-galactosidase: gene cloning, overexpression, purification and characterization. Appl Microbiol Biotechnol. 2010, 88: 155-165.
206. Prakash P, Jayalakshmi SK, Prakash B, Rubul M, Sreeramulu K: Production of alkaliphilic, halotolerent, thermostable cellulase free xylanase by *Bacillus halodurans* PPKS-2 using agro waste: single step purification and characterization. World J Microbiol Biotechnol. 2011, DOI: 10.1007/s11274-011-0807-2

Chapter 9

MARINE MICROBIAL BIODIVERSITY, BIOINFORMATICS AND BIOTECHNOLOGY (M2B3) DATA REPORTING AND SERVICE STANDARDS

Petra ten Hoopen, Stéphane Pesant, Renzo Kottmann, Anna Kopf, Mesude Bicak, Simon Claus, Klaas Deneudt, Catherine Borremans, Peter Thijsse, Stefanie Dekeyzer, Dick MA Schaap, Chris Bowler, Frank Oliver Glöckner and Guy Cochrane

European Nucleotide Archive, EMBL-EBI, Wellcome Trust Genome Campus Hinxton

ABSTRACT

Contextual data collected concurrently with molecular samples are critical to the use of metagenomics in the fields of marine biodiversity, bioinformatics and biotechnology. We present here Marine Microbial Biodiversity, Bioinformatics and Biotechnology (M2B3) standards for *"Reporting"* and *"Serving"* data. The M2B3 Reporting Standard (1) describes minimal mandatory and recommended contextual information for a marine microbial sample obtained in the epipelagic zone, (2) includes meaningful information for researchers in the oceanographic, biodiversity and molecular disciplines, and (3) can easily be adopted by any marine laboratory with minimum sampling resources. The M2B3 Service Standard defines a software interface through which these data can be discovered and explored in data repositories. The M2B3 Standards were developed by the European project Micro B3, funded under 7th Framework Programme "Ocean of Tomorrow", and were first used with the Ocean Sampling Day initiative. We believe that these standards have value in broader marine science.

BACKGROUND

An immense wealth of genetic, functional and morphological diversity in marine ecosystems remains unexplored, offering the potential for substantial scientific and biotechnological discoveries. Indeed, significant interest in this area has led to large-scale initiatives, such as Tara Oceans [1], the Global Ocean Survey [2] and Malaspina [3], that target the exploration of marine biodiversity on planetary scales. While the shared goal of such initiatives is the development of an understanding of the composition and ecology of marine microbial ecosystems, each focuses on different parts of the taxonomic breadth of ocean life and only a subset of ocean ecosystems, such as epi- meso- and bathypelagic systems. Ongoing and future marine survey projects will add value to these explorations and will continue to build a powerful marine data infrastructure from which ecosystems biology and biotechnology will derive benefit. Prerequisite for the successful exploitation of acquired data are standards that enable interoperability in the data infrastructure.

Just as marine studies span many disciplines (e.g. biological, oceanographic, molecular), use of data from marine studies requires approaches that traverse the many disciplines, asking questions, for example, of species distribution, physical oceanographic parameters, molecular biology and data licensing. Each discipline has established infrastructure and best practice for the dissemination of its data, including open data repositories, reporting and data standards and discovery and analysis portals. However, there remain major barriers when data are to be used across disciplines that relate to a lack of interoperability between standards and the lack of a consistent environment for the discovery and retrieval of data.

The Marine Microbial Biodiversity, Bioinformatics, Biotechnology Project (Micro B3) [4] unites intensive oceanographic monitoring, thorough biodiversity studies and high-throughput DNA sequencing of marine genomes, metagenomes and pan-genomes. The project addresses interdisciplinary needs in marine ecosystems biology and biotechnology by considering established best practice within the disciplines and deriving practical least-change means to align practices. Recognising that it is non-trivial to influence deeply-rooted working practices established over decades, we have delivered an extensive programme of workshop-based discussions amongst representatives of the disciplines [22,23].

This effort led to the development of two standards described here. First, the *M2B3 Reporting Standard* defines and describes fields of information to be made available with marine data sets. Second, the *M2B3 Service Standard* defines and describes a software interface through which hosts of marine data, such as the public data repositories, can present their marine data holdings.

The resulting standards were used by marine sampling stations and cruises participating in the Micro B3 sampling campaign, Ocean Sampling Day (OSD) [5], a simultaneous world-scale sample and contextual data collection to investigate dynamics and functions of marine microbial diversity. We believe that our work will also be of value to other marine surveys in the future.

M2B3 Reporting Standard

We have developed the M2B3 Reporting Standard to support data collection and dissemination for those involved in marine microbial sampling. The standard, shown in full in Tables 1,2,3,4,5,6,7 spans the biodiversity, molecular and oceanographic domains and adopts existing standards of each discipline with their mandatory, recommended and optional descriptors (fields of information) (see Figure 1). It represents a unique intersection of existing reporting requirements across all three domains.

Table 1: M2B3 Reporting Standard about an investigation effort

Descriptor name	Description of usage	Control vocabulary/ format/unit	Example
INVESTIGATION_ **Campaign**	Refers to a sampling activity that is either determined in time, repeated in time or continuous, e.g. a cruise, a mesocosm experiment, a time series, or live data streams	Free text	Micro B3-OSD2014
INVESTIGATION _Site	Refers to the unique identifier and name of the site/station where the sampling activity is performed.	Format: <Site ID from OSD Site Registry >, <Site name from OSD Site Registry>	OSD5, Poseidon-E1-M3A Time Series Station
INVESTIGATION _Platform	Refers to the specific unique stage from which the sampling device was deployed; includes the platform category and platform name.	Format: <Platform category from SDN:L06>,<Platform name>	research vessel, FILIA
INVESTIGATION _Authors	List of people who will appear in the citation of data publications. Please order the list according to authorship. The first author is the contact person.	Format: <LASTNAME>, <FirstName>, <Institution>, <email>	JONES, Peter, Institute1, pjones@institute1.eu; SMITH, Mary, Institute2, msmith@institute2.eu
INVESTIGATION _Project	Refers to the project that organised/funded the data/sample collection.	Free text	Micro B3

INVESTIGATION _Objective	Describes the scientific context/interest of the sampling activity. This information is useful to generate a short abstract as part of the data set citation.	Free text; 100-500 words	A short abstract

Mandatory information is in bold and other fields are recommended OSD Sites Registry is a controlled register for OSD sampling Sites (http://mb3is.megx.net/osd-registry). SDN:L06::XX is a controlled terms list describing "CATEGORIES" of platforms (http://seadatanet.maris2.nl/v_bodc_vocab_v2/search.asp?lib=L06).

Table 2: M2B3 Reporting Standard about a sampling event

Descriptor name	Description of usage	Control vocabulary/ format/unit	Example
EVENT_Date/Time	Date and time when the sampling event started and ended, e.g. each CTD cast, net tow, or bucket collection is a distinct event.	Date and time in UTC;	2013-06-21T14:05:00Z/
		Format: yyyy-mm-ddThh:mm:ssZ	2013-06-21T14:46:00Z
EVENT_Longitude	Longitude of the location where the sampling event started and ended, e.g. each CTD cast, net tow, or bucket collection is a distinct event.	Format: ###.######	035.666666
		Decimal degrees; East=+, West=-Format: Use WGS 84 for GPS data	035.670200
EVENT_Latitude	Latitude of the location where the sampling event started and ended, e.g. each CTD cast, net tow, or bucket collection is a distinct event.	Format: ##.######	−24.666666
		Decimal degrees; North=+, South=-Format: Use WGS 84 for GPS data	-24.664300
EVENT_Device	Refers to the instrument/gear used to collect the sample or the sensor used to measure environmental parameters.	Free text	10L-Niskins or 5L-Bucket
EVENT_Method	Refers to the standard deployment procedure of the Device.	Free text	12 Niskins were deployed on a Rosette

Marine Microbial Biodiversity, Bioinformatics and Biotechnology.... 191

| EVENT_Comment | Report any observation/deviation from the standard deployment procedure described in EVENT_Method | Free text | Lots of Jellyfish in the water |

Mandatory information is in bold and other fields are recommended.

Table 3: M2B3 Reporting Standard about a sample

Descriptor name	Description of usage	Control vocabulary/format/unit	Example
SAMPLE_**Title**	A short informative description of the sample. Must be unique for each sample, (i.e. for each filter generated during sampling).	Format: <OSD_SiteID>_<Month>_<Year>_<SiteName>_<Protocol_Label>_<SampleNo>_<Depth>	OSD3_06_14_Helgoland_NPL022_1_surface
SAMPLE_**Depth**	The distance below the surface of the water at which a measurement was made or a sample was collected.	Format: ##.#; Positive below the sea surface. SDN:P06:46:ULAA for m	1.5
SAMPLE_**Protocol_Label**	Identifies the protocol used to produce the sample, e.g. filtration and preservation.	Term list; See the *SAMPLE_Protocol_Label* in the OSD Protocols Section for details	NPL022
SAMPLE_Quantity	Refers to the quantity of environment that was sampled, most often with dimensions Length, Amount, Mass or Time.	Format : ###.### See the *SAMPLE_Quantity* in the OSD Protocols Section for details	20 Litres
SAMPLE_Container	Refers to the container in which the sample is stored prior to analysis.	Term list; See the *SAMPLE_Container* in the OSD Protocols Section for details	Sterivex cartridge
SAMPLE_Content	Refers to the content of the sample container. While the sample might target a specific organism (e.g. bacteria), the sample content might be a filter or a volume of water.	Term list; See the *SAMPLE_Content* in the OSD Protocols Section for details	Particulate matter on a 0.22 μm pore size filter

SAMPLE_Size-Fraction_Upper-Threshold	Refers to the mesh/pore size used to pre-filter/pre-sort the sample. Materials larger than the size threshold are excluded from the sample.	Term list;	no pre-filtration
		See the *SAMPLE_Size-Fraction_Upper-Threshold* in the OSD Protocols Section for details; in μm	
SAMPLE_Size-Fraction_Lower-Threshold	Refers to the mesh/pore size used to retain the sample. Materials smaller than the size threshold are exclude from the sample.	Term list;	0.22
		See the *SAMPLE_Size-Fraction_Lower-Threshold* in the OSD Protocols Section for details; in μm	
SAMPLE_Treatment_Chemicals	Refers to the chemicals (e.g. preservatives) added to the sample.	Terms list: ChEBI;	None
		See the *SAMPLE_Treatment_Chemicals* in the OSD Protocols Section for details	
SAMPLE_Treatment_Storage	Refers to the conditions in which the sample is stored, e.g. temperature, light conditions, time.	Term list;	−80 degrees Celsius
		See the *SAMPLE_Treatment_Storage* in the OSD Protocols Section for details	

Mandatory information is in bold and other fields are recommended. OSD Protocols are available at http://www.microb3.eu/sites/default/files/osd/OSD_Handbook_v2.0.pdf. ChEBI is an ontological classification and dictionary of small chemical compounds (http://www.ebi.ac.uk/chebi/init.do).

Table 4: M2B3 Reporting Standard about the sample environmental context

Descriptor name	Description of usage	Control vocabulary/format/unit	Example
ENVIRONMENT_Biome	Descriptor of the broad ecological context of a sample.	Terms list: EnvO	ENVO:01000023 for "marine pelagic biome"
ENVIRONMENT_**Feature**	Compared to biome, feature is a descriptor of a geographic aspect or a physical entity that strongly influences the more local environment of a sample.	Terms list: EnvO	ENVO:00000209 for "photic zone"
ENVIRONMENT_**Material**	Descriptor of the material that was displaced by the sampling activity, or material in which a sample was embedded, prior to the sampling event.	Terms list: EnvO	ENVO:00002149 for "sea water"

ENVIRONMENT_Temperature	Temperature of water at the time of taking the sample. Define the parameter according to Table 7.	Format: ##.#	16.2°C
		SDN:P02:75:TEMP	
		SDN:P06:46:UPAA for°C	
ENVIRONMENT_Salinity	Salinity of water at the time of taking the sample. Define the measurement according to Table 7.	Format: ##.#	39.1 psu
		SDN:P02:75:PSAL	
		SDN:P06:46:UGKG for PSU	
ENVIRONMENT_Marine_Region	It characterises the environment, based on the latitude and longitude, by reference to geographic, political, economic or ecological boundaries.	Terms list: Marine Regions	MRGID:21886 for Marine Ecoregion:South European Atlantic Shelf
ENVIRONMENT_Other_Parameters	Add as many fields as there are other environments parameters measured.		
	Define the measurement according to Table 7.		
	See the list of recommended environmental parameters in Table 5		

Mandatory information is in bold and other fields are recommended EnvO is the Environment Ontology (http://www.environmentontology.org/Browse-EnvO). SDN:P02:75:XXXX is a controlled terms list describing "WHAT" is measured (http://seadatanet.maris2.nl/v_bodc_vocab_v2/search.asp?lib=P02). SDN:P06:46:XXXX is a controlled terms list describing "UNITS" of measurements (http://seadatanet.maris2.nl/v_bodc_vocab_v2/search.asp?lib=P06). Marine Regions is a standard list of marine georeferenced place names (http://www.marineregions.org/).

Table 5: M2B3 Reporting Standard about environmental measurements

	Measurement	Description of usage	Control vocabulary/format/unit
General	Conductivity	Electrical conductivity of water	SDN:P02:75:CNDC
			SDN:P06:46:UECA for mS/cm
	Temperature	Temperature of water	SDN:P02:75:TEMP
			SDN:P06:46:UPAA for °C
	Depth (m)	Vertical spatial coordinates	SDN:P02:75:AHGT
			SDN:P06:46:ULAA for m
	Salinity	Salinity of water	SDN:P02:75:PSAL
			SDN:P06:46:UGKG for PSU
	Fluorescence	Raw (volts) or converted (mg Chla/m^3) fluorescence of the water	SDN:P02:75:FVLT
			SDN:P06:46:UVLT for volts

Nutrient status of a system	Nitrate	Nitrate concentration parameters in the water column	SDN:P02:75:NTRA
			SDN:P06:46:UPOX for μmol/L
	Nitrite	Nitrite concentration parameters in the water column	SDN:P02:75:NTRI
			SDN:P06:46:UPOX for μmol/L
	Phosphate	Phosphate concentration parameters in the water column	SDN:P02:75:PHOS
			SDN:P06:46:UPOX for μmol/L
	Silicate	Silicate concentration parameters in the water column	SDN:P02:75:SLCA
			SDN:P06:46:UPOX for μmol/L
	Ammonium	Ammonium concentration parameters in the water column	SDN:P02:75:AMON
			SDN:P06:46:UPOX for μmol/L
Chemical properties of a system	pH	Alkalinity, acidity and pH of the water column	SDN:P02:75:ALKY
	Dissolved oxygen concentration	Dissolved oxygen parameters in the water column	SDN:P02:75:DOXY
			SDN:P06:46:KGUM for μmol/kg
Optical properties of a system	Downward PAR	Visible waveband radiance and irradiance measurements in the water column	SDN:P02:75:VSRW
			SDN:P06:46:UMES for μE/m^2/s
	Turbidity	Transmittance and attenuance of the water column	SDN:P02:75:ATTN
			SDN:P06:46:USTU for FTU or NTU
Biogeo-chemistry (Amount or Mass)	Carbon organic particulate (POC)	Particulate organic carbon concentration in the water column	SDN:P02:75:CORG
			SDN:P06:46:UGPL for μg/L
	Nitrogen organic particulate (PON)	Particulate organic nitrogen concentration in the water column	SDN:P02:75:NTOT
			SDN:P06:46:UGPL for μg/L
	Carbon organic dissolved (DOC)	Dissolved organic carbon concentration in the water column	SDN:P02:75:DOCC
			SDN:P06:46:UPOX for μmol/L
	Nitrogen organic dissolved (DON)	Dissolved organic nitrogen concentration in the water column	SDN:P02:75:TDNT
			SDN:P06:46:UMGL for mg/L

Ecosystem trophic structure & biodiversity (Amount, Volume or Mass of organisms in the environment)	Pigment concentrations	Concentration of pigments (e.g. chlorophyll a) extracted and analysed by fluorometry or HPLC	SDN:P02:75:CPWC
			SDN:P06:46:UMMC for mg/m^3
	Picoplankton (Flow Cytometry)	Abundance of cells in the water column (+other avail. cell properties)	SDN:P02:75:BATX
			SDN:P06:46:UPMM for #/m^3
	Nano/Microplankton	Abundance of cells in the water column (+other avail. cell properties)	SDN:P02:75:MATX or PATX
			SDN:P06:46:UPMM for #/m^3
	Meso/Macroplankton	Abundance of individuals in the water column (+other avail. properties)	SDN:P02:75:ZATX
			SDN:P06:46:UPMM for #/m^3
Ecosystem trophic rates	Primary Production (isotope uptake)	Primary Production in the water column	SDN:P02:75:PPRD
			SDN:P06:46:UGDC for mg/m^3/d
	Primary Production (oxygen)	Primary Production in the water column	SDN:P02:75:PPRD
			SDN:P06:46:UGDC for mg/m^3/d
	Bacterial production (isotope uptake)	Bacterial production in the water column	SDN:P02:75:UPTH
			SDN:P06:46:UGDC for mg/m^3/d
	Bacterial production (respiration)	Bacterial production in the water column	SDN:P02:75:UPTH
			SDN:P06:46:UGDC for mg/m^3/d

Mandatory information is in bold and other fields are recommended.

Table 6: M2B3 Reporting Standard about organisms in a sample

Descriptor name	Description of usage	Control vocabulary/format/unit	Example
ORGANISM_Taxon_ID	An identifier for the nomenclatural (not taxonomic) details of a scientific name.	Terms list: WoRMS	urn:lsid:marinespecies.org:taxname: 345516
		Format: LSID	
ORGANISM_Taxon_Scientific_Name	The full name of the lowest level taxon.	Terms list: WoRMS	Prochlorococcus marinus
		Format: Taxon name	

ORGANISM_Sex	The sex of a specimen or collected/observed individual(s).	Terms list: M=Male; F=Female; H=Hermaphrodite; I=Indeterminate (examined but could not be determined; U=Unkown (not examined); T=Transitional (between sexes; useful for sequential hermaphrodites); B=Both Male and Female	M
ORGANISM_Life_Stage	Indicates the life stage present.	Free text	resting spores
ORGANISM_Size	Refers to size measurements that are made concurrently to the enumeration and identification of organisms.		
	Define the measurement according to Table 7.		
ORGANISM_Biovolume	Refers to volume measurements/calculations that are made concurrently to the enumeration and identification of organisms.		
	Define the measurement according to Table 7.		
ORGANISM_Biomass	Refers to biomass measurements/calculations that are made concurrently to the enumeration and identification of organisms.		
	Define the measurement according to Table 7.		

Mandatory information is in bold and other fields are recommended WoRMS is the World Register of Marine Species (http://www.marinespecies.org/aphia.php?p=search).

Table 7: M2B3 Reporting Standard about environmental measurement processes

Descriptor name	Description of usage	Control vocabulary/format/unit	Example
MEASUREMENT_ID	Unique ID from a controlled vocabulary.	SDN:P02:75:xxxx	SDN:P02:75:CORG for Particulate organic carbon concentration in the water column
MEASUREMENT_Name	Common name for the measurement.	Free text	POC
MEASUREMENT_Quantity	Describes the quantity measured using terms from the Système International of units.	Free text; SI of units	Mass concentration
MEASUREMENT_Dimensions	Describes the quantity measured using dimension terms from the Système International of units.	Free text; SI of units	$M^1 L^{-3}$
MEASUREMENT_Currency	May often refer to a TAXONOMY_ID or a CHEMICAL_ID.	Free text; Terms list: WoRMS; Terms list: ChEBI	Organic carbon

MEASURE-MENT_Units	Describes the units of the quantity measured using terms from the Système International of units.	SDN:P06:46:xxxx	SDN:P06:46:UGPL for µg/L
MEASURE-MENT_Method	Describes the measurement method used. Equivalent to methodological details provided in a paper.	Free text	Mass spectrometry
MEASURE-MENT_Comment	Any comment about the measurement.	Free text	Inorganic carbon removed by acidification

Mandatory information is in bold and other fields are recommended.

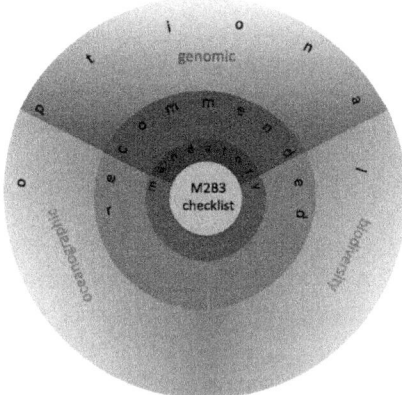

Figure 1: M2B3 Reporting Standard descriptors schematically depicted on the junction of three disciplines, adopting existing standards of each domain.

We have been strongly guided in this work by the existing standards MEDIN [6], MIxS [7] and Darwin Core [8], the expertise of the Tara Oceans project teams and the International Census of Marine Microbes (ICoMM) project [9], and knowledge of community-established reporting practice into public data archives bestowed by experts from the biodiversity, oceanographic and molecular domain.

The core of the M2B3 Reporting Standard is the M2B3 checklist, (see Figure 2). This core represents the minimal mandatory reporting requirement and consists of descriptors essential to oceanographic, biodiversity and molecular domains, representing research on microbial diversity and function in the marine environment. Marine scientists should be able to report this minimum contextual information about each marine microbial sample irrespective of their scientific expertise and resources available for the sampling.

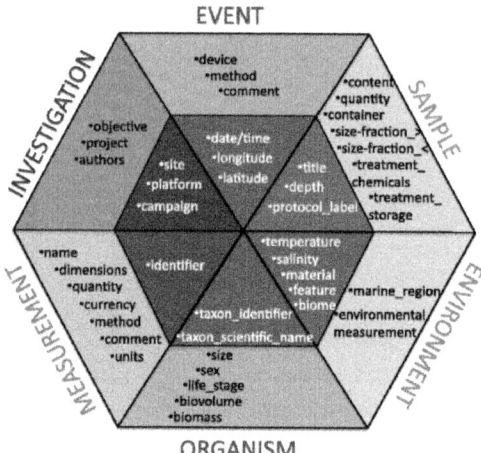

Figure 2: Mandatory and recommended information of the M2B3 Reporting Standard; descriptors are split into six categories represented by coloured triangles, where mandatory descriptors are in the dark-shaded area and recommended information elements are in the light-shaded area. Environmental measurements in the ENVIRONMENT section are further specified in Figure 3.

The M2B3 Reporting Standard includes a set of recommended descriptors (see Figure 2), provision of which brings each marine microbial sample into a rich environmental context and allows better ecological interpretation and experimental reproducibility. The standard's environmental parameters are recommended by the Micro B3 Consortium for description of the environmental landscape of each epipelagic microbial sample (see Figure 3). Here, we have taken an approach including descriptors that draw a balance between analysis requirements-driven methods and current reporting practice in marine microbial sampling. In the requirements-driven approach we analysed several use cases from the area of diatom biology and marine prokaryotic biodiversity. Collated environmental parameters, recorded and reported in these studies in order to answer the scientific questions posed in the studies, represent the optimal list of environmental variables to be measured at the time of microbial sample collection from the epipelagic zone. The current sampling practice-driven approach is the pragmatic counterpart, where environmental variables were identified based on current marine sampling practice surveys and consultations with experts from European marine stations with established long-term sea monitoring programs and a wealth of expertise, such as Western Channel Observatory in the UK, Station Biologique de Roscoff in France, the Stazione Zoologica in Naples, Italy, or the Biological Institute Helgoland (BAH) of the Alfred Wegener Institute, the Helmholtz Centre for Polar and Marine Research in Germany.

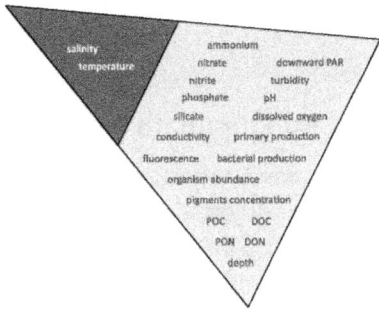

Figure 3: Mandatory (in the dark green area) and recommended (in the light green area) environmental measurements of the M2B3 Reporting Standard.

All mandatory and recommended information is described in detail in Tables 1,2,3,4,5,6,7 including specification relating to usage, formal requirements for structure, indication of appropriate units, where applicable, and an example. Descriptors are split for easy navigation into six categories: (1) the marine investigation effort, (2) the sample-taking event, (3) sample-specific details, (4a) the environmental context of the sample, (4b) environmental measurements, (5) marine species found in the sample and (6) description of environmental measurement processes. Descriptors of each conceptual category are prefixed with the category name. Table 4 specifies a broad and local environmental context of a sample including required minimum of measured environmental parameters. Table 5 focuses on specific environmental parameters that complement the fields in Table 4. Table 7 defines how environmental measurements are captured. The logical relationship between the environmental measurement, measurement description and measurement values is summarised in Figure 4.

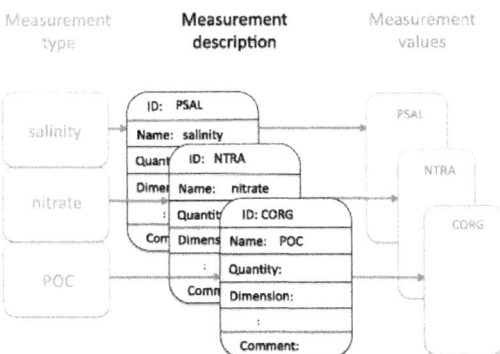

Figure 4: The logical connection between environmental measurements (Table 5), recording of the measurements (Table 7) and measured values, shown on the example

of three environmental parameters – salinity, nitrate and carbon organic particulate (POC).

M2B3 Reporting standard compliance

It is worth noting that if all mandatory descriptors from the M2B3 Reporting Standard are reported by a sampling station or a cruise, then a data management centre is frequently able to infer additional descriptors available from the public record. In one example, relating to OSD, the Micro B3 Information System (Micro B3 IS) [10] and the OSD coordinators are able to infer additional descriptors available from public data archives, such as the Data Publisher for Earth and Environmental Science (PANGAEA) for environmental data [11,12] and the European Nucleotide Archive (ENA) for molecular data [13,14]. The additional information can be added *post hoc* for all samples acquired within the OSD campaign since the campaign has standardised and published sampling protocols and a Registry of OSD stations and cruises [15]. The inferred descriptors include, for instance, a sample catalogue number and collection code assigned by the bio-archiving institution where the OSD samples will be centrally deposited. In a second example that applies very broadly across marine samples, remotely sensed data (such as cloud cover, air temperature and wind conditions) can be connected to appropriate records based upon geospatial fields.

Combining information compliant with the M2B3 Reporting Standard from marine sampling laboratories with inferred information has two major advantages: (1) it significantly reduces the reporting burden for the marine sampling laboratories and (2) it ensures that OSD data records created at the molecular data archive will be compliant with the MIxS molecular data standard, Version 4 [16], OSD data records created at the oceanographic data archive will be compliant with the oceanographic Common Data Index (CDI), Version 3 [17] and OSD data records created at the biodiversity data archive will be compliant with the biodiversity OBIS Schema, Version 1.1 [18].

M2B3 Service Standard

Six descriptors from the M2B3 Reporting Standard are central to data interoperability across disciplines. These descriptors provide the basis for connecting data points from one discipline to data points in another and are thus the indices upon which data resources providing services must present their data. The interoperability descriptors are: INVESTIGATION_Site, INVESTIGATION_Platform, EVENT_Date/Time, EVENT_Longitude, EVENT_Latitude and SAMPLE_Depth.

In order for users of marine data to discover and access data, there is a need for these fields of information to be made searchable in a single and consistent way across relevant data resources.

We define the M2B3 Service Standard as a standardised set of informatics methods through which marine data can be discovered in data resources. The six interoperability descriptors are presented by a compliant data resource using a programmatic service interface that follows Open Geospatial Consortium (OGC) standards, the Web Map Service (WMS), Web Feature Service (WFS) and/or the OpenSearch protocol. To date, the European Nucleotide Archive, European Ocean Biogeographic Information System (EurOBIS) [19,20], Micro B3 Information System, PANGAEA and SeaDataNet [21] have committed to supporting the M2B3 Service Standard for OSD data.

CONCLUSIONS

The M2B3 Reporting Standard combines reporting requirements of three disciplines. Compliance with the standard ensures that the collected data can be correctly directed to and stored in their respective domain-specific data archives, which are the ENA for molecular data and PANGAEA for environmental data and morphology-based biodiversity data. Compliance with the standard allows PANGAEA to create a condensed metadata summary and share it with pan-European oceanographic and biodiversity information networks, managed by SeaDataNet and EurOBIS, respectively. Micro B3 IS and other data resources compliant with the M2B3 Service standard can discover marine data compliant with the M2B3 Reporting Standard.

During its preparation, development of the M2B3 Reporting Standard and the M2B3 Service Standard allowed experts from the oceanographic, biodiversity and molecular disciplines to review current working practice, to extract and formulate what is essential and universal and to find common ground. Adoption of the M2B3 Reporting Standard will require a similar effort from the marine science community, as already started with the OSD sampling marine laboratories. The ultimate reward will be a unique collection of standardised marine data for the exploration of ecosystem biology and the advance of biotechnology.

Acknowledgements

We gratefully acknowledge contribution and comments from participants of the Micro B3 Use Case Workshop [22], Micro B3 Best Practice Workshop [23] and Micro B3 General Assembly. This project has received funding from the European Union's Seventh Framework Programme for research, technological development and demonstration (Joint Call OCEAN.2011-2: Marine microbial diversity – new insights into marine ecosystems functioning and its biotechnological potential) under grant agreement no 287589.

Competing interests

The authors declare that they have no competing interests.

Authors' contributions

PH coordinated the M2B3 standards development; SP contributed to the marine aspect of the M2B3 reporting standard, RK, GC, PH, MB and AK contributed to its molecular aspect and SC, KD CB and SD contributed to its biodiversity aspect; G.C DS, PT, RK and SC were leading the work on the M2B3 service standard. CB advised on the use case studies, GC and FOG provided overall guidance. PH wrote the manuscript with an editorial contribution of SP and GC and a revision by all co-authors. All authors read and approved the final manuscript.

REFERENCES

1. Karsenti E, Acinas SG, Bork P, Bowler C, De Vargas C, Raes J, et al. A holistic approach to marine eco-systems biology. PLoS Biol. 2011;9(10):e1001177.
2. Nealson KH, Venter JC. Metagenomics and the global ocean survey: what's in it for us, and why should we care? ISME J. 2007;1:185–7.
3. Malaspina. [http://www.expedicionmalaspina.es/]
4. Micro B3. [http://www.microb3.eu/]
5. OSD. [http://www.oceansamplingday.org]
6. MEDIN. [http://www.oceannet.org/marine_data_standards/]
7. Yilmaz P, Kottmann R, Field D, Knight R, Cole JR, Amaral-Zettler L, et al. Minimum information about a marker gene sequence (MIMARKS) and minimum information about any (x) sequence (MIxS) specifications. Nat Biotechnol. 2011;29:415–20.
8. Wieczorek J, Bloom D, Guralnick R, Blum S, Döring M, Giovanni R, et al. An Evolving Community-Developed Biodiversity Data Standard. PLoS One. 2012;7(1):e29715.
9. ICoMM. [http://news.coml.org/descrip/icomm.htm]
10. Micro B3 IS. [http://mb3is.megx.net/]
11. Schindler U, Diepenbroek M. Generic XML-based Framework for

Metadata Portals. Comput Geosci. 2008;34(12):1947–55. doi:10.1016/j.cageo.2008.02.023.

12. PANGAEA. [http://www.pangaea.de/]
13. Cochrane G, Alako B, Amid C, Bower L, Cerdeño-Tárraga A, Cleland I, et al. Facing growth in the European Nucleotide Archive. Nucleic Acids Res. 2013;41(D1):D30–5.
14. ENA. [http://www.ebi.ac.uk/ena]
15. OSD Registry. [http://mb3is.megx.net/osd-registry]
16. MIxS, v.4.0. [http://wiki.gensc.org/index.php?title=MIxS]
17. CDI, v.3.0. [http://www.seadatanet.org/Data-Access/Common-Data-Index-CDI]
18. OBIS, v.1.1. [http://www.iobis.org/node/304]
19. Vandepitte L, Hernandez F, Claus S, Vanhoorne B, De Hauwere N, Deneudt K, et al. Analysing the content of the European Ocean Biogeographic Information System (EurOBIS): available data, limitations, prospects and a look at the future. Hydrobiologia. 2011;667(1):1–14.
20. EurOBIS. [http://www.eurobis.org/]
21. SeaDataNet. [http://www.seadatanet.org]
22. Micro B3 Use Case Workshop. [http://www.microb3.eu/sites/default/files/deliverables/MB3_D4_1_PU.pdf]
23. Micro B3 Best Practice Workshop. [http://www.microb3.eu/sites/default/files/deliverables/MB3_D4_2_PU.pdf]

Chapter 10

ADVANCES IN BIOTECHNOLOGY AND GENOMICS OF SWITCHGRASS

Madhugiri Nageswara-Rao[1,2], Jaya R Soneji[2], Charles Kwit[1] and C Neal Stewart Jr[1,2]

[1]Department of Plant Sciences, The University of Tennessee, 252 Ellington Plant Sciences, 2431 Joe Johnson Dr., Knoxville, TN 37996, USA

[2]Department of Biological Sciences, Polk State College, Winter Haven, FL 33881, USA

ABSTRACT

Switchgrass (*Panicum virgatum* L.) is a C_4 perennial warm season grass indigenous to the North American tallgrass prairie. A number of its natural and agronomic traits, including adaptation to a wide geographical distribution, low nutrient requirements and production costs, high water use efficiency, high biomass potential, ease of harvesting, and potential for carbon storage, make it an attractive dedicated biomass crop for biofuel production. We believe that genetic improvements using biotechnology will be important to realize the potential of the biomass and biofuel-related uses of switchgrass. Tissue culture techniques aimed at rapid propagation of switchgrass and genetic transformation protocols have been developed. Rapid progress in genome sequencing and bioinformatics has provided efficient strategies to identify, tag, clone and manipulate many economically-important genes, including those related to higher biomass, saccharification efficiency, and lignin biosynthesis. Application of the best genetic tools should render improved switchgrass that will be more economically and environmentally sustainable as a lignocellulosic bioenergy feedstock.

INTRODUCTION

Resource consumption by humans continues to proceed at arguably unsustainable levels. In recent times, worldwide consumption of non-renewable fossil fuel reserves has increased drastically (U.S. Energy Information Administration;http://www.eia.gov/; Figure 1). With high rates of consumption anticipated and an ever-increasing population, a great challenge will be meeting the growing demand for energy for transportation, heating and industrial processes, and providing the raw industrial materials in a sustainable way [1]. Fossil fuels supply more than 80% of energy consumed globally and contribute to atmospheric greenhouse gases, declining water tables and climate change [2, 3]. All of these factors naturally lead to the development of renewable energy sources.

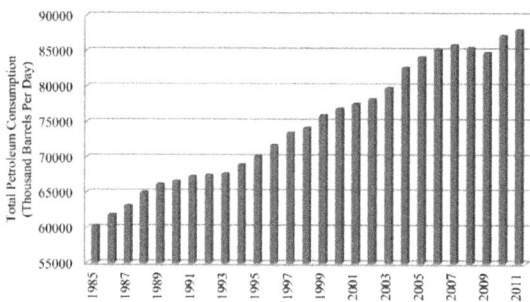

Figure 1: Total world petroleum consumption (thousand barrels/day) [Source: U.S. Energy Information Administration (EIA)].

Biomass and biomass-derived fuels may be able to provide a partial solution to today's energy challenges. In the last decade, there has been increased interest in dedicated biomass crops for biofuels [4]. It was considered that bioenergy provided by starch, sugar, and oils from plants would be crucial for accomplishing the goals of incremental substitution of petroleum-based transportation fuels in addition to reducing CO_2 emissions [5, 6]. However, first-generation biofuels were produced from traditional food and feed crops (e.g., sugarcane, corn, sugar beet), which may lead to supply shortages, and, in turn, to an increase in food prices [3, 7].

Even though plant-derived biofuels are renewable and, for the most part, carbon-neutral, they have been condemned for being associated with the loss of biological diversity and unfavorable consequences of changes in land use patterns [5]. These shortcomings led to a vision of developing second-generation lignocellulosic bioenergy crops, wherein stems, leaves, and/or husks of plants such as switchgrass, *Miscanthus*, jatropha, and poplar, may be used for the production of biofuels. In contrast to the easily-processed sugars and oils of first-generation bioenergy feedstocks, lignocellulosic biomass contains hard-to-digest matter from cell walls of grasses, crop residue, and woody biomass. One goal for the selection of second-generation bioenergy crops is that they should be able to grow on 'marginal' and low-cost land not suited for food crops, thus removing competition between the uses of land for food or fuel production [8]. Challenges remain in making second-generation bioenergy crops a reality. Among these are: (a) how to sustainably maximize the yield per hectare of biomass while minimizing agricultural inputs, (b) how to truly avoid competition with food and feed production, (c) how to increase the efficiency of biomass digestion by microbes and other processes [9], and (d) whether transgenic plants can be used [10].

Among all potential second-generation bioenergy crops, switchgrass (*Panicum virgatum* L.) has received, perhaps, the most attention as a dedicated lignocellulosic biofuel crop, beginning in the 1980s [11, 12] (Figure 2). Switchgrass is a member of the Paniceae tribe of grasses and belongs to the family Poaceae. It is native to North America and widely adapted; growing from 20°-60° north latitude and east of 100° west longitude [13, 14]. It exhibits tremendous diversity in its form and has been categorized into two ecotypes: upland and lowland [15, 16]. It can be grown on lands less-suitable for traditional agricultural crops for the production of biofuels, such as ethanol and butanol, from cellulose [17]. Switchgrass readily thrives on marginal land as a result of its deep-rooting habit, C_4 photosynthetic metabolism, among other traits [13]. Its perennial growth habit, wide adaptation, excellent conservation attributes, compatibility with conventional farming practices, ease of harvesting, handling, storage and amenability for being handled and stored both as wet or dry feedstock has made it a popular choice for biofuel feedstock crop [18–20]. Its high yielding potential on marginal lands and high yields across much of the eastern United States, especially the mid-South has set it apart from most other biofuel alternatives [12, 21, 22]. Switchgrass yields higher net energy than required to cultivate, harvest and convert it into cellulosic ethanol leading to much improved greenhouse gas balance compared with gasoline [23].

The importance of switchgrass as a bioenergy feedstock has increased interest in the generation of new cultivars optimized for energy production through breeding, biotechnology and management research. Improvement of biomass yield and nutritional quality should be amenable by conventional breeding. However, drastically better conversion of cell walls into fuels might not be possible by conventional breeding; genomics, biotechnology, systems- and synthetic biology tools might be required. Genomics and systems biology allow the identification and characterization of key genes that underlie critical fundamental processes. Overexpression of novel genes or knockdown of the expression of key endogenous genes can alter cell walls to dramatically improve fuel yield of switchgrass. The present scenario and the future prospects of the utilization of molecular and biotechnological tools for the genetic improvement of switchgrass have been emphasized in this review. While it is beyond the scope of this review, we envision that advanced biotechnology tools and synthetic biology will likely be required to optimize desired genetic improvements.

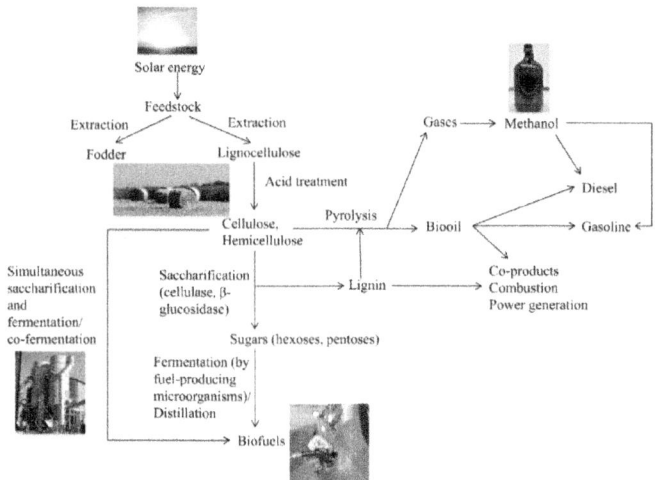

Figure 2: Flow chart of biofuel production in switchgrass [Photo credits: M Nageswara-Rao].

BIOTECHNOLOGICAL TOOLS FOR GENETIC IMPROVEMENT

Tissue culture

Efficient switchgrass cell and tissue culture is required for the production of transgenic plants as well as vegetative propagation. Prior to 1991, little switchgrass tissue culture research had been conducted. The initiation of US Bioenergy Feedstock Development Program enhanced opportunities for the long-term improvement of switchgrass [11]. Thus, in the 1990s, this program spurred research exploring explant types, tissue culture and regeneration of switchgrass with the ultimate goal of increasing the resource-base for developing transgenic lines. Switchgrass is amenable to regeneration after somatic embryogenesis and organogenesis.

Embryogenic callus

Somatic embryogenesis was used by Denchev and Conger [24] who reported high frequency plantlet regeneration. They used mature caryopses (seeds) and young leaf segments of the lowland cultivar 'Alamo' as explants to produce embryogenic callus on solidified Murashige and Skoog (MS) medium containing 2,4-dichlorophenoxyacetic acid (2,4-D) and 6-benzylaminopurine (BAP). The ease of handling and callus induction from mature caryopses made these valuable explants. When leaves were used as explants, there was a response gradient with regards to tissue age for callus initiation; young tissue is better than old tissue. Although somatic embryogenesis could be induced from embryogenic calli, regeneration of somatic embryos directly from the cells of the explants was not observed [24]. Somatic embryogenesis has also been reported from young infloresences of 'Alamo' [25, 26]. The cyclic production of plants from embryogenic callus renders this technique a viable option for rapid clonal propagation of switchgrass. However, compared with seed production, clonal propagation would be quite expensive and probably only used for the most valuable lines.

One disadvantage to the use of embryogenic callus- and seed-derived callus systems is that they generally have limited lifespans (months) of usefulness before they cease to be regenerable. Whereas the longevity of embryo viability can be only two months, the recently described switchgrass medium, LP9, increased the viability of callus and the ability to maintain it for a duration of over six months, making it more efficient for use in a transformation pipeline [27]. LP9 combined N_6 macroelements and B_5 microelements for the production and maintenance of switchgrass callus and its regeneration [27]. Also, the callus obtained was categorized as type II callus, which is more effective in grass transformation and regeneration [27] than type I callus obtained from previously described tissue culture systems [25, 26].

Cell Suspension Cultures

Cells divide faster in liquid suspension cultures compared with callus cells grown on solidified medium [28]. For large scale propagation, mutant selection, gene transfer and protoplast isolation, development of embryogenic cell suspension cultures would be advantageous. Cell suspension cultures were first obtained by Bob Conger's group that used young inflorescences of 'Alamo' as explants, which could directly yield embryogenic callus, which could be regenerated into plants [25]. This same group [26] showed that the utilization of osmotic pretreatment had a positive effect on the initiation and induction of somatic embryogenesis from suspension cultures derived from in vitro-cultured inflorescences of 'Alamo.' It was also observed that younger cultures gave a higher embryogenic response as compared with older cultures [26]. The HR8 line that was developed from a recurrent tissue culture selection of 'Alamo' had a higher seed germination capacity, and germinating seeds gave rise to higher percentage of somatic embryogenic callus [29]. Although this HR8 line, and indeed all improved Conger materials other than 'Alamo2' have been lost, the improved germplasm demonstrated very rapid propagation. These sorts of materials would have great use in breeding programs [11].

Cell suspensions are also excellent starting materials for the isolation of protoplasts. Protoplasts are useful in a wide range of applications including cell fusion and genetic manipulation [30]. Recently, Mazarei *et al.* reported protoplast isolation from switchgrass cell suspension cultures established from embryogenic callus [31]. They demonstrated that protoplast isolation efficiency was highly dependent on the type of cell suspension. Currently, our and other research groups are using cell suspension cultures for a variety of biotechnology-to-synthetic biology applications including deciphering the cell wall biology for improvments and high throughput multi-target genetic engineering and screening.

Organogenesis

Organogenesis illustrates a significant capability of plants to adapt to their altering environment; this process allows organ genesis from undifferentiated cells [32–34]. Switchgrass regeneration from organogenesis has been accomplished [24, 35]. Explants include mature caryopses, young leaf segments and young seedling explants and MS medium supplemented with auxins (2,4-D or picloram) and BAP is effective [24, 35]. The combination of 2,4-D and BAP induced a high regeneration frequency in both nonembryogenic and embryogenic calli derived from mature caryopses, while induction of shoots from young seedling explants was more effective when picloram was used in combination with BAP [35]. Protocols for high-throughput callus induction by plating whole dehusked caryopses and plant regeneration from new, higher yielding switchgrass cvs. 'NSL' and 'SL93' have been optimized [36]. Seed pretreatments, such as dehusking with sulfuric acid, chilling for two weeks at 4°C prior to plating, and sterilizing with sodium hypochlorite and ethanol, were found to have significant effect on callus induction and subsequent plant regeneration.

Micropropagation

As mentioned earlier, vegetative/micropropagation using tissue culture might be useful for valuable germplasm and also for research. Advanced regeneration techniques have been developed for switchgrass. For the efficient multiplication of switchgrass genotypes, micropropagation has been established using nodal explants especially the nodes below the top node [37]. Regardless of their position on the culm, all nodes exhibited shoot induction at a similar rate. It was also reported that 500 plantlets could be regenerated from a single parent plant in 12 weeks [37]. Clonal propagation can be used for scaling up the number of plants obtained from selected cultivars, for controlled pollination studies for use in breeding programs, in genetic transformation experiments, and also as an important explant source for additional in vitro culture initiation.

In switchgrass, the regeneration capacity is highly genotype-dependent [38, 39]. The recalcitrance of upland cultivars warranted the development of new efficient regeneration systems. Intact seedlings of both lowland ('Alamo') and upland ('Trailblazer' and 'Blackwell') cultivars exhibited multiple shoot regeneration on MS medium supplemented with various combinations of 2,4-D and thidiazuron (TDZ) [38]. This technique of inducing multiple shoots from intact seedlings was less labor intensive and more rapid, efficient and consistent across genotypes, and the shoots appeared to originate from enlarged shoot apice [38]. Since each caryopsis vary for genotype, owing to self-incompatibility and natural outcrossing that is inherent to switchgrass, this system did not have utility for clonal propagation.

Immature inflorescences are a significant resource for *in vitro* culture establishment. Young inflorescences of switchgrass have been utilized for callus induction and plant regeneration [40]. To reduce the damage caused by harvesting, endogenous or exogenous fungal and bacterial contamination, and toxicity of sterilization solutions on inflorescences, growth establishment in axenic cultures might be beneficial. A protocol for *in vitro* production of inflorescences from node cultures derived from greenhouse grown tillers of 'Alamo' has been reported [41]. These inflorescences, with completely developed spikelets and terminal florets, were used as axenic explants for callus induction and plant regeneration. This highly efficient procedure for the development of organ-specific differentiating tissues provides a vehicle for genetic transformation using microprojectile bombardment in switchgrass. *In vitro-* grown mature florets also provide an aseptic source of anthers for the production of haploids, and open up the possibilities for *in vitro* fertilization techniques to enhance breeding experiments between ecotypes that are naturally difficult to cross.

Genetic engineering

Genetic transformation is useful for gene discovery and characterization in plant biology. The commercial use of transformation is to introduce traits into plants that would not be possible by conventional breeding alone and also to increase trait development rate [42]. The main trait targets to address using genetic engineering in switchgrass include domestication, plant architecture, and especially reduced recalcitrance for cell wall conversion into biofuel and valuable bioproducts [6, 43]. The recent focus on the use of switchgrass as a biofuel crop has led to its large-scale production and genetic engineering (Table 1; Figure 3) for incorporating traits by overexpressing exotic genes and knocking down the expression of endogenous genes [44]. These genes may be for increasing the saccharification efficiency, modifying the cell wall structure and/or composition, enhancing biomass yields or affecting the growth and development of switchgrass plants [6, 9, 44, 45].

Table 1: Summary of genetic transformation of switchgrass

Cultivar used	Explant used	Method used	Gene(s) introduced	Reference
'Alamo'	Embryogenic calli	Particle bombardment	sgfp, bar	[46]
'Alamo'	Embryogenic calli, somatic embryos, mature caryopses, seedling segments	At-mediated, strain AGL1	uidA, bar	[47]
'Alamo', 'Alamo 2'	Protoplasts	PEG-mediated	GUS	[54]
'Alamo'	Embryogenic calli	At-mediated, strain AGL1	bar, phaA, phaB, phaC	[62]
'Alamo'	Harvested leaves	Agroinfiltration using Atstrain C58C1	uidA	[52]
'Alamo'	Embryogenic calli	At-mediated, strain EHA105	hpt, gusA, GUSPlus	[49]
'Alamo'	Germinating seedlings	At-mediated, strain AGL1	GUSPlus	[53]
'Alamo', 'Performer', 'Colony'	Embryogenic calli	At-mediated, strain EHA105	hpt, sgfp	[50]
'Alamo'	Embryogenic calli	At-mediated, strain AGL1	Cg1	[80]
'Alamo'	Embryogenic calli	At-mediated, strain EHA105	COMT	[68]
'ALBA4', 'ALBA22'	Embryogenic calli	At-mediated, strain AGL1	hpt, PviCAD2	[73]
'HR8'	Embryogenic calli	At-mediated, Strain C58C1	Pv4CL1	[67]
'Alamo'	Embryogenic calli	At-mediated, strain AGL1	hpt, Pre-OsmiR156b	[81]
'Alamo 2', 'ST1'	Embryogenic calli	At-mediated	bar, hpt, GUSPlus, pporRFP	[61]

| 'ST2' | Embryogenic calli | *At*-mediated | *PvMYB4* | [74] |
| 'Alamo', 'Cave-in-Rock' | Dehusked and husked seeds | *At*-mediated, strain EHA105 | *npt* II, *gusA*, *bar*, *hpt* | [51] |

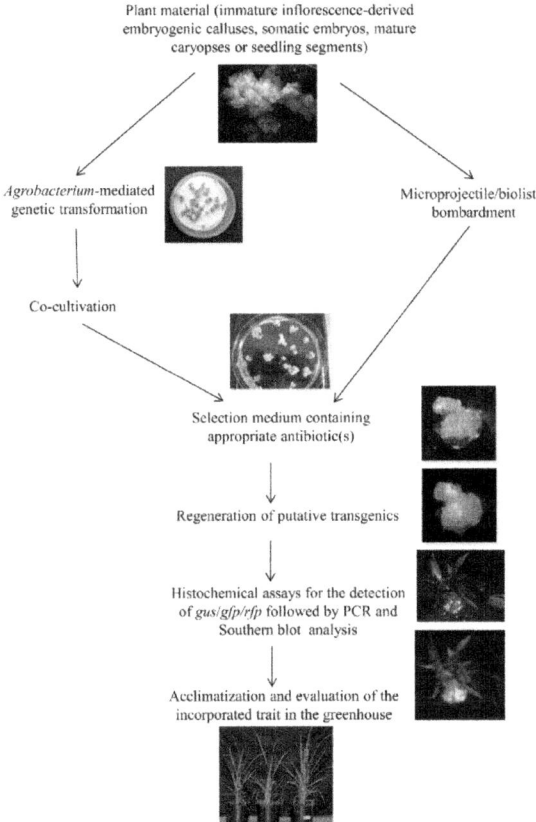

Figure 3: Flow chart of transgenic production in switchgrass [Photo credits: Wegi A. Wuddineh and M Nageswara-Rao].

The first transgenic switchgrass was obtained through bombardment of immature inflorescence-derived embryogenic calluses of 'Alamo' using a dual marker plasmid comprising the reporter gene *sgfp* (green fluorescent protein; GFP) driven by the rice actin (*Act1*) promoter and the selectable *bar* gene (conferring tolerance to the herbicide Basta) driven by the maize ubiquitin (*Ubi1*) promoter [46]. The leaf tissues and pollen of transgenic plants exhibited GFP and were also tolerant to Basta. T_1 seedlings from crosses between transgenic and non-transgenic control plants that inherited the *bar* transgene were also tolerant to Basta [46]. *Agrobacterium tumefaciens*-mediated transformation has been accomplished in switchgrass, and appears to be the most common method for switchgrass transformation. The hypervirulent *A. tumefaciens* strain AGL1 carrying the binary vector pDM805 containing the *bar* gene under the control of the *Ubi1*promoter and the *uidA* gene driven by *Act1* promoter was used for transforming four different explant types of which somatic embryos gave the highest transformation frequency [47]. This opened up new opportunities for genetic manipulation of switchgrass as *Agrobacterium*-mediated transformation is often the preferred method since it favors the integration of a low copy number of transgenes. Somleva et al. [48] was able to influence the transformation efficiency of switchgrass by manipulating explant type and genotype, pre-culture treatment of the explant, wounding of explants preceding infection, addition of acetosyringone during inoculation and cocultivation, and selection. These experiments have been valuable in making switchgrass transformation more routine.

Embryogenic calli derived from caryopses or inflorescences of 'Alamo' were transformed using *A. tumefaciens* strain EHA105 in combination with the binary vectors pCAMBIA 1301 (carrying a *gusA* from *E. coli*) and pCAMBIA 1305.2 (carrying a *GUSPlus* from *Staphylococcus* spp.) [49]. Since both binary vectors carried the *hygromycin phosphotransferase* gene (*hpt*) as a selectable marker, the transgenic plants were selected on medium supplemented with hygromycin. T_1 plants from crosses between transgenic and non-transgenic control plants that had multiple copies exhibited transgene silencing, whereas lines harboring only one insert expressed the transgene [49]. One of the largest sources, if not the largest source of efficiency improvement, has come from genotype. Highly regenerable and transformation-competent embryogenic calli developed from seeds of 'Alamo', 'Performer' and 'Colony' were used for genetic transformation using *A. tumefaciens* strain EHA105 containing the binary vectors pTOK47 (carrying a 20 kb *KpnI* fragment of Ti plasmid from pTiBo542, which contains*virB*, *virC* and *virG* virulence genes) and pJLU13 (a derivative of pCAMBIA 1301 containing *hpt* and *sgfp* genes) [50]. It appears that lines of 'Performer' are probably the best switchgrass for tissue culture and transformation. Application of vacuum during infection and dehydration at co-cultivation also enhanced the transformation efficiency, as did resting after infection and before culturing onto the selection medium [50]. Transformation efficiency can be improved by the optimization of the gene delivery system, and the appropriate selection and regeneration of transformed cells. Transformation efficiency was enhanced by utilizing the basal parts of 'Alamo' seedlings that had higher regeneration potential [51]. Genetic transformation of the type II callus derived from the inflorescences of switchgrass on LP9 medium [27] exhibited transformation efficiency of as high as 34% and also decreased the time taken for transgenic production by one month [52].

Though a number of procedures are well established for switchgrass plant transformation, evaluation of the transgene expression may take several weeks. To reduce this time required for testing gene constructs, transient transgene expression could be a rapid screen [53]. Inoculation of germinating 'Alamo' seedlings using an *Agrobacterium*-mediated transient gene expression system (agroinfiltration) was optimized using AGL1, C58, EHA105, and GV3101 strains, of which AGL1 showed the highest efficiency in gene delivery [54]. In another study, it was reported that EHA105 was more effective in gene delivery than LBA4404 or GV3101 [51]. To study the effects of agroinfiltration conditions such as mechanical wounding (bead beating, sonication or vortexing), concentration of the surfactant (Break-Thru S 240, Silwet L77 or Li700), and application of vacuum on transient β-glucuronidase expression, experiments were performed using harvested switchgrass leaves or seedlings [53, 54]. Though bead beating wounded the leaf surface, it did not have any effect on the transient β-glucuronidase expression [53]. On the other hand, utilization of sonication and vortexing with carborundum had a positive effect on the transient expression [54]. Use of 'Break-Thru S 240' under low vacuum application improved the transient expression [53] while Silwet L77 or Li700 had a negative effect [54]. Transient expression was also enhanced by increasing the vacuum application when surfactant concentration was low [53]. Incorporation of chemicals (L-cysteine and dithiothreitol), heat stress and separation by centrifugation also influenced transient transgene expression [54]. Agroinfiltration might provide a quick assay for overexpression studies in switchgrass.

Mazarei *et al.* [55] developed a protoplast system using leaves and roots of 'Alamo' and the 'Alamo2' clone followed by transient expression of polyethylene glycol (PEG) mediated DNA uptake in protoplasts [55]. GUS driven by either the CaMV 35S promoter or the maize *ubi1* promoter was utilized as the reporter gene. To develop a transformation system for upland cultivars, calli were induced from seedling segments of the upland octoploid cultivar 'Cave-in-Rock.' However, the callus was not amenable for regeneration and produced only roots [51]. Since the tissue culture and transformation systems have been developed for 'Alamo' or its derivatives, for a wide applicability across the species, there is a need to create more genotype-independent methodologies for switchgrass. It is also highly crucial to select the right candidate gene(s) for genetic transformation, and develop appropriate protocols for evaluation of transgenics with the non-transgenics [56]. Given the strong germplasm effects observed, this might be a difficult task. In addition, 'Alamo' and 'Performer' are both agronomically viable lowland cultivars. A wide variety of promoters have been used for monocot transformation [57–59], but only a few of these have been utilized in switchgrass [46, 47, 50]. Thus, attention has been given toward promoter testing and discovery for switchgrass genetic engineering [60, 61]. Two novel switchgrass ubiquitin gene (*PvUbi1* and *PvUbi2*) promoters have been tested [60]. Particle bombardment of callus using these two promoters exhibited expression patterns comparable to the maize Ubi1 promoter and much higher than that using the 35S promoter [60].

To rapidly screen transgenes in switchgrass, monocot-effective plant expression vectors are required. One such new vector set is pANIC, which uses a Gateway-compatible cassette for over-expression or RNAi of the target gene [62]. The set contains selectable marker and visible marker cassettes for *Agrobacterium*-mediated transformation as well as biolistic bombardment [62]. These vectors were designed especially for switchgrass and are being routinely used in several switchgrass transformation labs.

Production of bioproducts in transgenic switchgrass

Somleva *et al.* [63] demonstrated the amenability of transgenic switchgrass to synthesize polyhydroxybutyrate (PHB), a biodegradable polyhydroxyalkanoate biobased plastic, in which the pathway was engineered into switchgrass. PHB was accumulated to 3.72% and 1.23% (dry weight) in the leaves and whole tillers respectively. PHB production was stable in the next plant generation too. This study has shown the incorporation of a complex trait in switchgrass is possible for biomanufacturing.

Cell wall modification

Genetically modified feedstocks play an important role in scenarios for next-generation biofuel production [64]. Reducing lignin biosynthesis can lead to lower recalcitrance and higher saccharification efficiency, making lignin composition and amount an obvious target to change in lignocellulosic feedstocks [6]. Recalcitrance of cell walls conversion to biofuels is perhaps the greatest hurdle in realizing the economic potential of switchgrass and other lignocellulosic biofuel feedstocks [64, 65]. Currently, to enable efficient enzymatic degradation of cellulose, harsh physical or chemical pretreatment is required for the modification of the cell wall structures, removal of lignin and degradation of the hemicelluloses [66]. For augmenting the biofuel production from lignocellulosic feedstocks, changing lignin composition and amount are being performed [67, 68].

Fu *et al.* reported a reduction in lignin content, and increase (38%) in ethanol yield from transgenic switchgrass in which the endogenous *caffeic acid O-methyltransferase* (*COMT*) gene was down-regulated [69]. The syringyl:guaiacyl monolignol ratio was decreased and the transgenic plants required less pretreatment and enzymes to yield the same levels of ethanol using simultaneous saccharification and fermentation. As a result, there was also enhanced forage quality in the *COMT* down-regulated lines.

The last step in the biosynthesis of lignins is catalyzed by cinnamyl alcohol dehydrogenase (CAD) [70]. CAD deficiency modifies the lignin structure, reduces the lignin content, and augments the saccharification efficiency in grasses [71, 72].*Agrobacterium*-mediated transformation was utilized for RNAi of CAD in switchgrass [73, 74]. These two studies reported a reduction in lignin content and increased saccharification efficiency in the transgenic lines. Another important enzyme involved in the biosynthesis of lignin is 4-coumarate:coenzyme A ligase (4CL). Xu *et al.* carried out phylogenetic analysis and gene expression studies, and suggested the involvement of *Pv4CL1* in the biosynthesis of lignins [68]. *Pv4CL1* down-regulated transgenic switchgrass plants, obtained by *Agrobacterium*-mediated transformation, had normal biomass yields with reduced lignin content and increased saccharification efficiency [68].

In contrast to the above-mentioned approach in which endogenous lignin biosynthesis genes were down-regulated, Hui Shen and colleagues targeted the overexpression of a key transcription factor affecting the expression of many lignin biosynthesis genes [75]. A decrease in recalcitrance in transgenic switchgrass was observed when the repressor, *PvMYB4*was overexpressed [75]. The transgenic lines exhibited a drastic reduction in lignin, but no change in the S:G ratio. The plants were also morphologically affected, having more tillers and reduced height. The transgenics had increased cellulose and pectin contents, significantly reduced wall recalcitrance and phenolic fermentation inhibitors, and produced approximately 1.8-fold more ethanol using yeast based simultaneous saccharification and fermentation without pretreatment (Shen *et al.*, in review).

These efforts have highlighted the usefulness of lignin biosynthesis or lignin repressor gene targets for down-regulation, and these genetically engineered plants for reduced lignin may contain higher levels of free monolignols and other phenylpropanoids. The accessibility of cell wall carbohydrates for the production of biofuels is negatively correlated with the amount of lignin present [76, 77]. Decrease in lignin content or alteration in its composition alleviated the digestibility of the cellulose and hemicelluloses. This led to enhanced saccharification efficiency, reduction in the severity of the pretreatment, decrease in enzyme requirements and increase in the energy available to microorganisms for breaking down the carbohydrates [69, 76, 78]. To change the lignin content of the biomass, dwarfing might also be of use as it shifts the biomass allocation from the stem to the leaves [44]. Reduced lignin content during the vegetative phase in switchgrass might also delay flowering, which could also increase vegetative biomass [44, 79].

In is unclear whether the lignin biosynthetic pathway is perfectly conserved between widely-studied model species and switchgrass. There might be many more genes and transcription factors that have not been discovered in switchgrass and be manipulated for improved biofuel production. Other cell wall targets include cellulose, reducing the crystallinity of cellulose, hemicellulose, pectin, and their interactions with lignin. Research on the expression of cellulases, *in planta*, under extreme conditions and its thermal stability also needs to be carried out. The cost of lignocellulosic ethanol production may also be reduced by genetically modifying switchgrass to produce the enzymes that are required during fermentation. Devising strategies for recycling these enzymes will also lead to reduction in biofuel production cost.

Altering switchgrass development: microRNAs and other targets

Improvement in the rate of saccharification efficiency, which is inhibited by the complex structure of the plant cell wall, is an important objective in developing a competent and lucrative biofuel industry [80, 81]. Biomass yield could be enhanced by manipulating microRNAs (miRNAs) that regulate transcription factors controlling growth and development in plants [69,81–84]. The maize *Corngrass1* (*Cg1*) gene, which produces a miR156, targets the SQUAMOSA PROMOTER BINDING LIKE (SPL) family and reduces lignification while promoting juvenile characteristics in plants [85, 86]. To study how juvenile characters improve the biofuel potential of switchgrass, the *Cg1* gene was constitutively overexpressed in 'Alamo' [81]. A second miR156 study overexpressed the switchgrass *PvmiR156* in switchgrass, [82]. In both studies, the transgenic plants had delayed flowering, variant morphology, and improved sugar release. Transgene expression levels were sufficient to allow three morphology categories to be observed. Low expressers resembled non-transgenic switchgrass. Moderate expression levels rendered plants that were shorter and with more tillers. The plants had delayed flowering, which could be useful in bioconfinement of transgenes. High levels of miR156 accumulation induced severe dwarfism and reduced biomass accumulation [81, 82]. Thus, targeted overexpression of miR156 could not only make biofuel production more efficient but allow the production of switchgrass that is more suitable for production. These studies highlight the potential utility of this approach for the domestication of new switchgrass cultivars, and the lack or delay in flowering will have important implications for the limitation or prevention of transgene flow into native/wild relatives or non-transgenic agronomic plantings of switchgrass. Recently, it was demonstrated that the expression levels of miR156 and miR162 could be changed under drought conditions in switchgrass [87].

Genetic engineering can also be used to increase the biomass by modifying the plant growth regulators such as increasing the biosynthesis of gibberellins [88] to improve the growth and increase the biomass in switchgrass. Thus, early transgenic research in switchgrass has revealed that multiple targets for improvement have been reached. It appears that there could be a tradeoff between sugar release and plant growth, but results are promising with regards to increasing liters per hectare. To date, there has been no transgene stacking in switchgrass, which should be pursued. For example, it makes sense to hybridize miRI56 plants with those with greatly reduced lignin, such as *MYB4* overexpressers. In addition, tissue-specific and inducible expression of transgenes will also be valuable in decreasing off-target effects. Targeted expression is particularly needed for genes, such as those that are master regulators, to diminish or better control pleiotropic effects. The transgenic studies to date with switchgrass show the power of the technology, which is becoming increasingly routine.

GENETIC AND GENOMICS RESOURCES

Molecular markers

A number of DNA marker systems such as restriction fragment length polymorphism (RFLP), chloroplast DNA, randomly amplified polymorphic DNA (RAPD), amplified fragment length polymorphism (AFLP) and simple sequence repeats (SSRs) have been developed for the genetic diversity assessment and phylogenetic studies in switchgrass [20, 89–95]. Marker studies helped delineate upland and lowland variation and are useful in developing germplasm conservation and breeding programs [96]. Genetic linkage maps have been constructed using single dose restriction fragments (SDRFs), SSRs, sequence-tagged sites (STS) markers, expressed sequence tags (EST)-derived SSRs, gene-derived STS markers, and diversity array technology (DArT) markers [97–101]. Linkage maps will aid in the identification of quantitative trait loci linked with biomass yield, plant composition and other important agronomic traits, providing a genetic framework to facilitate marker-assisted breeding and genomics research in switchgrass.

Over the last few years, even though various technologies have emerged for whole genome sequencing, it is still technically difficult and expensive to completely sequence complex polyploid species such as switchgrass [102, 103]. Transcriptome sequencing of expressed sequence tags (ESTs) is amenable for any organism, including those for which *de novo* whole genome sequencing is difficult, thereby aiding in gene discovery and annotation [103–107]. ESTs have been successfully used for identification of molecular markers, analysis of tissue-specific patterns of expression or for comparative genomics [105, 108]. cDNA libraries derived from leaf, stem, crown, and callus of 'Kanlow' were utilized for generating 11,990 individual sequences of which 7,810 were unique gene clusters [105]. Sequence similarity and functional classification of these unique gene clusters was also performed. EST sequence information can also be mined for DNA sequence polymorphisms for single nucleotide polymorphisms (SNPs) and SSRs that can be used for genome characterization and genetic diversity assessment [94]. For developing SSR markers, Tobias *et al.* assessed the unique gene clusters and reported the occurrence of short tandem repeats in 3.8% of the ESTs tested [105]. ESTs were also produced by end-sequencing of callus, crown, and seedling tissue derived cDNA libraries of 'Kanlow,' and the assembled consensus sequences were aligned with the sorghum genome [108]. They observed that 3.3% of the sequences were similar to potential cell wall related genes. Millions of ESTs from tissue or xylem cell-specific EST libraries of 'Alamo' are also now available (http://compbio.dfci.harvard.edu/tgi/cgi-bin/tgi/gimain.pl?gudb=switchgrass) [56].

SSRs and EST-SSRs are significant resources for developing dense linkage maps and identifying economically important traits for utilization in molecular breeding programs intended to develop superior switchgrass cultivars [94]. EST-SSR markers were identified and assessed for the production of fragment length polymorphisms in the two individual parents of a mapping population [108]. To identify SSR sequences longer than 20 bp, available sequence data from switchgrass were assessed using the program SSRIT and approximately 32 genic di-, tri- and tetranucleotide repeat SSRs were characterized [109, 110]. When used to differentiate 'Alamo' and 'Kanlow' individuals, these SSRs exhibited a high degree of polymorphism consistent with their tetraploid, allogamous genome states [110]. Using genomic DNA of 'SL93 7 × 15', Wang *et al.* reported the construction of five genomic SSR-enriched libraries and identified 1,300 unique SSR-containing clones [94]. Given the power of genomics as described above, continued expansion of sequence availability, especially when assembled switchgrass genome is made available, will enable better understanding of switchgrass biology as well as facilitate genetic engineering.

BAC libraries and physical mapping

Efforts to map important traits for enhancing the breeding programs, and utilizing map-based cloning for the isolation of target genes are dependent on the availability of extensive physical and genetic maps; a switchgrass physical map is needed [111]. Genome assembly for switchgrass requires the genome structure information that can be obtained by sequencing bacterial artificial chromosome (BAC) libraries [96, 112, 113]. 'Alamo' has been extensively used in switchgrass breeding programs and is the parent of several mapping populations, therefore, it follows that the current whole-genome sequencing effort is focused on an 'Alamo' clone; the clone chosen was termed AP13. AP13 and all Alamo is a heterozygous tetraploid with two subgenomes [96]. Saski *et al.* assembled the first BAC library, by incomplete digestion of nuclear DNA of the 'Alamo'-derived genotype, SL93 2001–1 with EcoRI, which had approximately ten-fold coverage of the total nuclear content and five-fold of each of the two genomes based on a genome size of 3.2 gigabases (~1.6 Gb per genome) [111]. Since the study was restricted to a single locus and restriction enzyme, it warranted the need of additional libraries to attain fair and near-complete depiction for genome-wide studies. Recently, two (HindIII- and BstYI-fragmented) BAC libraries were constructed from AP13, which also aided in discoveries of SSRs [113]. Comparative analysis with other grass genomes such as foxtail millet, sorghum, rice, maize, and *Brachypodium* revealed high levels of homology with switchgrass exhibiting high microcolinearity with foxtail millet as compared with sorghum [114]. In addition, HudsonAlpha/Joint Genome Institute (JGI) has generated BAC-end sequences from a collection of BACs (http://genomicscience.energy.gov/). These studies provided a precise BAC-based physical platform that offers a definitive approach for sequencing and assembly of the switchgrass genome. They will also be able to give a precise estimate of the GC content, distribution of known, novel and repeat elements, and, thus, of the genome structure and composition of switchgrass.

Sub-organelle genome sequencing

Chloroplasts are invaluable for genetic and phylogenetic studies. Switchgrass chloroplasts are often maternally inherited and can be transformed, in several other plant species, to deliver high recombinant protein production [115, 116]. To differentiate genetic diversity in whole chloroplast genomes and a large number of nuclear loci in switchgrass, a unique strategy utilizing high-throughput sequencing of multiplexed restriction-digested reduced-representation libraries was used for the identification of SNPs [117]. The SNPs identified were able to characterize eight haplogroups. Switchgrass chloroplast genomes were also sequenced from individuals of the upland ('Summer Lin2') and lowland ('Kanlow Lin1') ecotypes giving an insight regarding the amount of variation within the two ecotypes, and facilitated comparisons within the ecotypes as well as among other sequenced plastid genomes [118]. These studies emphasize the use of chloroplast genome for comparing genetic variation between the upland and lowland ecotypes, are highly desirable for robust phylogenetic studies and can be used in differentiating mixed population into up- or lowland ecotypes. The complete chloroplast genome will facilitate the generation of species-specific transformation vectors [119] and will create an opportunity for the utilization of plastid genetic engineering in switchgrass.

Whole genome sequencing

Basic characterization of the switchgrass genome indicates that the tetraploid lowland cultivars have a nuclear DNA content of 3.07 ± 0.06 pg per nucleus [120], resulting in an effective genome size of ~1600 Mb for 'Alamo' derived genotypes, which is approximately twice that of sorghum and about three and a half times that of rice [111, 121]. Even with the availability of new and modern technologies, whole genome sequencing (WGS) of switchgrass would be difficult to achieve due to its large genome size and polyploidy. A practical solution to Sanger sequencing may be provided by pyrosequencing or other such next generation sequencing (NGS) technologies that offer quick and inexpensive technologies for transcriptomics by avoiding extensive and comparatively low throughput steps [122–124]. For *de novo* sequencing and transcriptomics of complex genomes, 454 pyrosequencing is the most extensively exploited NGS technology. GS FLX Titanium, the latest 454-sequencing platform, can produce a typical read length of approximately 330–700 bases [125, 126].

'Alamo' AP13 has been chosen for WGS by JGI (http://genome.jgi.doe.gov/genome-projects/). Sequencing of AP13 cDNA libraries produced from various tissues of switchgrass utilizing GS-FLX Titanium technology produced large number of reads for *de novo* assembly, and EST and SSR identification [103]. The accessibility to the foxtail millet draft genome also enhanced the switchgrass EST assembly and nearly doubled the EST information in the public domain. JGI also used a combination of Roche 454-based and Illumina-based sequencing to produce the switchgrass genomic sequence [96]. Initial investigations on assembly of switchgrass genome onto the foxtail millet framework led to the identification of paralogous assemblies from homoeologous assemblies [114]. However, autonomous assembly of both the subgenomes to achieve chromosome-scale contiguity for the reference is challenging [96]. Although dihaploid lines may simplify sequence assembly in switchgrass, they are not preferred for whole genome sequencing because of their elevated infertility and instability [126, 127]. The draft genome sequence of switchgrass is now available (http://www.phytozome.net/panicumvirgatum). The genome of switchgrass will help the biologists to determine the function and biotechnological potential of genes, especially those responsible for increasing the biofuel potential such as biomass yield, decreased lignin content and improved saccharification efficiency. Furthermore, comparative analysis of switchgrass with other sequenced grass genomes such as foxtail millet and sorghum will enable a more detailed annotation, and will play an important role in understanding how gene networks evolved and function (National Plant Genome Initiative:2009–2013; http://www.nsf.gov/bio/pubs/reports/). WGS also helps plant breeding efforts.

Gene expression studies

Information on the fundamental biology and the regulatory mechanisms of gene expression in switchgrass under abiotic stress conditions are required for determining the consequences of genetic improvements and for detection and manipulation of stress tolerance related gene candidates [87]. An Affymetrix microarray chip for switchgrass has been produced that contains representatives of most of its expressed genes that has been used to make a gene expression atlas (http://genomicscience.energy.gov/) [128] as well as the switchgrass relative *Panicum hallii*[129].

Of particular interest with regards to gene expression are miRNA studies. Mature miRNAs inhibit gene expression at the post-transcriptional levels by either targeting mRNAs for degradation or inhibiting protein translation [83, 130], which in turn can lead to transcriptional regulatory changes. Switchgrass traits of interest include cellulose biosynthesis, sucrose and fat metabolism, signal transduction, and plant development [131]. Investigations on the effect of salt and drought stress on the expression of miRNAs revealed an altered expression pattern of miRNAs in a dose-dependent manner [87]. Transgenic plants expressing the miR156 gene that exhibited severe morphological alterations was used to investigate the effects of miR156 overexpression on its downstream genes using Affymetrix microarray analysis [82]. The study discovered that transcript abundance reduced in eight SPL gene probe sets, leading to the expression analysis of the corresponding cDNA sequences, which showed that the highest miR156 expressers had the most reduction in *Pv* SPLs transcript abundance [82]. Such gene expression analyses will further augment the characterization and expression of genes controlling the biofuel traits, enhance the functional genomics studies and molecular breeding, and may further help in the assembly of the switchgrass genome.

Reverse genetics

To take advantage of the new DNA sequence information and to investigate the functions of specific genes, targeting induced local lesions in genomes (TILLING) was developed. TILLING is a non-transgenic technology that utilizes a reverse genetics approach for the production and detection of mutation [132]. EcoTILLING is a variation of TILLING that investigates the natural variation among cultivar/inbred line/accession when aligned with a sequenced reference genome for the identification of SNPs [133, 134]. In switchgrass, TILLING, EcoTILLING, or a permutation of both are being utilized [134]. This will lead to the identification of multiple SNPs within a target region of switchgrass accessions and when compared to a reference genome will be able to define the relatedness and differences among the target region. Traits such as biomass yield, saccharification efficiency, and flowering time may be potentially identified in switchgrass using these techniques. The limitations being that the mutations may be introduced randomly throughout the genome, and a large number of individuals need to be screened to identify the mutants having the trait of interest. It will also be difficult to identify recessive mutants due to the polyploidy nature of switchgrass.

DISCUSSION

Switchgrass has been the topic of important discoveries and relevance in genomics and biotechnology in the last decade [27, 46, 47, 63]. Significant trait improvement via biotechnology e.g. [69, 73, 74] with increased transformation efficiency [50] has been demonstrated in switchgrass. This suggests that genetic improvements of biofuel properties of switchgrass through expression and down-regulation of transgenes is a practical way to rapidly establish it as a viable bioenergy crop on a commercial level and will be achieved with growing reliability in the coming years [60, 81, 82]. Though transgenic approaches are considered imperative for the development of switchgrass and other biofuel crops, their cost-effectiveness will be dependent on their domestication, productivity and biofuel properties [44]. However, we speculate that a regulatory necessity, at least in the US, will likely be bioconfinement of transgenes [10, 135, 136].

Transgene escape has been considered as a major environmental, ecological and regulatory concern. Hence, for commercialization of transgenic switchgrass, efficient and reliable transgene bioconfinement strategies would be enabling, especially in US the geographic center of diversity of switchgrass. While transgenes can be vectored in pollen or seed and less commonly asexually, the prospective for long-distance pollination has made pollen-dispersed transgenes a major concern [137]. One strategy to control gene flow in switchgrass would be to introduce male sterility using transgene-encoded ribonucleases that inhibit pollen formation [138, 139]. With switchgrass being wind-pollinated, the excision of transgenes from the pollen genomes using site-specific recombination systems will also be desirable [140, 141]. Another strategy would be to use plastid (chloroplasts or mitochondria) transformation for the introduction of cytoplasmic male sterility into switchgrass, and thus developing plastid transformation for switchgrass would be helpful. Since the pollen of most plant species contain no chloroplasts, pollen spread will not introduce the foreign genes into wild or non-transgenic switchgrass populations [142, 143]. Thus, strategies for transgene bioconfinement and alleviation of gene flow and research that facilitates the utilization of information and proper regulatory guidelines for transgenic feedstocks are essential in developing the biofuel industry's infrastructure [10], including that for switchgrass [136]. The challenge is to generate efficient methods and procedures to accomplish elevated levels of agricultural productivity while conserving the environment and natural resources [7].

Recent advances in switchgrass genomics will further facilitate biotechnological interventions as well as its germplasm improvements via conventional and molecular breeding. The close colinearity of the switchgrass genome with other grasses will aid in the elucidation of gene function, regulation, and expression by leveraging off other resources. The application of new knowledge and tools developed from genomic resources such as identification of genes like those involved in the lignin pathway, saccharification efficiency, biomass yield, nutritional quality, and pest resistance will help geneticists and plant genetic improvement managers to overcome the limitations associated with conventional breeding, make sexual hybridization more efficient and manipulate various traits effectively. It is important to keep in mind, however, that the utility of new genetic combinations must be demonstrated ultimately by field trials and the value to consumers.

CONCLUSIONS

The development of switchgrass as a biofuel crop has the potential to contribute significantly to lignocellulosic ethanol production without competing with food and feed crops. Biotechnological advances made to genetically modify important biofuel related traits in switchgrass will play a key role in shaping the future of the switchgrass biofuel industry. Genomic information being generated for switchgrass will further enhance the breeding and biotechnological endeavors. Although plant biotechnology will play an important role to the successful generation of energy crops, it should be followed up with breeding programs aimed at sustaining or improving the significant agronomic attributes which made these plants imperative for biofuel generation to start with, namely resistance to abiotic and biotic factors, and low fertilization requirements [144]. The critical issue to be dealt with is how to improve the conversion efficiency from the solar energy to biofuel energy such that biofuels can meet anthropogenic energy consumption demands and be able to replace the fossil fuels.

DECLARATIONS

Acknowledgements

The authors would like to thank Mr. Wegi A. Wuddineh for providing photographs for Figure 3. This project was made possible through funding from USDA NIFA Biotechnology Risk Assessment Grants (BRAG) Program grant # 2010-39211-21699. C. Neal Stewart Jr. also received support from the BioEnergy Science Center, a Bioenergy Research Center supported by the Office of Biological and Environmental Research in the US Department of Energy Office of Science.

Competing interests

The authors declare that they have no competing interests.

Authors' contributions

MNR and JRS conceptualized, researched, wrote the manuscript and made the figures. CK and CNS conceptualized and critically revised the manuscript. All authors read and approved the final manuscript.

REFERENCES

1. Hahn-Hagerdal B, Galbe M, Gorwa-Grauslund MF, Liden G, Zacchi G: **Bio-ethanol – the fuel of tomorrow from the residues of today.** *Trends Biotechnol* 2006, **24:** 549-556. 10.1016/j.tibtech.2006.10.004

2. Brown LR: *World on the edge: How to prevent environmental and economic collapse*. New York: WW Norton & Company, Inc.; 2011.

3. Youngs H, Somerville C: **Growing better biofuel crops.** *Scientist* 2012. http://the-scientist.com/2012/07/01/growing-better-biofuel-crops/

4. Nageswara-Rao M, Kwit C, Stewart CN Jr: **Grass to solve global fuel crisis.** *BioSpectrum Asia* 2012. http://www.biospectrumasia.com/biospectrum/opinion/3439/grass-solve-global-petrol-scarcity

5. Kole C, Joshi CP, Shonnard DR: *Handbook of Bioenergy Crop Plants*. Boca Raton, London, New York: CRC Press, Taylor & Francis Group; 2012.

6. Yuan JS, Tiller KH, Al-Ahmad H, Stewart NR, Stewart CN Jr: **Plants to**

power: bioenergy to fuel the future. *Trends Plant Sci* 2008, **13**:421-429. 10.1016/j.tplants.2008.06.001

7. Herve G, Agneta F, Yves D: **Biofuels and world agricultural markets: outlook for 2020 and 2050.** In *Economic Effects of Biofuel Production*. Edited by: Bernardes MAS. Crotia: InTech Publishers; 2011:129-162.

8. Cai X, Zhang X, Wang D: **Land availability for biofuel production.** *Environ Sci Technol* 2011, **45:** 334-339. 10.1021/es103338e

9. Abramson M, Shoseyov O, Shani Z: **Plant cell wall reconstruction toward improved lignocellulosic production and processability.***Plant Sci* 2010, **178:** 61-72. 10.1016/j.plantsci.2009.11.003

10. Kausch AP, Hague J, Oliver M, Li Y, Daniell H, Mascia P, Watrud LS, Stewart CN Jr: **Transgenic perennial biofuel feedstocks and strategies for bioconfinement.** *Biofuels* 2010, **1:** 163-176.

11. McLaughlin SB, Kszos LA: **Development of switchgrass (*Panicum virgatum*) as a bioenergy feedstock in the United States.** *Biomass Bioenerg* 2005, **28:** 515-535. 10.1016/j.biombioe.2004.05.006

12. Wright L, Turhollow A: **Switchgrass selection as a "model" bioenergy crop: a history of the process.** *Biomass Bioenerg* 2010, **34:** 851-868. 10.1016/j.biombioe.2010.01.030

13. Moser LE, Vogel KP: **Switchgrass, big bluestem, and indiangrass.** In *Forages. Volume 1. An Introduction to Grassland Agriculture*. Edited by: Barnes RF, Miller DA, Nelson CJ. Ames: Iowa State University Press; 1995:409-420.

14. Vogel KP: **Switchgrass.** In *Warm-season (C4) grasses. Agronomy Monograph 45*. Edited by: Moser LE, Burson BL, Sollenberger LE. Madison: ASA, CSSA, and SSSA; 2004:561-588.

15. Parrish DJ, Fike JH: **The biology and agronomy of switchgrass for biofuels.** *Crit Rev Plant Sci* 2005, **24:** 423-459. 10.1080/07352680500316433

16. Porter CL Jr: **An analysis of variation between upland and lowland switchgrass, *Panicum virgatum* L., in central Oklahoma.** *Ecology*1966, **47:** 980-992. 10.2307/1935646

17. Narasimhamoorty B, Saha MC, Swaller T, Bouton JH: **Genetic diversity in switchgrass collections assessed by EST-SSR markers.***BioEnerg Res* 2008, **1:** 136-146. 10.1007/s12155-008-9011-0

18. McLaughlin SB, Bouton J, Bransby D, Conger B, Ocumpaugh W, Parrish D, Taliaferro C, Vogel K, Wullschleger S: **Developing switchgrass as a bioenergy feedstock.** In *Perspectives on New Crops and New Uses*.

Edited by: Janick J. Alexandria: ASHS Press; 1999:282-299.

19. Sokhansanj S, Mani S, Turhollow A, Kumar A, Bransby D, Lynd L, Laser M: **Large scale production, harvest and logistics of switchgrass (*Panicum virgatum* L.) - current technology and envisioning a mature technology.** *Biofuels Bioprod Bioref* 2009, **3**: 124-141. 10.1002/bbb.129

20. Todd J, Wu YQ, Wang Z, Samuels T: **Genetic diversity in tetraploid switchgrass revealed by AFLP marker polymorphisms.** *Genet Mol Res* 2011, **10**: 2976-2986. 10.4238/2011.November.29.8

21. Jager HI, Baskaran LM, Brandt CC, Davis EB, Gunderson CA, Wullschleger SD: **Empirical geographic modeling of switchgrass yields in the United States.** *Global Change Biol Bioenerg* 2010, **2**: 248-257. 10.1111/j.1757-1707.2010.01059.x

22. Wullschleger SD, Davis EB, Borsuk ME, Gunderson CA, Lynd LR: **Biomass production in switchgrass across the United States: database description and determinants of yield.** *Agron J* 2010, **102**: 1158-1168. 10.2134/agronj2010.0087

23. Schmer MR, Vogel KP, Mitchell RB, Perrin RK: **Net energy of cellulosic ethanol from switchgrass.** *Proc Natl Acad Sci USA* 2008, **105**:464-469. 10.1073/pnas.0704767105

24. Denchev PD, Conger BV: **Plant regeneration from callus cultures of switchgrass.** *Crop Sci* 1994, **34**: 1623-1637. 10.2135/cropsci1994.0011183X003400060036x

25. Dutta Gupta S, Conger BV: **Somatic embryogenesis and plant regeneration from suspension cultures of switchgrass.** *Crop Sci* 1999,**39**: 243-247. 10.2135/cropsci1999.0011183X003900010037x

26. Odjakova MK, Conger BV: **The influence of osmotic pretreatment and inoculum age on the initiation and regenerability of switchgrass suspension cultures.** *In Vitro Cell Dev Biol Plant* 1999, **35**: 442-444. 10.1007/s11627-999-0065-2

27. Burris JN, Mann DGJ, Joyce BL, Stewart CN Jr: **An improved tissue culture system for embryogenic callus production and plant regeneration in switchgrass.** *BioEnergy Res* 2009, **2**: 267-274. 10.1007/s12155-009-9048-8

28. Gurel S, Gurel E, Kaya Z: **Establishment of cell suspension cultures and plant regeneration in sugar beet (*Beta vulgaris* L.).** *Turk J Bot* 2002, **26**: 197-205.

29. Xu B, Huang L, Shen Z, Welbaum GE, Zhang X, Zhao B: **Selection and characterization of a new switchgrass (*Panicum virgatum* L.)**

line with high somatic embryogenic capacity for genetic transformation. *Scientia Hort* 2011, **129:** 854-861. 10.1016/j.scienta.2011.05.016

30. Hall RD: **The initiation and maintenance of plant cell suspension cultures.** In *Plant Tissue Culture Manual.* Edited by: Lindsey K. Dordrecht: Kluwer Academic Publishers; 1991:A3:1–21.

31. Mazarei M, Al-Ahmad H, Rudis MR, Joyce BL, Stewart CN Jr: **Switchgrass (*Panicum virgatum* L.) cell suspension cultures: Establishment, characterization, and application.** *Plant Sci* 2011, **181:** 712-715. 10.1016/j.plantsci.2010.12.010

32. Su WW: **Cell culture and regeneration of plant tissues.** In *Trangenic Plants and Crops.* Edited by: Khachatourians GG, McHughen A, Scorza R, Nip W, Hui YH. New York: Taylor & Francis Publishers; 2002:151-176.

33. Pernisova M, Klima P, Horak J, Valkova M, Malbeck J, Soucek P, Reichman P, Hoyerova K, Dubova J, Frimi J, Zazimalova E, Hejatko J: **Cytokinins modulate auxin-induced organogenesis in plants via regulation of the auxin efflux.** *Proc Natl Acad Sci USA* 2009, **106:**3609-3614. 10.1073/pnas.0811539106

34. Thomas E, Davey MR: *From single cells to plants.* London: Wykeham Publications; 1975.

35. Denchev PD, Conger BV: *In vitro* **culture of switchgrass: influence of 2,4-D and picloram in combination with benzyladenine on callus initiation and regeneration.** *Plant Cell Tiss Org Cult* 1995, **40:** 43-48. 10.1007/BF00041117

36. Foulk SM: *Tissue culture and recombination DNA technology: Developing protocols for potentially higher yielding switchgrass cultivars.* MS thesis: The University of Tennessee, Department of Plant Sciences; 2008.

37. Alexandrova KS, Denchev PD, Conger BV: **Micropropagation of switchgrass by node culture.** *Crop Sci* 1996, **36:** 1709-1711. 10.2135/cropsci1996.0011183X003600060049x

38. Dutta Gupta S, Conger BV: *In vitro* **differentiation of multiple shoot clumps from intact seedlings of switchgrass.** *In Vitro Cell Dev Biol Plant* 1998, **34:** 196-202. 10.1007/BF02822708

39. Seo M, Takahara M, Takamizo T: **Optimization of culture conditions for plant regeneration of *Panicum* spp. through somatic embryogenesis.** *Grassl Sci* 2010, **56:** 6-12. 10.1111/j.1744-697X.2009.00166.x

40. Chen CH, Sargent WA, Lo PF, Boe AA: **Plant regeneration and morphogenetic patterns in callus cultures derived from young inflorescences of switchgrass (Panicum virgatum L.) [abstract].** In *Proceedings of the VI Intern Cong Plant Tissue and Cell Cult.* Edited by: Somers DA. MN: St. Paul; 1986:227.

41. Alexandrova KS, Denchev PD, Conger BV: ***In vitro* development of inflorescences from switchgrass nodal segments.** *Crop Sci* 1996,**36:** 175-178. 10.2135/cropsci1996.0011183X003600010031x

42. Wang Z-Y, Ge Y: **Recent advances in genetic transformation of forage and turf grasses.** *In Vitro Cell Dev Biol Plant* 2006, **42:** 1-18.

43. Himmel ME: **Biomass recalcitrance: engineering plants and enzymes for biofuels production.** *Science* 2007, **315:** 804-807. 10.1126/science.1137016

44. Gressel J: **Transgenics are imperative for biofuel crops.** *Plant Sci* 2008, **174:** 246-263. 10.1016/j.plantsci.2007.11.009

45. Rubin EM: **Genomics of cellulosic biofuels.** *Nature* 2008, **454:** 841-845. 10.1038/nature07190

46. Richards HA, Rudas VA, Sun H, McDaniel JK, Tomaszewski Z, Conger BV: **Construction of a GFP-BAR plasmid and its use for switchgrass transformation.** *Plant Cell Rep* 2001, **20:** 48-54. 10.1007/s002990000274

47. Somleva MN, Tomaszewski Z, Conger BV: ***Agrobacterium* -mediated genetic transformation of switchgrass.** *Crop Sci* 2002, **42:** 2080-2087. 10.2135/cropsci2002.2080

48. Somleva MN: **Switchgrass (Panicum virgatum L.).** In *Methods in Molecular Biology. Volume 344: Agrobacterium Protocols.* Edited by: Wang K. Totowa: Humana Press Inc; 2006:65-74.

49. Xi Y, Fu C, Ge Y, Nandakumar R, Hisano H, Bouton J, Wang Z-Y: ***Agrobacterium* -mediated transformation of switchgrass and inheritance of the transgenes.** *BioEnerg Res* 2009, **2:** 275-283. 10.1007/s12155-009-9049-7

50. Li R, Qu R: **High throughput *Agrobacterium* -mediated switchgrass transformation.** *Biomass Bioenerg* 2010, **35:** 1046-1054.

51. Song G, Walworth A, Hancock JF: **Factors influencing *Agrobacterium* -mediated transformation of switchgrass cultivars.** *Plant Cell Tiss Org Cult* 2012, **108:** 445-453. 10.1007/s11240-011-0056-y

52. Burris JN **MS thesis.** In *An improved tissue culture and transformation*

system for switchgrass (Panicum virgatum L.). University of Tennessee, Department of Plant Sciences; 2010.

53. VanderGheynst JS, Guo H, Simmons CW: **Response surface studies that elucidate the role of infiltration conditions on*Agrobacterium tumefaciens* -mediated transient transgene expression in harvested switchgrass (*Panicum virgatum*).** *Biomass Bioenerg* 2008, **32:** 372-379.

54. Chen X, Equi R, Baxter H, Berk K, Han J, Agarwal S, Zale J: **A high-throughput transient gene expression system for switchgrass (*Panicum virgatum L.*) seedlings.** *Biotechnol Biofuels* 2010, **3:** 9. 10.1186/1754-6834-3-9

55. Mazarei M, Al-Ahmad H, Rudis MR, Stewart CN Jr: **Protoplast isolation and transient gene expression in switchgrass,** *Panicum virgatum* **L.** *Biotechnol J* 2008, **3:** 354-359. 10.1002/biot.200700189

56. Shen H, Fu C, Xiao X, Ray T, Tang Y, Wang Z, Chen F: **Developmental control of lignifications in stems of lowland switchgrass variety 'Alamo' and the effects on saccharification efficiency.** *BioEnerg Res* 2009, **2:** 233-245. 10.1007/s12155-009-9058-6

57. Christensen AH, Sharrock RA, Quail PH: **Maize polyubiquitin genes: structure, thermal perturbation of expression and transcript splicing, and promoter activity following transfer to protoplasts by electroporation.** *Plant Mol Biol* 1992, **18:** 675-689. 10.1007/BF00020010

58. McElroy D, Zhang W, Cao J, Wu R: **Isolation of an efficient actin promoter for use in rice transformation.** *Plant Cell* 1990, **2:** 163-171.

59. Wang J, Jiang J, Oard JH: **Structure, expression and promoter activity of two polyubiquitin genes from rice (*Oryza sativa* L.).** *Plant Sci* 2000, **156:** 201-211. 10.1016/S0168-9452(00)00255-7

60. Mann DGJ, King ZR, Liu W, Joyce BL, Percifield RJ, Hawkins JS, LaFayette PR, Artelt BJ, Burris JN, Mazarei M, Bennetzen JL, Parrott WA, Stewart CN Jr: **Switchgrass (*Panicum virgatum L.*) ubiquitin gene (*PvUbi1* and *PvUbi2*) promoters for use in plant transformation.** *BMC Biotechnol* 2011, **11:** 74. 10.1186/1472-6750-11-74

61. Peremarti A, Twyman R, Gómez-Galera S, Naqvi S, Farré G, Sabalza M, Miralpeix B, Dashevskaya S, Yuan D, Ramessar K, Christou P, Zhu C, Bassie L, Capell T: **Promoter diversity in multigene transformation.** *Plant Mol Biol* 2010, **73:** 363-378. 10.1007/s11103-010-9628-1

62. Mann DGJ, LaFayette PR, Abercombie LL, King ZR, Mazarei M, Halter

MC, Poovaiah CR, Baxter H, Shen H, Dixon RA, Parrott WA, Stewart CN Jr: **Gateway-compatible vectors for high-throughput gene functional analysis in switchgrass (*Panicum virgatum* L.) and other monocot species.** *Plant Biotechnol J* 2012, **10:** 226-236. 10.1111/j.1467-7652.2011.00658.x

63. Somleva M, Snell K, Beaulieu J, Peoples O, Garrison B, Patterson N: **Production of polyhydroxybutyrate in switchgrass, a value-added co-product in an important lignocellulosic biomass crop.** *Plant Biotechnol J* 2008, **6:** 663-678. 10.1111/j.1467-7652.2008.00350.x

64. Lynd LR: **Overview and evaluation of fuel ethanol from cellulosic biomass: technology, economics, the environment, and policy.***Annu Rev Energ Evn* 1996, **21:** 403-465. 10.1146/annurev.energy.21.1.403

65. Mosier N, Wyman C, Dale B, Elander R, Lee YY, Holtzapple M, Ladisch M: **Features of promising technologies for pretreatment of lignocellulosic biomass.** *Biores Technol* 2005, **96:** 673-686. 10.1016/j.biortech.2004.06.025

66. Liu C, Sun C: **The future crops for biofuels.** In *Economic Effects of Biofuel Production*. Edited by: Bernardes MAS. Crotia: InTech Publishers; 2011:25-38.

67. Joyce BL, Stewart CN Jr: **Designing the perfect plant feedstock for biofuel production: using the whole buffalo to diversify fuels and products.** *Biotechnol Adv* 2012, **30:** 1011-1022.

68. Xu B, Escamilla-Treviño LL, Sathitsuksanoh N, Shen Z, Shen H, Zhang YHP, Dixon RA, Zhao B: **Silencing of 4-coumarate:coenzyme A ligase in switchgrass leads to reduced lignin content and improved fermentable sugar yields for biofuel production.** *New Phytol* 2011, **192:** 611-625. 10.1111/j.1469-8137.2011.03830.x

69. Fu C, Mielenz JR, Xiao X, Ge Y, Hamilton CY, Rodriguez M Jr, Chen F, Foston M, Ragauskas A, Bouton J, Dixon RA, Wang ZY: **Genetic manipulation of lignin reduces recalcitrance and improves ethanol production from switchgrass.** *Proc Natl Acad Sci USA* 2011, **108:**3803-3808. 10.1073/pnas.1100310108

70. Li X, Weng J, Chapple C: **Improvement of biomass through lignin modification.** *Plant J* 2008, **54:** 569-581. 10.1111/j.1365-313X.2008.03457.x

71. Chen L, Auh CK, Dowling P, Bell J, Chen F, Hopkins A, Dixon RA, Wang ZY: **Improved forage digestibility of tall fescue (*Festuca arundinacea*) by transgenic down-regulation of cinnamyl alcohol dehydrogenase.** *Plant Biotechnol J* 2003, **1:** 437-449. 10.1046/j.1467-

7652.2003.00040.x

72. Dien BS, Sarath G, Pedersen JF, Sattler SE, Chen H, Funnell-Harris DL, Nichols NN, Cotta MA: **Improved sugar conversion and ethanol yield for forage sorghum (*Sorghum bicolor* L. Moench) lines with reduced lignin contents.** *BioEnerg Res* 2009, **2:** 153-164. 10.1007/s12155-009-9041-2

73. Fu C, Xiao X, Xi Y, Ge Y, Chen F, Bouton J, Dixon RA, Wang ZY: **Downregulation of *cinnamyl alcohol dehydrogenase* (CAD) leads to improved saccharification efficiency in switchgrass.** *BioEnerg Res* 2011, **4:** 153-164. 10.1007/s12155-010-9109-z

74. Saathoff AJ, Sarath G, Chow EK, Dien BS, Tobias CM: **Down-regulation of *cinnamyl-alcohol dehydrogenase* in switchgrass by RNA silencing results in enhanced glucose release after cellulase treatment.** *PLoS One* 2011, **6:** e16416. 10.1371/journal.pone.0016416

75. Shen H, He X, Poovaiah CR, Wuddineh WA, Ma J, Mann DGJ, Wang H, Jackson L, Tang Y, Stewart CN Jr, Chen F, Dixon RA: **Functional characterization of the switchgrass (*Panicum virgatum*) R2R3-MYB transcription factor *PvMYB4* for improvement of lignocellulosic feedstocks.** *New Phyto* 2012, **193:** 121-136. 10.1111/j.1469-8137.2011.03922.x

76. Casler MD: **Switchgrass breeding, genetics and genomics.** In *Switchgrass, Green Energy and Technology*. Edited by: Monti A. London: Springer-Verlag; 2012:29-53.

77. Vogel KP, Jung HJG: **Genetic modification of herbaceous plants for feed and fuel.** *Crit Rev Plant Sci* 2001, **20:** 15-49.

78. Bouton J: **Improvements of switchgrass as a bioenergy crop.** In *Genetic Improvement of Bioenergy Crops*. Edited by: Vermerris W. Berlin Heidelberg: Springer; 2008:295-308.

79. Jung HJG, Vogel KP: **Lignification of switchgrass (*Panicum virgatum*) and big bluestem (*Andropogon gerardii*) plant-parts during maturation and its effect on fiber degradability.** *J Sci Food Agric* 1992, **59:** 169-176. 10.1002/jsfa.2740590206

80. Carroll A, Somerville C: **Cellulosic biofuels.** *Annu Rev Plant Biol* 2009, **60:** 165-182. 10.1146/annurev.arplant.043008.092125

81. Chuck GS, Tobias C, Sun L, Kraemer F, Li C, Dibble D, Arora R, Bragg JN, Vogel JP, Singh S, Simmons BA, Pauly M, Hake S:**Overexpression of the maize *Corngrass1* microRNA prevents flowering, improves digestibility, and increases starch content of switchgrass.** *Proc Natl Acad Sci USA* 2011, **108:** 17550-17555. 10.1073/pnas.1113971108

82. Fu C, Sunkar R, Zhou C, Shen H, Zhang J, Matts J, Wolf J, Mann DGJ, Stewart CN Jr, Tang Y, Wang ZY: **Overexpression of miR156 in switchgrass (*Panicum virgatum* L.) results in various morphological alterations and leads to improved biomass production.** *Plant Biotechnol J* 2012, **10:** 443-452. 10.1111/j.1467-7652.2011.00677.x

83. Zhang B, Pan X, Cobb GP, Anderson TA: **Plant microRNA: a small regulatory molecule with big impact.** *Dev Biol* 2006, **289:** 3-16. 10.1016/j.ydbio.2005.10.036

84. Poethig RS: **Phase change and the regulation of shoot morphogenesis in plants.** *Science* 1990, **250:** 923-930. 10.1126/science.250.4983.923

85. Chuck G, Cigan AM, Saeteurn K, Hake S: **The heterochronic maize mutant *Corngrass1* results from overexpression of a tandem microRNA.** *Nat Genet* 2007, **39:** 544-549. 10.1038/ng2001

86. Rhoades MW, Reinhart BJ, Lim LP, Burge CB, Bartel B, Bartel DP: **Prediction of plant microRNA targets.** *Cell* 2002, **110:** 513-520. 10.1016/S0092-8674(02)00863-2

87. Sun G, Stewart CN Jr, Xiao P, Zhang B: **MicroRNA expression analysis in the cellulosic biofuel crop switchgrass (*Panicum virgatum*) under abiotic stress.** *PLoS One* 2012, **7:** e32017. 10.1371/journal.pone.0032017

88. Eriksson ME, Israelsson M, Olsson O, Moritiz T: **Increased gibberellin biosynthesis in transgenic trees promotes growth, biomass production and xylem fiber length.** *Nat Biotechnol* 2000, **18:** 784-788. 10.1038/77355

89. Casler MD, Stendal CA, Kapich L, Vogel KP: **Genetic diversity, plant adaptation regions, and gene pools for switchgrass.** *Crop Sci* 2007, **47:** 2261-2273. 10.2135/cropsci2006.12.0797

90. Cortese LM, Honig J, Miller C, Bonos SA: **Genetic diversity of twelve switchgrass populations using molecular and morphological markers.** *Bioenerg Res* 2010, **3:** 262-271. 10.1007/s12155-010-9078-2

91. Gunter LE, Tuskan GA, Wullschleger SD: **Diversity among populations of switchgrass based on RAPD markers.** *Crop Sci* 1996, **36:** 1017-1022. 10.2135/cropsci1996.0011183X003600040034x

92. Missaoui AM, Paterson AH, Bouton JH: **Molecular markers for the classification of switchgrass (*Panicum virgatum* L.) germplasm and to assess genetic diversity in three synthetic switchgrass populations.** *Genet Resour Crop Evol* 2006, **53:** 1291-1302. 10.1007/s10722-005-3878-9

93. Nageswara-Rao M, Stewart CN Jr, Kwit C: **Genetic diversity and

structure of natural and cultivated switchgrass (*Panicum virgatum*L.) populations. *Genet Resour Crop Evol* 2013, **60:** 1057-1068. 10.1007/s10722-012-9903-x

94. Wang YW, Samuels TD, Wu YQ: **Development of 1,030 genomic SSR markers in switchgrass.** *Theor Appl Genet* 2011, **122:** 677-686. 10.1007/s00122-010-1477-4

95. Zalapa JE, Price DL, Kaeppler SM, Tobias CM, Okada M, Casler MD: **Hierarchical classification of switchgrass using SSR and chloroplast sequences: ecotypes, ploidies, gene pools, and cultivars.** *Theor Appl Genet* 2011, **122:** 805-817. 10.1007/s00122-010-1488-1

96. Casler MD, Tobias CM, Kaeppler SM, Buell CR, Wang Z-Y, Cao P, Schmutz J, Ronald P: **The switchgrass genome: tools and strategies.** *The Plant Genome* 2011, **4:** 273-282. 10.3835/plantgenome2011.10.0026

97. Liu L, Wu Y, Wang Y, Samuels T: **A high-density simple sequence repeat-based genetic linkage map of switchgrass.** *Genes Genom Genet* 2012, **2:** 357-370.

98. Missaoui AM, Paterson AH, Bouton JH: **Investigation of genomic organization in switchgrass (*Panicum virgatum* L.) using DNA markers.** *Theor Appl Genet* 2005, **110:** 1372-1383. 10.1007/s00122-005-1935-6

99. Okada M, Lanzatella C, Saha MC, Bouton J, Wu R, Tobias C: **Complete switchgrass genetic maps reveal subgenome collinearity, preferential pairing and multilocus interactions.** *Genetics* 2010, **185:** 745-760. 10.1534/genetics.110.113910

100. Serba DD, Dhanasekaran V, Saha MC, Bouton JH: **Mapping of QTLs for biomass, plant composition, and agronomic traits in switchgrass.** In *Proceedings of the Plant and Animal Genomes XIX Conference*. San Diego: ; 2011:365.

101. Serba D, Ziebell A, Bahri BA, Sykes R, Devos K, Brummer C, Bouton JH, Saha MC: **Identification of putative genomic regions controlling recalcitrance in AP13 x VS16 switchgrass population.** In *Proceedings of the Plant and Animal Genomes XX conference*. San Diego: ; 2012:0746.

102. Soneji JR, Nageswara-Rao M, Sudarshana P, Panigrahi J, Kole C: **Current status on on-going genome initiatives.** In *Principles and Practices of Plant Genomics. Volume 3: Advanced Genomics*. Edited by: Enfield, New Hampshire, Edenbridge Ltd, Channel Islands, Kole C, Abbott AG. British Isles: Science Publishers, Inc; 2010:305-353.

103. Wang Y, Zeng X, Iyer NJ, Bryant DW, Mockler TC, Mahalingam

R: **Exploring the switchgrass transcriptome using second-generation sequencing technology.** *PLoS One* 2012, **7:** e34225. 10.1371/journal.pone.0034225

104. Adams M, Kelley J, Gocayne J, Dubnick M, Polymeropoulos M, Xiao H, Merril CR, Wu A, Olde B, Moreno RF, Kerlavage AR, McCombie R, Venter JC: **Complementary DNA sequencing: expressed sequence tags and human genome project.** *Science* 1991, **252:** 1651-1656. 10.1126/science.2047873

105. Tobias C, Twigg P, Hayden DM, Vogel KP, Mitchell RM, Lazo GR, Chow EK, Sarath G: **Analysis of expressed sequence tags and the identification of associated short tandem repeats in switchgrass.** *Theor Appl Genet* 2005, **111:** 956-964. 10.1007/s00122-005-0030-3

106. Andersen JR, Lubberstedt T: **Functional markers in plants.** *Trends Plant Sci* 2003, **8:** 554-560. 10.1016/j.tplants.2003.09.010

107. Emrich SJ, Barbazuk WB, Li L, Schnable PS: **Gene discovery and annotation using LCM-454 transcriptome sequencing.** *Genom Res* 2007, **17:** 69-73.

108. Tobias CM, Sarath G, Twigg P, Lindquist E, Pangilinan J, Penning BW, McCann MC, Carpita NC, Lazo GR: **Comparative genomics in switchgrass using 61,585 high-quality expressed sequence tags.** *The Plant Genom* 2008, **1:** 111-124. 10.3835/plantgenome2008.08.0003

109. Temnykh S, DeClerck G, Lukashova A, Lipovich L, Cartinhour S, McCouch S: **Computational and experimental analysis of microsatellites in rice (*Oryza sativa* L.): frequency, length variation, transposon associations, and genetic marker potential.** *Genom Res* 2001, **11:** 1441-1452. 10.1101/gr.184001

110. Tobias CM, Hayden DM, Twigg P, Sarath G: **Genic microsatellite markers derived from EST sequences of switchgrass (*Panicum virgatum* L.).** *Mol Ecol Notes* 2006, **6:** 185-187. 10.1111/j.1471-8286.2006.01187.x

111. Saski CA, Li Z, Feltus FA, Luo H: **New genomic resources for switchgrass: a BAC library and comparative analysis of homoeologous genomic regions harboring bioenergy traits.** *BMC Genomics* 2011, **12:** 369. 10.1186/1471-2164-12-369

112. Venter JC, Smith HO, Hood L: **A new strategy for genome sequencing.** *Nature* 1996, **381:** 364-366. 10.1038/381364a0

113. Sharma MK, Sharma R, Cao P, Jenkins J, Bartley LE, Qualls M, Grimwood J, Schmutz J, Rokhsar D, Ronald PC: **A genome-wide survey of switchgrass genome structure and organization.** *PLoS*

One 2012, **7:** e33892. 10.1371/journal.pone.0033892

114. Bennetzen JL, Schmutz J, Wang H, Percifield R, Hawkins J, Pontaroli AC, Estep M, Feng L, Vaughn JN, Grimwood J, Jenkins J, Barry K, Lindquist E, Hellsten U, Deshpande S, Wang X, Wu X, Mitros T, Li P, Sharma M, Sharma R, Ronald PC, Panaud O, Kellogg EA, Brutnell TP, Doust AN, Tuskan GA, Rokhsar D, Devos KN: **Reference genome sequence of the model plant *Setaria*** . *Nat Biotechnol* 2012, **30:** 555-561. 10.1038/nbt.2196

115. Bock R: **Structure, function, and inheritance of plastid genomes.** In *Cell and Molecular Biology of Plastids. Volume 19*. Edited by: Bock R. Berlin Heidelberg: Springer; 2007:1610-2096.

116. Raubeson L, Jansen R: **Chloroplast genomes of plants.** In *Plant Diversity and Evolution: Genotypic and Phenotypic Variation in Higher Plants*. Edited by: Henry R. Cambridge: CABI Publishing; 2005:45-68.

117. Morris GP, Grabowski PP, Borevitz JO: **Genomic diversity in switchgrass (*Panicum virgatum* L.): from the continental scale to a dune landscape.** *Mol Ecol* 2011, **20:** 4938-4952. 10.1111/j.1365-294X.2011.05335.x

118. Young HA, Lanzatella CL, Sarath G, Tobias CM: **Chloroplast genome variation in upland and lowland switchgrass.** *PLoS One* 2011, **6:**e23980. 10.1371/journal.pone.0023980

119. Grevich J, Daniell H: **Chloroplast genetic engineering: recent advances and future perspectives.** *Crit Rev Plant Sci* 2005, **24:** 83-107. 10.1080/07352680590935387

120. Lu K, Kaeppler SM, Vogel KP, Arumuganathan K, Lee DJ: **Nuclear DNA content and chromosome numbers in switchgrass.** *Great Plains Res* 1998, **8:** 269-280.

121. Hultquist SJ, Vogel KP, Lee DJ, Arumuganathan K, Kaeppler S: **Chloroplast DNA and nuclear DNA content variations among cultivars of switchgrass, *Panicum virgatum* L.** *Crop Sci* 1996, **36:** 1049-1052. 10.2135/cropsci1996.0011183X003600040039x

122. Mardis ER: **The impact of next-generation sequencing technology on genetics.** *Trends Genet* 2008, **24:** 133-141. 10.1016/j.tig.2007.12.007

123. Morozova O, Marra MA: **Applications of next-generation sequencing technologies in functional genomics.** *Genom* 2008, **92:** 255-264. 10.1016/j.ygeno.2008.07.001

124. Wang W, Wang Y, Zhang Q, Qi Y, Guo D: **Global characterization of *Artemisia annua* glandular trichome transcriptome using 454**

pyrosequencing. *BMC Genomics* 2009, **10:** 465. 10.1186/1471-2164-10-465

125. Metzker ML: **Sequencing technologies - the next generation.** *Nat Rev Genet* 2010, **11:** 31-46. 10.1038/nrg2626

126. Liu L, Li Y, Li S, Hu N, He Y, Pong R, Lin D, Lu L, Law M: **Comparison of next-generation sequencing systems.** *J Biomed Biotechnol* 2012. 10.1155/2012/251364

127. Young HA, Hernlem BJ, Anderton AL, Lanzatella CL, Tobias CM: **Dihaploid stocks of switchgrass isolated by a screening approach.** *BioEnerg Res* 2010, **3:** 305-313. 10.1007/s12155-010-9081-7

128. Zhang J, Lee Y, Torres-Jerez I, Wang M, Yin Y, Chou W, Je J, Shen H, Srivastava AC, Pennacchio C, Lindquist E, Grimwood J, Schmutz J, Xu Y, Sharma M, Sharma R, Bartley LE, Ronald PC, Saha MC, Dixon RA, Tang Y, Udvardi MK: **Development of an integrated transcript sequence database and a gene expression atlas for gene discovery and analysis in switchgrass (*Panicum virgatum* L.).** *Plant J* 2013. 10.1111/tpj.12104

129. Meyer E, Logan TL, Juenger TE: **Transcriptome analysis and gene expression atlas for *Panicum hallii* var. *filipes* , a diploid model for biofuel research.** *Plant J* 2012, **70:** 879-890. 10.1111/j.1365-313X.2012.04938.x

130. Bartel DP: **MicroRNAs: Genomics, biogenesis, mechanism, and function.** *Cell* 2004, **116:** 281-297. 10.1016/S0092-8674(04)00045-5

131. Xie F, Frazier TP, Zhang B: **Identification and characterization of microRNAs and their targets in the bioenergy plant switchgrass (*Panicum virgatum*).** *Planta* 2010, **232:** 417-434. 10.1007/s00425-010-1182-1

132. Henikoff S, Till BJ, Comai L: **TILLING: traditional mutagenesis meets functional genomics.** *Plant Physiol* 2004, **135:** 630-636. 10.1104/pp.104.041061

133. Comai L, Young K, Till BJ, Reynolds SH, Greene EA, Codomo CA, Enns LC, Johnson JE, Burtner C, Odden AR, Henikoff S: **Efficient discovery of DNA polymorphisms in natural populations by Ecotilling.** *Plant J* 2004, **37:** 778-786. 10.1111/j.0960-7412.2003.01999.x

134. Weil C: **TILLING in grass species.** *Plant Physiol* 2009, **149:** 158-164. 10.1104/pp.108.128785

135. Stewart CN Jr: **Biofuels and biocontainment.** *Nat Biotechnol* 2007, **25:** 283-284. 10.1038/nbt0307-283

136. Kwit C, Stewart CN Jr: **Geneflow matters in switchgrass (*Panicum virgatum* L.), a potential widespread biofuel feedstock.** *Ecol Appl* 2012, **22:** 3-7. 10.1890/11-1516.1

137. Rieger MA, Lamond M, Preston C, Powles SB, Roush RT: **Pollen-mediated movement of herbicide resistance between commercial canola fields.** *Science* 2002, **296:** 2386-2388. 10.1126/science.1071682

138. Daniell H: **Molecular strategies for gene containment in transgenic crops.** *Nat Biotechnol* 2002, **20:** 581-586.

139. Mariani C, DeBeuckeleer M, Trueltner J, Leemans J, Goldberg RB: **Induction of male sterility in plants by a chimeric ribonuclease gene.** *Nature* 1990, **347:** 737-741. 10.1038/347737a0

140. Luo K, Duan H, Zhao D, Zheng X, Deng W, Chen Y, Stewart CN Jr, McAvoy R, Jiang X, Wu Y, He A, Pei Y, Li Y: **'GM-gene-deletor': fused loxP-FRT recognition sequences dramatically improve the efficiency of FLP or CRE recombinase on transgene excision from pollen and seed of tobacco plants.** *Plant Biotechnol J* 2007, **5:** 263-274. 10.1111/j.1467-7652.2006.00237.x

141. Moon HS, Abercombie LL, Eda S, Blanvillain R, Thonson JG, Ow D, Stewart CN Jr: **Transgene excision in pollen using a codon optimized serine resolvase CinH- *RS* 2 site-specific recombination system.** *Plant Mol Biol* 2011, **75:** 621-631. 10.1007/s11103-011-9756-2

142. Daniell H, Datta R, Varma S, Gray S, Lee SB: **Containment of herbicide resistance through genetic engineering of the chloroplast genome.** *Nat Biotechnol* 1998, **16:** 345-348. 10.1038/nbt0498-345

143. Hagemann R, Schroeder M: **The cytological basis of plastid inheritance in angiosperms.** *Protoplasma* 1989, **152:** 57-64. 10.1007/BF01323062

144. Vega-Sanchez ME, Ronald PC: **Genetic and biotechnological approaches for biofuel crop improvement.** *Curr Opi Biotechnol* 2010, **21:** 218-224. 10.1016/j.copbio.2010.02.002

Chapter 11

A NOVEL, HIGHLY SELECTIVE RT-QPCR METHOD FOR QUANTIFICATION OF MSRV USING PNA CLAMPING SYNCYTIN-1 (ERVWE1)

Grzegorz Machnik[1], Estera Skudrzyk[1], Łukasz Bułdak[1], Krzysztof Łabuzek[1], Jarosław Ruczyn´ski[2], Magdalena Alenowicz[2], Piotr Rekowski[2], Piotr Jan Nowak[3], Bogusław Okopien´[*1]

[1]Department of Pharmacology, Medical University of Silesia, Medyko´w 18, 40-752 Katowice, Poland

[2]Faculty of Chemistry, University of Gdan´sk, Wita Stwosza 63, 80-308 Gdan´sk, Poland

[3]Department of Nephrology, Hypertension, and Kidney Transplantation, Medical University of Łódź, Pomorska 251, 92-213 Łódź, Poland

ABSTRACT

HERV-W is a multi-locus family of human endogenous retroviruses (HERVs) that has been found to play an important role in human physiology and pathology. Two particular members of HERV-W family are of special interests: ERVWE1 (coding syncytin-1, which is a glycoprotein essential in the formation of the placenta) and MSRV (multiple sclerosis-associated retrovirus that is thought to play a significant role in human pathology as a result of its increased expression in the brain tissue and blood cells derived from patients with multiple sclerosis (MS)). Both ERVWE1 and MSRV mRNA share high level of similarity and hence a method that allows to exclusively quantify the MSRV expression in clinical samples would be desirable. We developed a quantitative polymerase chain reaction (QPCR) technique for the detection and quantification of the multiple sclerosis-associated retrovirus. The assay utilises fluorescently labelled oligonucleotide probe, which is complementary to the conservative fragment of MSRV *env* gene and a peptide nucleic acid (PNA) probe, fully complementary to the ERVWE1 sequence fragment that efficiently blocks the polymerase action on ERVWE1 templates. The PNA molecule, if used parallel with hydrolysis probe in QPCR analysis, greatly facilitates the detection efficiency of MSRV even if ERVWE1 is present abundantly in respect to MSRV in the analysed sample. We achieved a wide and measurable range from $1 \times 10 \text{ e}^2$ to $1 \times 10 \text{ e}^8$ copies/reaction; the linearity of the technique was maintained even at the low MSRV level of 1 % in respect to ERVWE1. Using our newly developed method we confirmed that the expression of MSRV takes place in normal human astrocytes and in human umbilical vein endothelial cells in vitro. We also found that the stimulation of human monocytes did not influence the specific expression of MSRV but it caused changes in mRNA level of distinct HERV-W templates.

INTRODUCTION

Human endogenous retroviruses of the HERV-W family are broadly dispersed thorough the human genome; it is estimated that HERV-W are represented by more than one hundred copies per haploid genome [1]. According to Pavlicek et al. [2] and their in silico evaluation, this retroviral group comprises 77 proviral copies with complete or, at least, partial internal (non-LTR) sequences, and of 343 long terminal repeats without any internal sequences (solo-LTRs). Despite the huge number of HERV copies, only a few intact open reading frames (ORFs) survived, avoiding the gradual accumulation of compromising mutations and/or deletions. The role of retroviral sequences in human genome has been broadly investigated recently, as there is numerous evidence of their adverse effect upon (auto-) immunologic disorders. Some gene expression products of human endogenous retroviruses have been detected as transcripts and as proteins in the central nervous system and they have been frequently connected with neuroinflammation. The HERV-W family has received substantial attention in large part because of its associations with diverse syndromes including multiple sclerosis (MS) and several psychiatric disorders, such as schizophrenia [3–5]. It has been shown that both HERV-W and HERV-H/F were specifically activated in the circulation and in the central nervous system (CNS) in the majority of MS patients and, particularly, the envelopes (*env* mRNA and *env*proteins) appear to be strongly associated with disease activity and progression [6]. Brudek et al. found a significantly higher expression of HERV-H and HERV-W *env* epitopes on B cells and monocytes from patients with active MS compared with patients with stable MS or control individuals [7]. Another known member of the HERV-W family is ERVWE1, which is located on chromosome 7, coding syncytin-1. This glycoprotein has highly fusogenic properties that are essential during the development of syncytiotrophoblast in placenta [8]. Both multiple sclerosis-associated retrovirus (MRSV) and ERVWE1 mRNA are very similar to each other. MSRV and ERVWE1 *env* sequences belonging to the HERV-W family are distinct from other HERV families, showing only 38–47 % sequence similarity with HERV-FRD/syncytin-2 or HERV-H and HERV-K and also differ from exogenous human retroviruses, such as human T-lymphotropic virus 1 and human immunodeficiency virus (20–25 % similarity) [9]. On the other hand MSRV and ERVWE1 pol region share 92 % of sequence while their *env* genes are identical in 81 % [10, 11].

Because an increased level of HERV-W RNA has been reported in brain tissue from MS patients [12], a precise technique that allows differentiating between MSRV and syncytin-1 (from ERVWE1 locus) mRNA would be very desirable. Indeed, an increased level of ERVWE1, but not MSRV mRNA in MS brain samples has been reported [11]. The most popular method that was also utilised by researchers from Prof. Power's group [13], is to use allele-specific oligonucleotides (ASO) as primers in the amplification reaction of HERV-W template. This technique allows the selective amplification of mutant or wild-type gene fragment and the principle of the method is based on the phenomenon that DNA polymerase cannot bind to the 3'end of the primer if there is a mismatch between oligonucleotide and the template [14]. However, as shown by Kwok et al., many 3' mismatches between a primer and its template do not significantly impair the PCR reaction [15]. Moreover, Yoshida et al. noted that if they used hydrolysis probes in real-time PCR settings, it was difficult to measure the amounts of hepatitis-B virus mutants accurately, especially when the target strain was only a minor component of the mixed population [16]. Some analogy is observed in the case of HERV-W family where the target i.e. MSRV mRNA level may comprise only a small portion of abundantly expressed ERVWE1 mRNA.

Peptide nucleic acid (PNA) is a DNA mimicking synthetic oligonucleotide with peptide back-bones instead of sugar groups and was first reported in 1991 by Nielsen et al. from the University of Copenhagen [17]. By comparison with nucleic acid, PNA is much more stable chemically and biologically and, what is particularly noteworthy is, it is resistant against nucleases [18]. It has been documented that PNA sequence (probe) recognises and binds to the complementary DNA template with higher affinity and specificity than DNA-to-DNA does, because of the lack of electrostatic repulsion. On the other hand, high affinity and binding strength of PNA/DNA duplexes dropped drastically if even a single nucleotide mismatch occurs in the target DNA sequence. This phenomenon presented the opportunity to utilise PNA oligomers in single nucleotide polymorphism (SNP) analysis [17]. Peptide nucleic acid cannot be extended by the DNA polymerase, therefore it cannot serve as a primer in polymerase chain reaction (PCR) as well. It can, instead, be useful as a PCR blocking (clamping) molecule that effectively prevents primer hybridisation and subsequent amplification of undesired (wild type) template. If standard oligonucleotide PCR primers are designed to overlap the SNP site in the target (mutated) sequence and a PNA oligomer that is fully complementary to the other (wild type) DNA is added into the reaction, PNA preferentially hybridises (binds) to the wild type but not to the mutated site DNA. This, in turn, strongly reduces the amplification of the wild type relative to the mutant DNA. In such a setting, only the amplification of the sequence of interest is allowed.

As mentioned above, PNA-mediated PCR clamping method is especially desired if DNA of interest is present in a very low amount relative to the wild type one [19]. This situation can occur, for example, when only a few mutant cancer cells have to be detected along with a large excess of healthy cells in a tissue biopsy or for the seeking of a drug-resistant mutant among heterogeneous viral population [20]. As shown by Kim et al., this method allows SNP detection even when the wild-type DNA is present in up to 20,000-fold excess relative to the mutant [19]. This unique feature characterising the clamping PCR assay is especially important in the analysis of HERV-W family because ERVWE1, being physiologically active in many cells, expresses abundantly and may overshadow possible discrete differences of MSRV level in investigated samples.

In our previous report, we described a real-Time polymerase chain resaction/PNA-mediated methodology that allows the specific quantitative amplification of multiple sclerosis- associated retrovirus (MSRV). The assay is based on two sequence-specific oligonucleotides (PCR primers) that are common for both MSRV and ERVWE1 gene fragment, an ERVWE1- specific PNA probe and SYBR Green 1 intercalating dye that was essential in the quantitative assessment. Those results showed that it is possible to estimate the MSRV level in a biological sample if the ERVWE1 template is excluded from the reaction by means of PNA-mediated PCR clamping. According to our results with the PNA inhibition of ERVWE1, we showed that the overall expression of HERV-W was similar in samples treated with PNA or without PNA clamping, therefore we concluded that the expression of MSRV is predominant over syncytin-1 in human astrocytic U-87 MG cell line. Because the calculated HERV-W *env* mRNA copy number differed in respect of the presence or absence of the ERVWE1-specific PNA, therefore we concluded that MSRV is present predominantly over syncytin-1 in human astrocytic U-87 MG cell line. Additionally, compared to the intercalating dyes (e.g. SYBR Green I), our novel method incorporated the specific fluorescent probe that is much more suitable for virology/microbiology experiments, due to higher pathogen-specificity of the procedure primers, the utilisation of a third sequence-specific element, such as oligonucleotide probe is preferred over an intercalating dye in the virology/microbiology fields due to higher pathogen-specificity of the method. In the probe-based chemistry, both primers and probe hybridise to the common target, thus as much as three events must occur independently to give a specific fluorescent signal [21].

MATERIALS AND METHODS

In Silico Analysis of HERV-W Sequences

HERV-W *env* sequences were derived from the GenBank database according to previously published data [10, 11]. We compared three *env* sequences denoted as MSRV with four *env* sequences denoted as coding for syncytin-1. Accession numbers of the investigated entries were as follows: AF331500, AF127229, AF072498 for MSRV and AY101582, AF513360, AF208161 and NM_014590 for syncytin-1. In both sequence types, a consensus sequence was built using CONS software from the EMBOSS package [22]. Whole design was then focused only on a highly conserved common fragment for all MSRV sequences and for syncytin-1 sequences, which was located between nucleotide position 1187 and 1456 in respect to the reference sequence (NM_014590). We paid attention on this region as it bears several point mutations that are specific and conservative for MSRV or syncytin-1 sequences only and therefore provides an appropriate place for the hybridisation of sequence-specific PNA and hydrolysis probe.

Peptide nucleic acids, 15-nucleotides in length, were synthesised and purified at the Department of Chemistry, University of Gdańsk, Poland (please see "Synthesis of peptide nucleic acids complementary to ERVWE1" caption for details). Lyophilised PNA was resolved in 5 % DMSO and left for 20 min at room temperature (RT) to obtain a stock concentration of 10 μM. PNA was then aliquoted and stored at −20 °C.

Oligonucleotide primers for polymerase chain reaction were designed according to previously published guidelines [23]. Three reverse and one common forward primer were chosen for further analysis. Reverse primers were located in different positions in respect to the target region for PNA and for the probe. The reciprocal primers, PNA and probe hybridisation regions in MSRV and ERVWE1 *env* gene sequence are shown in Fig. 1. All oligonucleotides that were used in our work are also specified in details in Table 1.

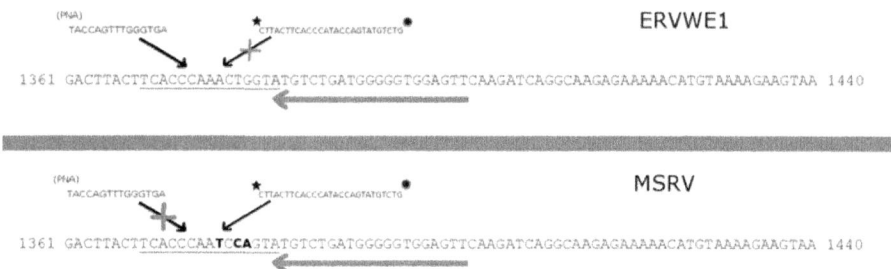

Figure 1: A principle of the method behind the quantitation of MSRV expression using PNA-mediated quantitative polymerase chain reaction. Peptide nucleic acid (PNA) probe (on the *left-hand side* of the figure) binds to the ERVWE1 template with high affinity due to 100 % complementarity of their sequences and prevents the hybridisation of fluorescently labelled probe (on the *right*). The probe remains intact and therefore no signal is emitted. If MSRV serves as template, PNA probe cannot bind tightly to the DNA due to three mismatches between PNA and template that diminish drastically its affinity. On the contrary, fluorescent probe now is allowed to bind to the MSRV template and afterwards it is cleaved by the polymerase during the elongation phase and emits a signal as normally seen in probe-based QPCR

Table 1: Oligonucleotide primers used in the work—a summary

Oligonucleotide name	Sequence	Remarks
Syncytin_out_F	5'-TGCCCCATCGTATAG-GAGTC-3'	For the cloning procedure, PNA-mediated PCR and megaprimer mutagenesis
Syncytin_out_R	5'-CGGGTGAGTTGGGA-GATTAC-3'	For the cloning procedure
sync1_REV/-2[*)]	5'-ATTCCACCCCCAT-CAGACATA-3'	[*)] Reverse primers used in the PNA-mediated PCR clamping together with Syncytin_out_F as forward primer
sync1_REV/-1[*)]	5'-AATTCCACCCCCAT-CAGACAT-3'	As above
sync1_REV/0[*)]	5'-GAATTCCACCCCCAT-CAGACA-3'	As above
Mutagenic MRSV	5'-CAGACATACTGG-TATGGGTGAAGT-3'	For use in the megaprimer mutagenesis technique
PNA ERVWE1	N'-TACCAGTTTGGGTG-C'	For the PNA-mediated PCR clamping of ERVWE1

PNA ERVWE1-T	N'-TACCAGTTTGGGT-GA-C'	For the PNA-mediated PCR clamping of ERVWE1
MSRV Probe	FAM 5'-CT-TACTTCACCCATACCAG-TATGTCTGATG- 3' BHQ1	For the quantitation of MSRV sequences by means of QPCR

Extraction of Ribonucleic Acids and Reverse Transcription Reaction

Total RNA was extracted from normal human astrocytes (NHA) cell line (Lonza Ltd, Warsaw, Poland), human umbilical vein endothelial cells (HUVECs) (Sigma-Aldrich, Poznań, Poland and from human monocytes/macrophages. The study was accepted by the Bioethical Committee of the Medical University of Silesia in Katowice, Poland. The investigation conformed to the principles of the Declaration of Helsinki. Human whole-blood samples were taken from antecubital veins of healthy volunteers (aged 18–40 years old). Peripheral blood monocytic cells (PBMCs) were obtained after centrifugation on blood in the Histopaque (Sigma-Aldrich Co., Poznań, Poland). Afterwards monocytes were isolated from the PBMCs by negative immunomagnetic separation using Pan-T and Pan-B Dynabeads (Dynal, Oslo, Norway) [24]. Monocyte-derived macrophages were then stimulated with medium supplemented with 1 μg/ml of lipopolysaccharide (LPS, Escherichia coli serotype 0111:B4, Sigma, Germany) as described previously [25]. Control cells were maintained at the same conditions except the addition of LPS-containing medium.

For NHA and HUVECs, 3 ml of TRI reagent (MRC Inc., USA) was used to lyse cells from one 10 mm in diameter tissue culture plate after the >80 %-confluence was achieved. Before TRI reagent was added, culture medium was removed but without washing the cells with PBS. Monocyte cultures were lysed using one ml of TRI reagent, which was added directly into wells of a 24-well culture plate after removing the culture medium. Three wells were then pooled together. After cell lysis and complete disruption of nucleoprotein complexes, total RNA was precipitated and further purified according to the one-step extraction method described by Chomczynski et al. [26]. Finally, ribonucleic acids were resolved in 50 μl of deionised formamide (Formazol, MRC Inc., Cincinnati, Ohio, USA). RNA samples were examined and quantified spectrophotometrically at wavelengths of 260 and 280 nm. One μg of total RNA was diluted in nuclease-free water and treated with 1.2 U of Recombinant DNase I, RNase-free (Ambion-Life Technologies Corp., Carlsbad, California, USA) in order to remove any residual DNA present in RNA extracts. This step was crucial for further interpretation of results as

HERV-W amplicons of both cDNA and genomic DNA origins are of the same length, and it would not be possible to utilise intron/exon spanning primers. After DNase treatment we performed additional quantitative PCR assays of RNA and DNA samples without reverse transcriptase (RT negative), to exclude the possibility of RNA contamination by DNA. After the QPCR, the reaction mixtures were loaded onto agarose gel. Neither the QPCR plots nor theagarose gel analysis gave positive signals for that samples. In the next step, 500 ng of DNA-free RNA was reverse-transcribed using GoScript Reverse Transcription System (Promega Corp. USA) in the final volume of 20 μl. Reverse transcription (RT) reaction contained of 1×Reverse Transcription buffer, 4.0 μl of 25 mM $MgCl_2$, 2.0 μl of 10 mM dNTP mixture, 0.5 μl of RNasin Ribonuclease Inhibitor, 15 units of AMV Reverse Transcriptase and 0.5 μg of Random hexamers. Reaction mixture was incubated for 10 min at room temperature and then for 30 min at 42 °C followed by thermal inactivation of reverse transcriptase at 95 °C for 5 min. According to the manufacturer's recommendations, reverse transcription reaction mixture was further diluted 1:4 with nuclease-free water before the use in QPCR. Five μl of diluted RT reaction mixture (i.e. an equivalent of about 25 ng of total RNA) was then used as a template for subsequent QPCR analysis.

The Choice of a Proper Priming in Reverse Transcription Reaction

In our assay, we used random hexamers as primers for the reverse transcriptase enzyme. As it is stated in Promega's instruction for use of their product # A3500 "Choose Oligo(dT)15 when priming at the 3'poly(A) region is desired. Choose Random Primers when priming throughout the length of the RNA is desired".

It should be emphasised that the choice of optimal RT priming is crucial to get the most representative cDNA pool. Human endogenous retroviral-derived particles make even more complex issue because they are broadly dispersed thorough the genome and are mostly truncated; e.g. a significant part of HERV-W possesses a poly-A tail (such as syncytin-1) while other retroviral members does not. If so, a mixture of $oligo(dT)_{15}$ and random primers may, in theory, improve the quality of cDNA pool. However, the use of $oligo(dT)_{15}$ concomitantly with random hexamers/nonamers is problematic because both solutions require slightly different thermal conditions and finally the resulting cDNA library may not be more representative than when using random primers alone.

For cloning purposes, DNA extract obtained from human placenta was used (a kind gift from Prof. dr Urszula Mazurek, Dept. of Molecular Biology, Medical University of Silesia, Sosnowiec, Poland). The Amplification of Syncytin-1 (ERVWE1) Gene Fragment and Site-Directed Mutagenesis

Polymerase chain reaction was performed using Qiagen Taq PCR Master Mix kit (Qiagen, Germany). The reaction mixture in a total volume of 25 µl consisted of 12.5 µl of 2× Qiagen PCR Master Mix, 200 nM of each Syncytin_out_F and Syncytin_out_R primer and 0.5 µl of placental DNA extract as a template. After reaction, amplicons were resolved by means of agarose gel electrophoresis, excised from the gel and extracted/purified using silica-based spin columns (Invisorb Fragment CleanUp, InviTec, Germany). Four separately purified amplimers were then sequenced (Genomed SA, Warsaw, Poland). BLAST analysis of obtained sequences confirmed that the specific amplification of ERVWE1 locus took place, giving a 100 % similarity to the reference sequence (GenBank: NM_014590). For cloning purposes, PCR reaction was repeated at identical conditions as described above, but less error-prone Pfu polymerase was used instead of less accurate Taq polymerase (Thermo Scientific, Lithuania). Furthermore, Pfu polymerase generated blunt-ended amplicons that were essential for the subsequent blunt-end cloning procedure.

In the next step, site-directed mutagenesis was made in order to introduce three point mutations into appropriate positions in the ERWVE1-specific PCR product making this region identical with that of MRSV (Fig. 1). To introduce point mutations, we utilised a megaprimer mutagenesis technique as described previously by Brons-Poulsen et al. [27]. The megaprimer method consists of two steps of polymerase chain reaction, where the first round of PCR generates a fragment containing the desired mutation. For the first PCR, Pfu polymerase was used (Thermo Scientific, Lithuania) and the reaction mixture consisted of 1xPfu Buffer, 2 mM of $MgSO_4$, 200 nM of each deoxyribonucleotide, 1 U of Pfu polymerase and of 25 pmol of each syncytin_mut_F and a mutagenic_MRSV_R primer (Table 1). The reaction was performed in total volume of 25 µl under the following thermal conditions: 95 °C/5 min, 35 repeats of (95 °C/30 s., 58 °C/30 s, 72 °C/40 s.) and 72 °C/10 min. After amplification, the reaction product (megaprimer) was resolved by agarose gel electrophoresis and purified by means of silica-based spin columns. Obtained megaprimer's amplimers were semi-quantified using Ethidium bromide-containing agarose plates and lambda DNA as a quantitative standard of known DNA copy number (Thermo Scientific, Lithuania) [28]. Reaction components of the second polymerase chain reaction were identical as in the first PCR, but one pmol of syncytin_mut_R primer and one microlitre of megaprimer solution were used as primers for Pfu polymerase (i.e. both megaprimer and reverse primer were in similar concentrations in the reaction mixture). Thermal settings of second PCR were adopted from the work of Brons-Poulsen et al.: 94 °C/2 min, 23 repeats of (94 °C/45 s, 58 °C/2 min, 72 °C/40 s) and 72 °C/10 min. Again, reaction products were resolved in agarose gel and a specific band of appropriate molecular weight (i.e. of 270 bp) was excised from the gel, purified as described previously and used directly for cloning parallel with ERVWE1 fragment.

Construction and Preparation of MSRV and ERVWE1 Plasmids Employed as a Reference and for the Optimization of PNA-Mediated PCR Clamping

Both ERVWE1–derived and mutated MSRV-like PCR products were cloned into pSC-B-amp/kan plasmid using StrataClone Ultra Blunt PCR Cloning Kit according to the manufacturer's guide (Agilent Technologies, USA). Mini preps were prepared in 3.5 ml of Luria Broth medium and were purified by NucleoSpin Plasmid kit (Macherey–Nagel, Germany). Obtained plasmids were given an initial examination to confirm the presence of the insert (regardless of their orientation) by PCR with a plasmid-specific universal T3/T7 primer pair. Finally, pSC-B-MSRV and pSC-B-ERVWE1 positive clones were sequenced in both directions by means of dideoxy-sequencing with the use of insert-specific primers, i.e. Syncytin_out_F and Syncytin_out_R (Genomed SA, Warsaw, Poland). The mini-prep extracts were quantitatively measured using SYBR Green 1 fluorescent dye (Life Technologies, Warsaw, Poland) against the known amounts of lambda DNA as a reference (Thermo Fermentas, Vilnius, Lithuania) in a fluorometer (Fluoroscan microplate reader, Labsystems, Finland). The excitation wavelength of the dye was 495 nm, and emission was filtered using a 520-nm barrier filter. Plasmid DNA copy number was then calculated using dsDNA copy number calculator (http://cels.uri.edu/gsc/cndna.html).

The Synthesis of Complementary to ERVWE1 Peptide Nucleic Acids

All reagents and solvents were of analytical or HPLC grade. Solutions were freshly prepared with distilled, deionised water using a Milli-Q Millipore system (Millipore Corporation, Billerica, Massachusetts, USA) and filtered with a 0.22-µm syringe filter before use. Fmoc-XAL-PEG-PS resin for the PNA synthesis was obtained from Merck KGaA (Darmstadt, Germany). Fmoc/Bhoc-protected PNA monomers were purchased from Panagene (Billingham Cleveland, United Kingdom). DNA oligonucleotides were obtained from Future Synthesis (Poznań, Poland). The other reagents and solvents were obtained from Sigma-Aldrich Co. (Poznań, Poland).

PNA sequence: TACCAGTTTGGGTGA-amide complementary to ERVWE1 was synthesised using a Labortec AG SP-650 Peptide Synthesizer on a 3.8 µmol scale using a 2.5-fold molar excess of the Fmoc/Bhoc-protected monomers and Fmoc-XAL-PEG-PS resin (amine groups loading of 0.19 mmol/g) [29]. Monomers were activated with the use of a HATU/NMM/2,6-lutidine (0.7:1:1.5) mixture and coupled for 30 min as active derivatives. Fmoc deprotection was performed with the use of 20 % piperidine in DMF (2 × 2 min). Deprotection of the Bhoc group and cleavage of PNA from resin was carried out with the use of a TFA/m-cresol (95:5) mixture for 30 min.

The crude oligomer obtained was lyophilised and purified by semi-preparative reverse phase, high performance liquid chromatography RP-HPLC using a Knauer system with a Kromasil C8 column (20 × 250 mm, 5 µm particle size) (Sigma-Aldrich Co., Poznań, Poland). The mobile phase consisted of 0.08 % TFA in acetonitrile (solvent A) and 0.1 % TFA in water (solvent B). The gradient profile for the mobile phase was as 0–30 % of solvent A held for 90 min. The column was maintained at ambient temperature. The flow rate was 4.5 ml/min, and the eluate was monitored with a UV detector at 254 nm. Fractions, which had purity greater than 98 % were collected and lyophilised.

All analytical RP-HPLC analyses were performed on a Phenomenex Kinetex XB-C18 column (Phenomenex- Shim-Pol, Izabelin, Poland) (4.6 × 150 mm, 5 µm particle size) using a Beckman Gold System. The mobile phase consisted of 0.08 % TFA in acetonitrile (solvent A) and 0.1 % TFA in water (solvent B). The gradient profile for the mobile phase was as 0–30 % of solvent A held for 30 min. The column was maintained at ambient temperature. The flow rate was 1 ml/min, and the eluate was monitored with a UV detector at 254 nm.

Finally, PNA was characterised by MALDI-TOF mass spectrometry (Bruker, BIFLEX III), which confirmed the identity of the product synthesised.

The Capillary Electrophoresis of PNA and Its DNA Homologues

The specificity of PNA hybridisation to the complementary ERVWE1 ssDNA fragment and not complementary MRSV ssDNA fragment was analyzed and confirmed by capillary electrophoresis [30]. Separations were performed using a P/ACE System MDQ (Beckman Instruments, Fullerton, CA, USA) controlled by Karat software. A coated eCAP DNA capillary (Beckman Coulter, USA) of 39.5 (29.5 to detector) cm × 100 µm thermostated at 35 °C was used. Analyses were performed at 7.8 kV using a background electrolyte (BGE) of dsDNA 1000 Gel Buffer (Beckman Coulter, USA) with reverse electrodes polarisation (anode at the detector end). Samples were introduced to the capillary at its anodic end by electrokinetic injection at 5 kV for 5 s. The capillary was rinsed with a new portion of a separation buffer between runs for 5 min. PNA/DNA hybrids were monitored with a UV detector at 254 nm. All experiments were performed in triplicate.

Polymerase Chain Reaction and Its Inhibition by ERWVE1 (syncytin-1)-Specific PNA

To optimise the proper reaction conditions for an efficient, PNA-mediated PCR clamping, 1×10^5 copies of pSC-B-ERVWE1 and of pSC-B-MRSV plasmids were subjected, in parallel, as the template into polymerase chain reactions. Reaction mixtures contained 1× Qiagen PCR Master Mix, 200 nM of Syncytin_out_F and 200nM of an appropriate reverse primer (sync1_REV/-2 or sync1_REV/-1 or sync1_REV/0, respectively, see Table 1 for details).

In our previous experiment, we found that the addition of PNA annealing step to a standard 3-step PCR in PNA-mediated, end-point PCR clamping assay cycle gave excellent results, much better as those in 3-step PCR [31]. After optimization, the reaction's thermal conditions were finally set as follows: 95 °C/5 min, 26 repeats of (95 °C/30 s, 63 °C/50 s, 58 °C/30 s, 72 °C/40 s) and 72 °C/10 min. In the next step, the smallest but still effective amount of ERWVE1- complementary PNA was estimated. Two types of peptide nucleic acid molecules were taken into consideration that differ by only one monomer at the C'-end. Final concentrations in the range of 0.05–5 µM PNA in PCR mixture were examined. Control reaction was supplemented with an appropriate amount of 5 % dimethyl sulfoxide (DMSO) instead of PNA. All reactions were performed in duplicate and analysed in 1.5 % agarose gel stained with ethidium bromide (0.5 µg/ml).

Real-Time PCR

Real-time PCR with the hydrolysis probe and/or with PNA was performed using a Roche real-time PCR system; model LightCycler 480 (Roche Diagnostics Polska, Warsaw, Poland). A hydrolysis probe, complementary to MSRV but not to ERVWE-1 gene fragment was designed using EPRIMER3 software available from the EMBOSS package [22]. The target sequence for the probe hybridisation was overlapped with that for PNAs.

The hydrolysis probe used in the assay possessed a reporter dye (6-carboxyfluorescein, FAM) and a non-fluorescent quencher (Black Hole Quencher®-1, BHQ-1) at the 5' and 3' ends, respectively. The reporter dye's emission is quenched when the probe remains intact. Only if the probe hybridised with its complementary target (i.e. MSRV *env* gene), the fluorescence of a reporter dye becomes detectable due to 5' → 3' exonuclease activity of Taq polymerase. In the case of MSRV, the hybridisation process of the probe is not impeded by PNA because of its non-complementarity to MSRV sequence (Fig. 1). On the contrary, the synthesis on the ERVWE1 locus/template cannot take place for two reasons: first, because the fluorescent probe is complementary to the MSRV but not to the ERVWE1; second, because PNA molecules hybridise with high affinity to the ERVWE1 template and thus prevent hybridisation of the hydrolysis probe. Thermal conditions for the quantitative analysis with the use of hydrolysis probe differ from that used in the end-point PCR. This results from the fact that the probe is hydrolysed during the primers' annealing step. Finally, the conditions of the PCR were as follows: 10 min at 95 °C for the initial denaturation, followed by 50 cycles at 95 °C for 30 s for denaturation and 61 °C for 1 min for primer extension (with signal acquisition at this step). Reaction mixtures consisted of 1× Qiagen PCR Master Mix (containing the 1.5 mM of $MgCl_2$), 2.5 mM $MgCl_2$, 200 nM of each Syncytin_out_F forward- and sync1_REV/-1 reverse primer, 400nM of MSRV-complementary hydrolysis probe, 300nM of PNA (or an appropriate volume of 5 % DMSO).

Preparation of Standard Curve for the Measurement of MSRV Using a Hydrolysis Probe

PSC-B-MSRV plasmids that had been quantitatively adjusted using an fluorometer (Fluoroscan microplate reader, Labsystems, Finland) beforehand were used for tenfold serial dilutions from $1 \times 10\ e^9$ copies/reaction to $1 \times 10\ e^2$ copies/reaction; from these, the standard curves were generated (Fig. 2).

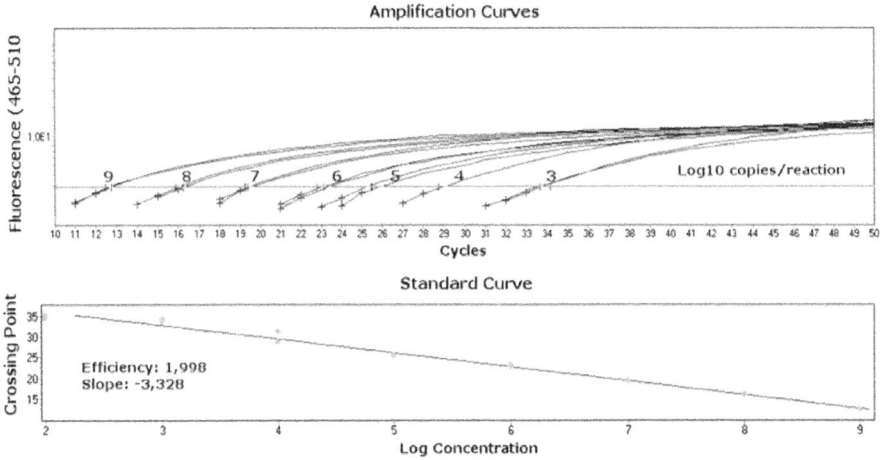

Figure 2: Standard curve prepared for the quantitation of MSRV. Plasmids contained MSRV env gene fragment (pSC-B-MSRV) that had been quantitatively adjusted were

used for sevenfold serial dilutions from 1×10 e^{9}copies/reaction to 1×10 e^3 copies/reaction; from these, the standard curves were generated. A strong inverse correlation between the logarithmic concentration of samples and CT values was obtained, and linearity was confirmed within a whole range of standard concentrations. The efficiency of the reaction reaches a value of 1.998

RESULTS

PNA-Mediated PCR Clamping of ERWVE1 Plasmid DNA Template

In our previous experiments, we have documented that the amplification of ERVWE1 gene fragment by means of polymerase chain reaction (PCR) can be specifically inhibited by complementary peptide nucleic acids (PNA) while no inhibition occurred when MSRV clones were used as the template. Furthermore, we also found that slight differences in PNA design or in primers localization altered drastically the efficiency of PCR blocking. A PNA probe, 14-monomers in length that hybridise at the position 1370–1383 in respect to the reference gene (GenBank: NM_014590) efficiently clamped the amplification of ERVWE1 at the concentration of at least 4 μM while we noted that its longer analogue (15-monomers, hybridised at positions 1369–1383) worked effectively even at tenfold lower concentration of 400nM or less (Fig. 3).

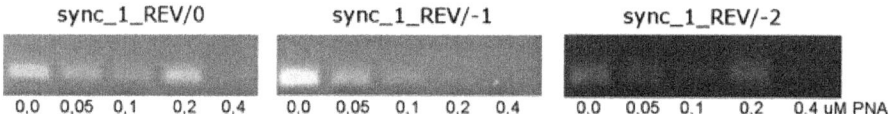

Figure 3: The optimization of reaction conditions of PNA-mediated PCR clamping. The specific inhibition of ERVWE1 by an assay relies on exact reciprocal probe-to-primer localization and on the optimal concentration of the PNA. The best results were obtained if a PNA probe shared only one base with the adjacent reverse primer (sync_1_REV/-1)

In theory, PCR clamping operates by three possible mechanisms of action, i.e.: (1) by the competition for a common target site between a PNA and one of the PCR primers; (2) by the interference of primer elongation and/or (3) by the product elongation arrest. According to our observations, primer elongation stoppage was the only efficient way to inhibit the reaction. Neither the competition between PNA and one of the PCR primer for a common target nor the product elongation arrest was efficient to inhibit the reaction even at the final PNA concentration of as much as 10 μM. Even in the case of an inhibition caused by primer elongation stoppage, the efficiency of the clamping was strongly influenced by the exact reciprocal primer-to-PNA localization (Figs. 1, 3).

Real-Time PCR of MSRV Template

Standard curve for the quantitative measurement of MSRV template in the sample was prepared using serial dilutions of the pSC-B-MSRV plasmid stock from $1 \times 10\,e^9$ copies/reaction to $1 \times 10\,e^2$ copies/reaction. A strong inverse correlation between the logarithmic concentration of samples and Ct value was obtained, and linearity was confirmed within a range from $1 \times 10\,e^3$ to $1 \times 10\,e^9$ copies/reaction (Efficiency: 1.998; Slope: −3.328, Fig. 2). All standard concentrations were run in duplicates.

Measurement of a Target MSRV in the Presence of ERVWE1 Sequence

The accuracy of quantification of MSRV DNA was determined in a mixed MSRV/ERVWE1 template by real-time PCR with a MRSV-specific probe with or without PNA. Briefly, $1 \times 10\,e^4$ molecules of the pSC-B-ERVWE1 plasmid DNA were serially diluted with the pSC-B-MSRV plasmid DNA of the same stock concentration. Proportions of pSC-B-MRSV were adjusted to 100, 80, 60, 40, 20, 10 and 1 % (pSC-B-MSRV to pSC-B-ERVWE1 ratio). If PNA was not present in the reaction, the target MSRV sequence was correctly measured only at the highest concentration of the MSRV compared to ERVWE1 and an estimated molecule number was similar to that theoretically calculated only in these experimental conditions. Measurements were underestimated when the concentration of MSRV template was lower (Fig. 4). On the contrary, the addition of ERVWE1-specific PNA probe to the reaction improved greatly the accuracy of the method and allowed to estimate the MSRV copy number even if this template was in minority in respect to ERVWE1. Additionally, a linearity of the assay was maintained.

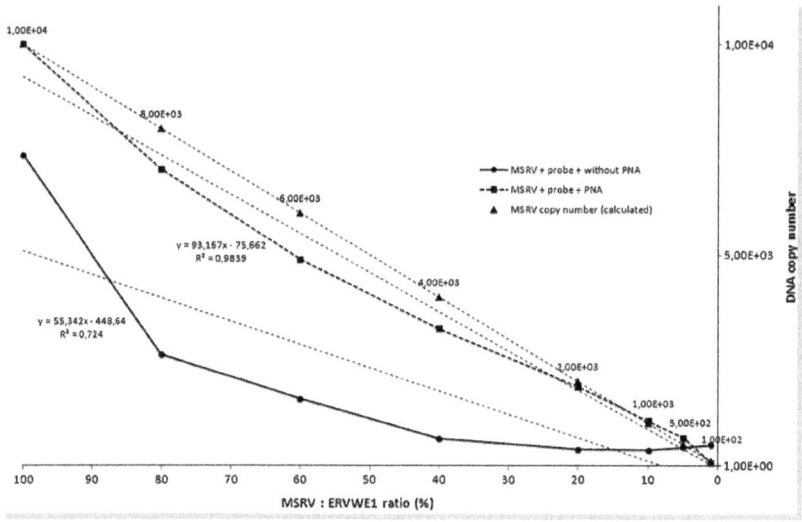

Fig. 4

The measurement of a target MSRV sequence in mixed templates. Serially diluted standard templates were amplified in the presence or absence of ERVWE1-specific PNA during probe-based real-time PCR to measure the MSRV expression level. An amount of $1 \times 10 \text{ e}^4$ copies/reaction of plasmid DNA containing MSRV fragment (pSC-B-MSRV) was used as template and 0.4 µM of ERVWE1-specific PNA was added into the reaction mixture in this assay. The pSC-B-MSRV was then serially diluted with plasmid DNA that contained ERVWE1 sequence (pSC-B-ERVWE1). The proportions of pSC-B-MRSV were adjusted to 100, 80, 60, 40, 20, 10, 5 and 1 % (pSC-B-MSRV:pSC-B-ERVWE1 ratio). Control samples that did not contain PNA were analysed parallelly. In PNA-containing samples, a linearity of the plot was maintained in the whole range of MSRV concentration values but if no PNA was added into the reaction, a correct quantitation of MSRV was possible only if an amount of MSRV DNA was equal to 100 % in respect to ERVWE1 but not at the lower MSRV-to-ERVWE1 values when the measurement was not linear. The accuracy of the method was evaluated according to the calculated plasmid copy number (depicted as *triangles* at the plot). All samples were run in duplicates

Evaluation of the Method Using Biological Samples

In the next step we examined the level of MSRV mRNA copy number in human cells using both hydrolysis probe and PNA. As a target for the assay we chose those cell lines, in which specific blocking of ERVWE1 amplification may be crucial to monitor the discrete expression changes of MRSV irrespective of the ERVWE1 transcription. It was documented that various human tissues and cell lines exhibit an overall expression of HERV-W *env* elements but their highest mRNA level was observed in placenta, glial cells and in testes [32]. Regardless the cell type, an altered expression of human endogenous retroviral W family was observed in numerous pathological conditions, such as neuroinflammation [33], burn-caused injury [34] or exogenous viral infection [35, 36].

Based on our previous observations, we checked if the expression of MSRV was detectable and could be measured quantitatively in normal human astrocytes (NHA) and in human umbilical vein endothelial cells (HUVECs) cultured in vitro. We also examined whether the MSRV expression level changed after the lipopolysaccharide (LPS) - induced activation of human monocytes in vitro.

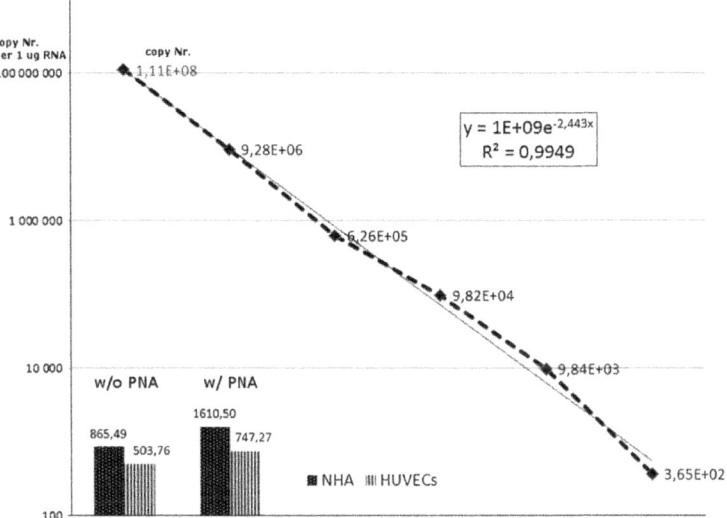

Figure 5: The quantitative analysis of MSRV mRNA in cell lines. An amount of MSRV mRNA was estimated in normal human astrocytes (NHA) and in human umbilical vein endothelial cells (HUVECs) in in vitro cultures. The addition of PNA into the reaction increased the calculated MSRV copy number in both NHA and HUVECs when compared with the quantitative analysis without the use of PNA. These results indicate that an inhibitory effect caused by ERVWE1 present in the sample may be eliminated by the use of ERVWE1-specific PNA and that ERVWE1 is more abundantly expressed in NHA than in HUVECs (1.86- and 0.67-fold increase in calculated mRNA copies number after addition of PNA, respectively). For the quantitative analysis of MSRV in cells, an individual, in-run standard curve was generated in the range from $1 \times 10 \text{ e}^8$ copies/reaction to $1 \times 10 \text{ e}^2$ copies/reaction (*dotted line* with data labels)

We confirmed that the expression of MSRV was present in investigated cell types albeit its level was relatively low in NHA and HUVECs $1.61 \times 10\,e^3 \pm 2.88 \times 10\,e^2$ and $7.47 \times 10\,e^2 \pm 3.4 \times 10\,e^1$ mRNA copies number/µg RNA, respectively. For beta actin gene expression, the results were as follow: $1.29 \times 10\,e^5 \pm 1.54 \times 10\,e^4$ and $2.52 \times 10\,e^4 \pm 1.4 \times 10\,e^3$ mRNA copies per one microgram of total RNA, respectively. As we expected, the measurement values of MSRV mRNA copy number change if ERVWE1- specific PNA was added into the QPCR reaction mixture simultaneously with MSRV-specific fluorescent probe. In the control reactions where no PNA was added but only probe was present, the obtained results were lower in comparison with that after the addition of PNA (NHA: $8.65 \times 10\,e^2 \pm 1.6 \times 10\,e^1$ mRNA copy number/µg RNA; HUVECs: $5.03 \times 10\,e^2 \pm 2.6 \times 10\,e^1$ mRNA copy number/µg RNA) (Fig. 5) indicating that the possible reaction inhibition by an abundant ERVWE1 expression was done.

In human monocytes, the stimulation by lipopolysaccharide (LPS) did not alter the expression of MSRV because no differences in HERV-W expression were observed after addition of PNA into the reaction ($1.198 \times 10\,e^3 \pm 1.83 \times 10\,e^2$ vs. $1.16 \times 10\,e^3 \pm 3.2 \times 10\,e^1$ mRNA copy number/µg RNA). Interestingly, an inhibitory effect of LPS on the expression of HERV-W was observed in samples without PNA against ERVWE1 ($9.64 \times 10\,e^3 \pm 1.93 \times 10\,e^2$; vs. $5.22 \times 10\,e^3 \pm 6.0 \times 10\,e^0$ mRNA copy number), indicating that LPS may influence the ratio between MSRV and HERV-W *env* expression in human monocytes, especially by the diminishing of non-MSRV-derived HERV-W copy number (Fig. 6).

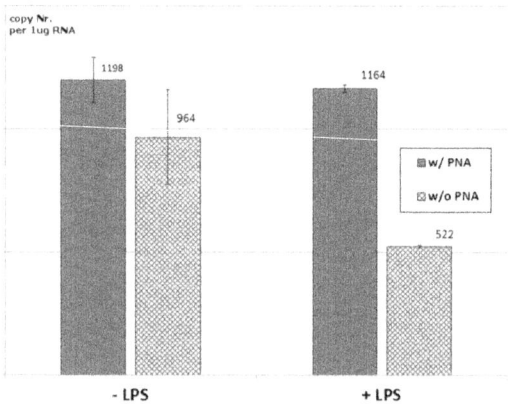

Figure 6: The use of PNA-mediated, Real-Time PCR for the quantitation of MSRV mRNA copy number in human monocytes after (LPS)-induced activa-

tion. The calculated MSRV mRNA copy number did not change in the QPCR quantitation method that utilised PNA regardless the stimulation by LPS or not. On the contrary, in standard probe-based QPCR without PNA, a strong decrease of MSRV expression level was noted, indicating that LPS may influence the ratio between MSRV and HERV-W *env* expression in human monocytes especially by the diminishing of non-MSRV-derived HERV-W copy number

DISCUSSION

The endogenous retrovirus HERV-W family is broadly represented in human genome; its sequences have been detected on almost all chromosomes. However, currently, only two loci are thought to play a specific role in both human physiology and pathology. All the explored HERV-W sequences are closely related and are difficult to distinguish [10].

Phylogenetic study of HERV-W sequences derived from brain and peripheral blood mononuclear cells of multiple sclerosis (MS) patients provided by Mameli et al. showed that envelope gene intracellular domain of MSRV and ERVWE1 (syncytin-1) RNA share >89 % similarity while whole envelope gene showed >93 % similarity in both MSRV and ERVWE1.

The homology between MSRV *env* and syncytin-1 gave rise to speculations regarding their possible physiologic and/or pathogenic effects. A major difference is that only MSRV has been detected as an extracellular virus (visualised by electron microscopy, with a polyA (+) RNA containing terminal repeats, gag, pol and *env* sequences) while the syncytin-1 protein has been found only intracellularly or on the plasma membrane [37].

As a result of high similarity at the protein level, no specific antibody for unique HERV-W family members has been identified so far and it is not possible to discriminate between MSRV *env* and syncytin-1 on the protein level. Thus, any results based on immunostaining or Western blot could reflect either MSRV *env* or syncytin-1 expression, or both.

Another approach to distinguish between the expression of MSRV *env* and syncytin-1 in research assays is to utilise PCR, reverse transcriptase (RT) PCR and/or real-time PCR assays that could selectively amplify either MSRV or syncytin-1 mRNA sequences. Reliability of the assays with respect to specificity, sensitivity and inter- and intra-assay variations has been published previously [11]. With respect to ERVWE1, MSRV*env* sequences harbour a 12-nucleotide insertion in the transmembrane moiety. Based on this insertion, discriminatory real-time PCR assay was developed that can selectively amplify either MSRV *env* or ERVWE1. The selectivity of the method was achieved by the use of allele-specific primer pairs and appropriate probes, TaqMan-type for MSRV and ERVWE1 and locked nucleic acid (LNA) probe for the generic HERV-W *env* [10].

A similar approach was demonstrated by Antony et al., where authors employed a real-time PCR approach using MRSV or syncytin-1-selective oligonucleotide primers and fluorescent, intercalating SYBR Green I dye which allowed the quantitative analysis of gene expression in brain tissue samples, peripheral blood leukocytes (PBLs) and monocyte-derived macrophages (MDM) from multiple sclerosis (MS) and non-MS patients as well as in an astrocytic U373 cell line. Analyses provided by Antony et al. revealed that ERVWE1 *env*-encoding DNA and RNA exhibited statistically significant increased detection and expression in the brains of MS patients in comparison with control. Moreover, ERVWE1 *env* transcripts were inducible in glial cells as well. Conversely, such changes in MSRV expression were not reported in that work [13].

Due to an error-prone nature of Taq polymerase, a reverse transcriptase-quantitative polymerase chain reaction based on intercalating dyes is not recommended for the detection and quantification of mutant nucleic acids, especially when slight differences among investigated sequences are expected (such as in the case of single-nucleotide mutations, SNPs). Any suboptimal reaction conditions (temperature fluctuations at the primers annealing, the overload of an enzyme co-factor $MgCl_2$ etc.) may result in possible primers-to-template mishybridisation and, subsequently, to the synthesis of unspecific amplimers [38]. Therefore, in the allele-specific amplification PCR (ASO) where allele- specific primers are used, many of 3' mismatches between a primer and its template do not efficiently impair the PCR reaction of the wild-type template and may giving false-positive signals [15].

In order to improve the specificity of the assay, the use of SYBR Green I has to be replaced by a sequence-specific fluorescent probe. Until now, SYBR Green 1 dye has been commonly used because it is simple to use, sequence-independent (there is no need to design any sequence-specific probe) and generates fluorescent signal, which theoretically is proportional to the number of double-stranded DNA molecules that are produced during PCR. This dye is suitable in most typical, quantitative, nucleic acid- based assays. On the contrary, if the reaction conditions are not optimal and/or if any double-stranded DNA molecules other than that from desired template arise in the reaction mixture, SYBR Green I gives inappropriate, false-positive results. Indeed, in respect to HERV-W research, Garson et al. in his Letter to the Editor of "AIDS Research and Human Retroviruses" journal ascertain some technical flaws in the real-time PCRs previously employed in the studies performed by Antony et al. [13, 39]. Thus, in our opinion, in MSRV analysis the best way to obtain much more accurate results is the use of PNA concomitantly with fluorescent probe and perform a direct quantification method instead of relative quantification based on intercalating dyes. In our assay, the addition of a sequence-specific fluorescently labelled oligonucleotide probe enhances greatly the selectivity of the reaction, as there are three independent factors that must occur simultaneously to give specific level of fluorescence.

Peptide nucleic acid (PNA) is an artificially synthesised polymer that has properties of both nucleic acids and proteins. Importantly, PNA binds to the complementary sequence of nucleic acid with much stronger binding capacity than DNA does, because of the lack of electrostatic repulsion.

Thus, in our assay, PNA oligomer is intended to bind specifically to the cDNA of the ERVWE1 locus and prevent them from being amplified by polymerase action. Our methodology allows the detection and measurement of discrete MSRV expression level due to MSRV-specific fluorescent oligonucleotide. What is more, any ERVWE1- derived templates are excluded because of their specific clamping by PNA. These factors improve greatly the analysis of MSRV quantity even if it is present in minority in respect to other similar molecules. In our previously published data, we found that SYBR Green I-based Real-Time PCR and its specific clamping by ERVWE-1-specific PNA favour the phenomenon in which virtually all "non-ERVWE1-derived" amplimers could be synthesised. Because only two HERV-W members, namely MSRV and ERVWE1 comprise the majority of HERV-W expression products at all, the rest of amplimers that arose and that are quantified in course of the Real-Time PCR are, in theory, considered as being of MSRV origin. However, any possible expression from distinct HERV-W loci cannot be excluded. The principle of an assay changes if non-specific SYBR Green-I dye is replaced by a MSRV-specific oligonucleotide probe. Thus, the fluorescent signal source for the quantification of the results is derived exclusively from the amplimers synthesised on MSRV template due to the reciprocal, independent hybridisation of both primers and the probe. To sum up, PNA molecules play a role as "scavenger inhibitors" that alleviate the impact of substantial ERVWE-1 overload that could interfere with the reaction efficiency (see Fig. 4, dotted line vs. solid line).

CONCLUSIONS

We developed a novel, real-time PCR technique that uses a syncytin-1 (ERVWE-1)-specific PNA oligomer and MSRV-specific fluorescent probe. This technique is intended for the discrete detection and measurement of MSRV (multiple sclerosis-associated retrovirus) expression level in biological samples regardless of syncytin-1 expression from ERVWE-1 locus.

ACKNOWLEDGMENTS

The authors would like to thank Mrs. Halina Klimas for her excellent technical cooperation. In developing the ideas presented here, I have received helpful input from Mrs. Ewa Batko. The research for this paper was financially supported by the Medical University of Silesia, Katowice, Poland; Grant No. KNW-1-097/N/4/0.

REFERENCES

1. Paces J, Pavlicek A, Paces V. HERVd: Database of human endogenous retroviruses. Nucleic Acids Research. 2002;30(1):205–206. doi: 10.1093/nar/30.1.205.
2. Pavlicek A, Paces J, Elleder D, Hejnar J. Processed pseudogenes of human endogenous retroviruses generated by lines: Their integration, stability, and distribution. Genome Research. 2002;12(3):391–399. doi: 10.1101/gr.216902.ArticlepublishedonlinebeforeprintinFebruary2002.
3. Antony JM, Ellestad KK, Hammond R, Imaizumi K, Mallet F, Warren KG, Power C. The human endogenous retrovirus envelope glycoprotein, syncytin-1, regulates neuroinflammation and its receptor expression in multiple sclerosis: A role for endoplasmic reticulum chaperones in astrocytes. Journal of Immunology. 2007;179(2):1210–1224. doi: 10.4049/jimmunol.179.2.1210.
4. Emmer A, Staege MS, Kornhuber ME. The retrovirus/superantigen hypothesis of multiple sclerosis.Cellular and Molecular Neurobiology. 2014;34(8):1087–1096. doi: 10.1007/s10571-014-0100-7.
5. Yao Y, Schröder J, Nellåker C, Bottmer C, Bachmann S, Yolken RH, Karlsson H. Elevated levels of human endogenous retrovirus-w transcripts in blood cells from patients with first episode schizophrenia.Genes Brain and Behavior. 2008;7(1):103–112.
6. Christensen T. HERVs in neuropathogenesis. Journal of Neuroimmune Pharmacology. 2010;5(3):326–335. doi: 10.1007/s11481-010-9214-y.
7. Brudek T, Christensen T, Aagaard L, Petersen T, Hansen HJ, Møller-Larsen A. B cells and monocytes from patients with active multiple sclerosis exhibit increased surface expression of both HERV-H *env* and HERV-W *env*, accompanied by increased seroreactivity. Retrovirology. 2009;6:104. doi: 10.1186/1742-4690-6-104.
8. Vargas A, Zhou S, Éthier-Chiasson M, Flipo D, Lafond J, Gilbert C, Barbeau B. Syncytin proteins incorporated in placenta exosomes are important for

cell uptake and show variation in abundance in serum exosomes from patients with preeclampsia. The FASEB Journal. 2014;28(8):3703–3719. doi: 10.1096/fj.13-239053.

9. Mameli G, Astone V, Arru G, Marconi S, Lovato L, Serra C, Sotgiu S, Bonetti B, Dolei A. Brains and peripheral blood mononuclear cells of multiple sclerosis (MS) patients hyperexpress MS-associated retrovirus/ HERV-W endogenous retrovirus, but not human herpesvirus 6. Journal of General Virology.2007;88(Pt 1):264–274. doi: 10.1099/vir.0.81890-0.

10. Mameli G, Poddighe L, Astone V, Delogu G, Arru G, Sotgiu S, Serra C, Dolei A. Novel reliable real-time PCR for differential detection of MSRV *env* and syncytin-1 in RNA and DNA from patients with multiple sclerosis. Journal of Virological Methods. 2009;161(1):98–106. doi: 10.1016/j.jviromet.2009.05.024.

11. Antony JM, Zhu Y, Izad M, Warren KG, Vodjgani M, Mallet F, Power C. Comparative expression of human endogenous retrovirus-W genes in multiple sclerosis. AIDS Research and Human Retroviruses.2007;23(10):1251–1256. doi: 10.1089/aid.2006.0274.

12. Schmitt K, Richter C, Backes C, Meese E, Ruprecht K, Mayer J. Comprehensive analysis of human endogenous retrovirus group HERV-W locus transcription in multiple sclerosis brain lesions by high-throughput amplicon sequencing. Journal of Virology. 2013;87(24):13837–13852. doi: 10.1128/JVI.02388-13.

13. Antony JM, Izad M, Bar-Or A, Warren KG, Vodjgani M, Mallet F, Power C. Quantitative analysis of human endogenous retrovirus-W *env* in neuroinflammatory diseases. AIDS Research and Human Retroviruses. 2006;22(12):1253–1259. doi: 10.1089/aid.2006.22.1253.

14. Gassen, H.G., & Minol, K. (1996). *Gentechnik: Einfuehrung in Prinzipien und Methoden*. Gustav Fischer Verlag.

15. Kwok S, Kellogg DE, McKinney N, Spasic D, Goda L, Levenson C, Sninsky JJ. Effects of primer-template mismatches on the polymerase chain reaction: Human immunodeficiency virus type 1 model studies. Nucleic Acids Research. 1990;18(4):999–1005. doi: 10.1093/nar/18.4.999.

16. Yoshida S, Hige S, Yoshida M, Yamashita N, Fujisawa S-I, Sato K, Kitamura T, Nishimura M, Chuma M, Asaka M, Chiba H. Quantification of lamivudine-resistant hepatitis B virus mutants by type-specific TaqMan minor groove binder probe assay in patients with chronic hepatitis b. Annals of Clinical Biochemistry. 2008;45(Pt 1):59–64. doi: 10.1258/acb.2007.006219.

17. Nielsen PE, Egholm M, Berg RH, Buchardt O. Sequence-selective recognition of dna by strand displacement with a thymine-substituted polyamide. Science. 1991;254(5037):1497–1500. doi: 10.1126/science.1962210.

18. Ray A, Nordén B. Peptide nucleic acid (PNA): Its medical and biotechnical applications and promise for the future. The FASEB Journal. 2000;14(9):1041–1060.

19. Kim HJ, Lee KY, Kim Y-C, Kim K-S, Lee SY, Jang TW, Lee MK, Shin K-C, Lee GH, Lee JC, Lee JE, Kim SY. Detection and comparison of peptide nucleic acid-mediated real-time polymerase chain reaction clamping and direct gene sequencing for epidermal growth factor receptor mutations in patients with non-small cell lung cancer. Lung Cancer. 2012;75(3):321–325. doi: 10.1016/j.lungcan.2011.08.005.

20. Hige S, Yamamoto Y, Yoshida S, Kobayashi T, Horimoto H, Yamamoto K, Sho T, Natsuizaka M, Nakanishi M, Chuma M, Asaka M. Sensitive assay for quantification of hepatitis B virus mutants by use of a minor groove binder probe and peptide nucleic acids. Journal of Clinical Microbiology. 2010;48(12):4487–4494. doi: 10.1128/JCM.00731-10.

21. Mackay IM, Arden KE, Nitsche A. Real-time PCR in virology. Nucleic Acids Research.2002;30(6):1292–1305. doi: 10.1093/nar/30.6.1292.

22. Rice P, Longden I, Bleasby A. EMBOSS: The european molecular biology open software suite. Trends in Genetics. 2000;16(6):276–277. doi: 10.1016/S0168-9525(00)02024-2.

23. Orum H. PCR clamping. Current Issues in Molecular Biology. 2000;2(1):27–30.

24. Łabuzek K, Liber S, Bułdak Ł, Machnik G, Liber J, Okopień B. Eplerenone promotes alternative activation in human monocyte-derived macrophages. Pharmacological Reports. 2013;65(1):226–234. doi: 10.1016/S1734-1140(13)70983-6.

25. Suchy D, Łabuzek K, Bułdak Ł, Szkudlapski D, Okopień B. Comparison of chosen activation markers of human monocytes/macrophages isolated from the peripheral blood of young and elderly volunteers.Pharmacological Reports. 2014;66(5):759–765. doi: 10.1016/j.pharep.2014.04.008.

26. Chomczynski P, Sacchi N. Single-step method of RNA isolation by acid guanidinium thiocyanate-phenol-chloroform extraction. Analytical Biochemistry. 1987;162(1):156–159. doi: 10.1016/0003-2697(87)90021-2.

27. Brøns-Poulsen J, Petersen NE, Hørder M, Kristiansen K. An improved PCR-based method for site directed mutagenesis using

megaprimers. Molecular and Cellular Probes. 1998;12(6):345–348. doi: 10.1006/mcpr.1998.0187.

28. Christen AA, Montalbano B. An ethidium bromide-agarose plate assay for the nonradioactive detection of cDNA synthesis. Analytical Biochemistry. 1989;178(2):269–272. doi: 10.1016/0003-2697(89)90637-4.

29. Nielsen PE. Gene targeting and expression modulation by peptide nucleic acids (PNA) Current Pharmaceutical Design. 2010;16(28):3118–3123. doi: 10.2174/138161210793292546.

30. Basile A, Giuliani A, Pirri G, Chiari M. Use of peptide nucleic acid probes for detecting dna single-base mutations by capillary electrophoresis. Electrophoresis. 2002;23(6):926–929. doi: 10.1002/1522-2683(200203)23:6<926::AID-ELPS926>3.0.CO;2-J.

31. Machnik G, Łabuzek K, Skudrzyk E, Rekowski P, Ruczyński J, Wojciechowska M, Mucha P, Giri S, Okopień B. A peptide nucleic acid (PNA)-mediated polymerase chain reaction clamping allows the selective inhibition of the ERVWE1 gene amplification. Molecular and Cellular Probes. 2014;28(5–6):237–241. doi: 10.1016/j.mcp.2014.04.003.

32. Yi J-M, Kim H-M, Kim H-S. Expression of the human endogenous retrovirus HERV-W family in various human tissues and cancer cells. Journal of General Virology. 2004;85(Pt 5):1203–1210. doi: 10.1099/vir.0.79791-0.

33. Perron H, Lang A. The human endogenous retrovirus link between genes and environment in multiple sclerosis and in multifactorial diseases associating neuroinflammation. Clinical Reviews in Allergy and Immunology. 2010;39(1):51–61. doi: 10.1007/s12016-009-8170-x.

34. Lee K-H, Rah H, Green T, Lee Y-K, Lim D, Nemzek J, Wahl W, Greenhalgh D, Cho K. Divergent and dynamic activity of endogenous retroviruses in burn patients and their inflammatory potential. Experimental and Molecular Pathology. 2014;96(2):178–187. doi: 10.1016/j.yexmp.2014.02.001.

35. Mameli G, Madeddu G, Mei A, Uleri E, Poddighe L, Delogu LG, Maida I, Babudieri S, Serra C, Manetti R, Mura MS, Dolei A. Activation of MSRV-type endogenous retroviruses during infectious mononucleosis and Epstein-Barr virus latency: the missing link with multiple sclerosis? PLoS ONE.2013;8(11):e78474. doi: 10.1371/journal.pone.0078474.

36. Machnik G, Klimacka-Nawrot E, Sypniewski D, Matczyńska D, Gałka S, Bednarek I, Okopień B. Porcine endogenous retrovirus (PERV) infection of HEK-293 cell line alters expression of human endogenous retrovirus

(HERV-W) sequences. Folia Biologica (Praha) 2014;60(1):35–46.

37. Dolei A, Perron H. The multiple sclerosis-associated retrovirus and its HERV-W endogenous family: A biological interface between virology, genetics, and immunology in human physiology and disease. The Journal of NeuroVirology. 2009;15(1):4–13. doi: 10.1080/13550280802448451.

38. Bracho MA, Moya A, Barrio E. Contribution of Taq polymerase-induced errors to the estimation of RNA virus diversity. Journal of General Virology. 1998;79(Pt 12):2921–2928.

39. Garson, J. A., Huggett, J. F., Bustin, S. A., Pfaffl, M. W., Benes, V., Vandesompele, J., and Shipley, G. L. (2009). Unreliable real-time PCR analysis of human endogenous retrovirus-W (HERV-W) RNA expression and DNA copy number in multiple sclerosis. *AIDS Research and Human Retroviruses*, *25*:377–3778, author reply 379–81.

Chapter 12

MICROFLUIDICS IN BIOTECHNOLOGY

Richard Barrycorresponding[1] and Dimitri Ivanov[2]

[1]School of Biological Sciences Royal Holloway, University of London Egham, Surrey TW20 0EX United Kingdom

[2]"Laboratoire de Physique des Polymères, CP223 Université Libre de Bruxelles" B-1050 Brussels Belgium

ABSTRACT

Microfluidics enables biotechnological processes to proceed on a scale (microns) at which physical processes such as osmotic movement, electrophoretic-motility and surface interactions become enhanced. At the microscale sample volumes and assay times are reduced, and procedural costs are lowered. The versatility of microfluidic devices allows interfacing with current methods and technologies. Microfluidics has been applied to DNA analysis methods and shown to accelerate DNA microarray assay hybridisation times. The linking of microfluidics to protein analysis techologies, e.g. mass spectrometry, enables picomole amounts of peptide to be analysed within a controlled microenvironment. The flexibility of microfluidics will facilitate its exploitation in assay development across multiple biotechnological disciplines.

BACKGROUND

Current analytical techniques in biotechnology can potentially benefit from an integrated reduction in scale through lowered production and operating costs, and via the specific dynamics of flowing fluids occurring at the mico-scale, which enable the generation of accurate quantitative assays. Microfluidics combines multiple disciplines including biotechnology, microtechnology, physics, and analytical chemistry and has flourished as a research field. The processes involved in biotechnology and microfluidics technologies take place on a very small scale (microns to millimetres) where some physical processes can become enhanced, e.g. osmotic movement, electrophoretic motility and surface interactions. Microfluidics technology has essentially taken advantage of the inherent properties of liquids and gases at the microscale and combined this with semiconductor technology in order to build singular devices using a streamlined manufacturing process.

COMMERCIAL PRODUCTS/TECHNOLOGIES

In general, microfluidic devices can offer a number of advantages over more conventional systems, e.g. their compact size, disposable nature, increased utility and a prerequisite for reduced concentrations of sample reagents. Miniaturised assemblies can be designed to perform a wide range of tasks that range from detecting airborne toxins to analysing DNA and protein sequences. Therefore, microfluidics systems provide a real potential for improving the efficiency of techniques applied in drug discovery and diagnostics. In order for microfluidic technology to interface with, and provide improvements for, current assaying techniques it needs to be adaptable. Some commercial microfluidics systems illustrate their suitability to biotechnological applications.

Typical devices include passive flow systems, such as the Passive Fluid Control (PFC™) micro fluid analysis system by BioMicro Systems http://www.biomicro.com. PFC incorporates 'building block-like' components into circuit designs in order to carry out sample processing, e.g. immobilisation, mixing, incubation. Essentially, PFC utilises hydrophobicity and 'passive valves' (a narrowing of capillaries) to control the movement of small volumes of fluids (< 1 µl) within a network of channels. Incorporation of active or passive pumps can also be used to control the movement of fluids in microfluidic systems, e.g. Nanostream's Snap-n-Flow™ system http://www.nanostream.com. Modules are 'snapped' together to construct a completely integrated and versatile system. A further setup by Gyros http://www.gyrosmicro.com has integrated a CD element with the movement and control of nanolitre volumes. When the CD is set spinning centrifugal forces are created allowing the device to be used to produce a controlled passage of samples through ‹microfabricated units› on the surface of the CD. This technology can be applied to sample preparation for maldi-mass spectrometric analysis.

Microfluidics systems capable of assaying ‹unprocessed› biological samples, e.g. blood, have been developed therefore eliminating the requirement for sample preparation, e.g. Micronics http://www.micronics.net. Micronics› MicroFlow™ system can be used to extract analytes directly from whole blood and other particulate suspensions (5–200 µl volumes). The system utilises disposable ‹lab cards›, e.g. the ActiveH™ card can be used for sample preparation and isolation whereas the ActiveT™ card can be used in immunoassays.

DNA APPLICATIONS

Some specific microfluidic systems have been developed that are capable of a range of DNA-type analyses. A microfluidic integrated system, which minimises sample processing and handling, has been developed for PCR analysis. Here DNA typing is achieved from whole blood samples using capillary microfluidics and capillary array electrophoresis [1], see Figure 1, whereby blood is used directly as the sample template for a PCR amplification analysis.

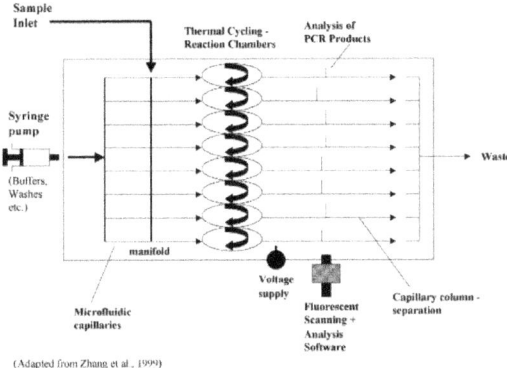

Figure 1: Capillary flow direct PCR analysis. Whole blood samples are used for direct PCR analysis. Samples are manipulated within microfluidic channels.

Microfluidics technology has also illustrated a potential to be allied with the detection of very low numbers of DNA molecules, i.e. potentially individual molecules. Foquet *et al.* [2] have shown that the construction of fluidic channels of <1 μm enables the detection and relative proportions of mixtures of DNA molecules to be measured. In addition, using an electrical field to control the flow rates analysis times of only several milliseconds per DNA molecule become achievable.

Electrophoretic mobility shift assays for the detection of DNA-protein interactions have also been carried out in a microfluidic chip environment [3]. Some of the benefits achieved are reduced sample volumes, an avoidance of labelling procedures and decreased analysis times.

The application of DNA microarrays revolutionised the analysis of gene expression studies. However, the technique generally relies on passive diffusion of the sample volume, containing the target analytes, towards the immobilised probe elements and this can result in long hybridisation times (normally hours). A method of accelerating the hybridisation time for DNA arrays using plastic microfluidic chips, comprising networks of microfluidics channels plus an integrated pump, have been developed [4]. It has been shown that 'high initial hybridisation velocities' can be attained and that equilibrium, in terms of bound versus free analyte, is quickly achieved and so negates the requirement for such long hybridisation events. The assembly of arrays into microfluidic channels in order to improve the kinetics associated with hybridisation has also been shown by other researchers. A low-density array, generated within microfluidics channels, has been used to detect gene fragments (K-Ras) carrying a point mutation [5]. Again it was found that microfluidics reduced the hybridisation time in this assay from hours, i.e. the time required in conventional static hybridisations, down to less than 1 minute. An alternative method of reducing array hybridisation times based on cavitation microstreaming has also been shown [6]. Essentially cavitation microstreaming involves the use of a sound field to induce the vibration of air-bubbles (at a solid surface) present within a fluid. This ultimately causes a circulatory flow within the fluid and so mixing times become reduced from hours to seconds. Hybridisation signals and kinetics are also reported to increase by approximately 5-fold.

Protein applications

Microfluidic technology has also been incorporated into the analysis of proteins / peptides [7, 8]. In particular, microfluidics can be linked with a mass spectrometric analysis of proteins or peptides. Thus, peptides can be adsorbed onto hydrophobic membranes, desalted, and through the use of microfluidics eluted in a controlled manner to allow the direct mass spectrometric analysis of picomole amounts of peptides by electrospray ionisation mass spectrometry procedures [9], Figure 2.

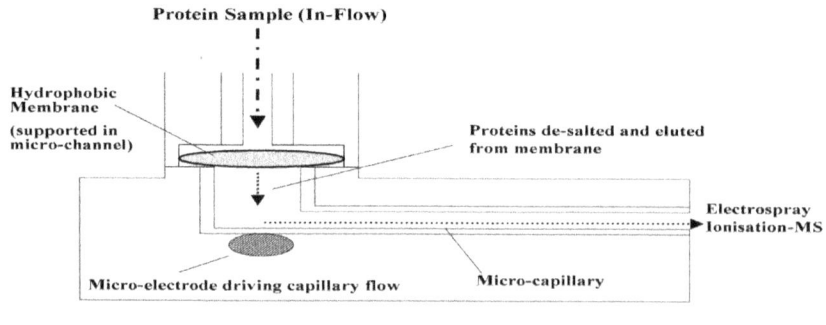

(Adapted from Lion et al. 2003)

Figure 2: Microfluidic mass spectrometric protein analysis. Proteins are applied directly to a membrane, desalted and directed by microfluidic channel to mass spectrometric analysis.

The recently reported combinatorial peptidomics approach [10] is also perfectly suited for use with integrated microfluidic systems and in principle allows identification of tryptic peptides directly from the crude proteolytic digest. Combinatorial peptidomics initially utilises peptidomics where a protein sample is proteolytically digested prior to assaying, and combines it with a combinatorial depletion of the digest (peptide pool) by chemical cross-linking via amino acid side chains to allow a subsequent profiling of the resulting sample, Figure 3.

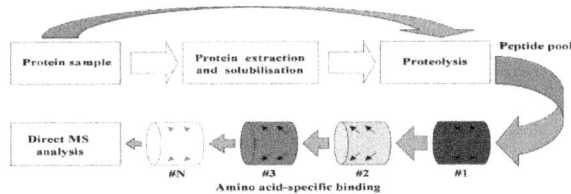

Figure 3: Combinatorial peptidomics. Sample solubilisation and protein purification are not necessary, since proteolyric digection may be carried on native cells/tissues (dashed lines). The amino acid filtering (depletion) step may be repeated using combinations of up to 6 amino acid "filters", i.e. chemically reactive surfaces (*e.g.* derivatised beads) able to covalently cross-link particular amino-acids. Chemical depletion reduces the complexity of the peptide pool to a sufficient degree to make it compatible with direct MS detection.

Other protein analysis methods have utilised microfluidics channels linked to membranes imprinted with trypsin. This allows the amount of protein delivered to the membrane, the reaction temperature within the device and the reaction time to be directly controlled for optimal digestion [11]. Thus, using microfluidics the sample can be supplied directly from upstream processing procedures, e.g. purification products from cell lysates. The peptide mixture can subsequently be analysed by electrospray ionisation mass spectrometry. Therefore, protein identification can be achieved in minutes using nanograms of sample.

The development of protein microarray methods [12–14], analogous to DNA microarray technologies, for protein / peptide analysis has the potential to hasten the discovery of proteins of pharmacological value. As is the case with DNA microarrays it is important that sample volumes required for analysis are low, the sensitivity of the assay is high (particularly for low-abundance proteins), and hybridisation times are kept to a minimum in order to produce an efficient assay. A system incorporating protein microarrays, fluorescent detection and integrated microfluidics using planar waveguide technology has been developed [15]. In combination these components enable quantitative measurements for protein profiling to be carried out with high sensitivity and also require shorter analysis times than static hybridisations.

FUTURE PROSPECTS

Finally, more novel uses for microfluidic technology at a cellular level include the handling of mammalian embryos [16], the manipulation of embryos and oocytes in assisted reproduction [17] and even the isolation of motile spermatozoa [18]. It is evident that the inherent flexibility of microfluidic systems will allow them to permeate and advance the development of assays in multiple biological, chemical and physical disciplines. Thus, microfluidics should ultimately reduce the cost of running assays, decrease procedural times and limit the required concentration and hands-on manipulation of samples.

Acknowledgements

RB acknowledges the support from the Research Strategy Fund, Royal Holloway Univesity of London.

REFERENCES

1. Zhang N, Tan H, Yeung ES: Automated and integrated system for high-throughput DNA genotyping directly from blood. Anal Chem. 1999, 71: 1138-1145. 10.1021/ac981139j.
2. Foquet M, Korlach J, Zipfel W, Webb WW, Craighead HG: DNA fragment sizing by single molecule detection in submicrometer-sized closed fluidic channels. Anal Chem. 2002, 74: 1415-1422. 10.1021/ac011076w.
3. Clark J, Shevchuk T, Swiderski PM, Dabur R, Crocitto LE, Buryanov YI, Smith SS: Mobility-shift analysis with microfluidics chips. Biotechniques. 2003, 35: 548-554.
4. Lenigk R, Liu RH, Athavale M, Chen Z, Ganser D, Yang J, Rauch C, Liu Y, Chan B, Yu H, Ray M, Marrero R, Grodzinski P: Plastic biochannel hybridization devices: a new concept for microfluidic DNA arrays. Anal Biochem. 2002, 311: 40-49. 10.1016/S0003-2697(02)00391-3.
5. Wang Y, Vaidya B, Farquar HD, Stryjewski W, Hammer RP, McCarley RL, Soper SA, Cheng YW, Barany F: Microarrays assembled in microfluidic chips fabricated from poly(methyl methacrylate) for the detection of low-abundant DNA mutations. Anal Chem. 2003, 75: 1130-1140. 10.1021/

ac020683w.

6. Liu RH, Lenigk R, Druyor-Sanchez RL, Yang J, Grodzinski P: Hybridization enhancement using cavitation microstreaming. Anal Chem. 2003, 75: 1911-1917. 10.1021/ac026267t.

7. Figeys D, Gygi SP, McKinnon G, Aebersold R: An integrated microfluidics-tandem mass spectrometry system for automated protein analysis. Anal Chem. 1998, 70: 3728-3734. 10.1021/ac980320p.

8. Figeys D, Aebersold R: High sensitivity analysis of proteins and peptides by capillary electrophoresis-tandem mass spectrometry: recent developments in technology and applications. Electrophoresis. 1998, 19: 885-892.

9. Lion N, Gellon JO, Jensen H, Girault HH: On-chip protein sample desalting and preparation for direct coupling with electrospray ionization mass spectrometry. J Chromatogr A. 2003, 1003: 11-19. 10.1016/S0021-9673(03)00771-4.

10. Soloviev M, Barry R, Scrivener E, Terrett J: Combinatorial peptidomics: a generic approach for protein expression profiling. J Nanobiotechnology. 2003, 1: 4-10.1186/1477-3155-1-4.

11. Gao J, Xu J, Locascio LE, Lee CS: Integrated microfluidic system enabling protein digestion, peptide separation, and protein identification. Anal Chem. 2001, 73: 2648-2655. 10.1021/ac001126h.

12. Barry R, Scrivener E, Soloviev M, Terrett J: Chip-Based Proteomics Technologies. Int Genomic / Proteomic Technology. 2002, 14-22.

13. Scrivener E, Barry R, Platt A, Calvert R, Masih G, Hextall P, Soloviev M, Terrett J: Peptidomics: A new approach to affinity protein microarrays. Proteomics. 2003, 3: 122-128. 10.1002/pmic.200390020.

14. Barry R, Diggle T, Terrett J, Soloviev M: Competitive assay formats for high-throughput affinity arrays. J Biomol Screen. 2003, 8: 257-263. 10.1177/1087057103008003003.

15. Pawlak M, Schick E, Bopp MA, Schneider MJ, Oroszlan P, Ehrat M: Zeptosens' protein microarrays: a novel high performance microarray platform for low abundance protein analysis. Proteomics. 2002, 2: 383-393. 10.1002/1615-9861(200204)2:4<383::AID-PROT383>3.0.CO;2-E.

16. Glasgow IK, Zeringue HC, Beebe DJ, Choi SJ, Lyman JT, Chan NG, Wheeler MB: Handling individual mammalian embryos using microfluidics. IEEE Trans Biomed Eng. 2001, 48: 570-578. 10.1109/10.918596.

17. Beebe D, Wheeler M, Zeringue H, Walters E, Raty S: Microfluidic

technology for assisted reproduction. Theriogenology. 2002, 57: 125-135. 10.1016/S0093-691X(01)00662-8.
18. Schuster TG, Cho B, Keller LM, Takayama S, Smith GD: Isolation of motile spermatozoa from semen samples using microfluidics. Reprod Biomed Online. 2003, 7: 75-81.

CITATION

CHAPTER 1
Lázaro-Silva, D., De Mattos, J., Castro, H., Alves, G. and Amorim, L. (2015) The Use of DNA Extraction for Molecular Biology and Biotechnology Training: A Practical and Alternative Approach. Creative Education, 6, 762-772. doi: 10.4236/ce.2015.68079.

CHAPTER 2
Chinnathambi, V., Balasubramanium, M., Gurusamy, R. and Paramasamy, G. (2015) Molecular Cloning and Expression of a Family 6 Cellobiohydrolase Gene cbhII from Penicillium funiculosum NCL1. Advances in Bioscience and Biotechnology, 6, 213-222. doi: 10.4236/abb.2015.63021.

CHAPTER 3
Matsunaga, G., Karasuda, S., Nishino, R., Fukushima, H. and Matsumiya, M. (2016) Molecular Cloning of a Chitinase Gene from the Ovotestis of Kuroda's Sea Hare Aplysia kurodai. Advances in Bioscience and Biotechnology, 7, 38-46. doi: 10.4236/abb.2016.71005.

CHAPTER 4
H Christina Fan, Jianbin Wang, Anastasia Potanina, and Stephen R Quake (2011). Whole-genome molecular haplotyping of single cells, Nature Biotechnology 29, 51–57 (2011), doi:10.1038/nbt.1739.

CHAPTER 5
Farahnaz Movahedzadeh, Ryan Patwell, Jenna E. Rieker, and Trinidad Gonzalez, "Project-Based Learning to Promote Effective Learning in Biotechnology Courses,"

Education Research International, vol. 2012, Article ID 536024, 8 pages, 2012. doi:10.1155/2012/536024

CHAPTER 6

Montserrat Rodrigo-Baños, Inés Garbayo, Carlos Vílchez, María José Bonete, and Rosa María Martínez-Espinosa (2015). Carotenoids from Haloarchaea and Their Potential in Biotechnology, Mar. Drugs 2015, 13(9), 5508-5532; doi:10.3390/md13095508

CHAPTER 7

Lutz Herrmann, Ulrike Bockau, Arno Tiedtke, Marcus WW Hartmann and Thomas Weide, The bifunctional dihydrofolate reductase thymidylate synthase of Tetrahymena thermophila provides a tool for molecular and biotechnology applications, doi:10.1186/1472-6750-6-21.

CHAPTER 8

Ram Karan, Melinda D Capes and Shiladitya DasSarma, Function and biotechnology of extremophilic enzymes in low water activity, DOI: 10.1186/2046-9063-8-4.

CHAPTER 9

Petra ten Hoopen, Stéphane Pesant, Renzo Kottmann, Anna Kopf, Mesude Bicak, Simon Claus, Klaas Deneudt, Catherine Borremans, Peter Thijsse, Stefanie Dekeyzer, Dick MA Schaap, Chris Bowler, Frank Oliver Glöckner and Guy Cochrane, Marine microbial biodiversity, bioinformatics and biotechnology (M2B3) data reporting and service standards, DOI 10.1186/s40793-015-0001-5.

CHAPTER 10

Madhugiri Nageswara-Rao, Jaya R Soneji, Charles Kwit and C Neal Stewart Jr, Advances in biotechnology and genomics of switchgrass, DOI: 10.1186/1754-6834-6-77.

CHAPTER 11

Grzegorz Machnik, Estera Skudrzyk, Łukasz Bułdak, Krzysztof Łabuzek, Jarosław Ruczyński, Magdalena Alenowicz,Piotr Rekowski, Piotr Jan Nowak, and Bogusław Okopień, A Novel, Highly Selective RT-QPCR Method for Quantification of MSRV Using PNA Clamping Syncytin-1 (ERVWE1), doi: 10.1007/s12033-015-9873-2.

CHAPTER 12

Richard Barry and Dimitri Ivanov, Microfluidics in biotechnology, DOI: 10.1186/1477-3155-2-2.

INDEX

A

abscisic acid (ABA) 101
Agarose gel electrophoresis 4, 7, 12
allele-specific oligonucleotides (ASO) 243

B

Biological Sciences and Biotechnology 1
biotechnology 205, 207, 210, 224, 225, 276
Bromophenol blue 7

C

cellobiohydrolase 19, 20, 22, 23, 24, 26, 27, 28, 29, 30, 31
Cellulose 20, 31, 32
central nervous system (CNS) 242
chemical defined medium (CDM) 137
chitin binding domains (CBDs) 41
cinnamyl alcohol dehydrogenase (CAD) 217, 233
collaboration 82, 84, 91, 92
Common Data Index (CDI) 200
copy number variants (CNVs) 59
Crystal structure 152, 153, 154, 155, 156, 157, 173, 175, 180, 184

D

dihydrofolate reductase (DHFR) 132, 133, 157
Dihydrofolate reductase (DHFR) 131
direct deterministic phasing (DDP) 51
DNA polymerases 150
Dunaliella salina 98, 115, 119

E

ethidium bromide 5, 7, 12
Ethylenediaminetetraacetic acid 5
expressed sequence tags (ESTs) 219

F

fluorescence resonance energy transfer (FRET) 159

G

glycoside hydrolase (GH) 35

H

haloarchaea 97, 98, 99, 103, 105, 107, 108, 110, 111, 112, 114, 115, 116, 127
Haloferax mediterranei 97, 110, 125, 126
Harold Washington College (HWC) 82

hemicelluloses 20
high-density lipoproteins (HDL) 102
high performance liquid chromatography-mass spectrometry (HPLC-MS) 98
human endogenous retroviruses (HERVs) 241
human leukocyte antigen (HLA) 50
hydrolase enzyme 19

L

Lignocellulosic biomass 20
low-density lipoproteins (LDL) 102

M

macronucleus 132, 141
Marine Microbial Biodiversity, Bioinformatics and Biotechnology (M2B3) 187
micronucleus 132, 141
microorganisms 149, 150, 151, 156, 174, 176
micropropagation 210
multiple sclerosis- associated retrovirus (MSRV) 244
multiple sclerosis (MS) 241, 242, 259, 260, 263

N

nitrogen oxidative species (NOS) 100
nuclear magnetic resonance (NMR) 98

O

Ocean Sampling Day (OSD) 188
Open Geospatial Consortium (OGC) 201
open reading frame (ORF) 39
open reading frames (ORFs) 242
Organogenesis 210

P

pedagogy 82

Penicillium funiculosum 19, 20, 30, 31, 275
Peptide nucleic acid (PNA) 243, 246, 261, 264
photoprotectors 101
polydimethylsiloxane 62
polyethylene glycol (PEG) 215
polyhydroxybutyrate (PHB) 216
project-based learning (PBL) 81
proliferating cell nuclear antigen (PCNA) 157

Q

quantitative polymerase chain reaction (QPCR) 241

R

reactive oxygen species (ROS) 100
restriction fragment length polymorphism (RFLP) 219
reverse transcription-polymerase chain reaction (RT-PCR) 35

S

sacrificing educational quality 81
science, technology, engineering, and mathematics (STEM) 81
sequence-tagged sites (STS) 219
Single-nucleotide polymorphism (SNP) 49
sodium dodecyl sulfate-polyacrylamide gel electrophoresis (SDS-PAGE) 39
synthetic biology 207, 210

T

thin layer chromatography (TLC) 98
thymidylate reductase (TS) 133
thymidylate synthase (TS) 131, 132

U

UV transilluminator 3, 5